Die Grundlehren der mathematischen Wissenschaften

in Einzeldarstellungen
mit besonderer Berücksichtigung
der Anwendungsgebiete

Band 203

B. Schoeneberg

Elliptic Modular Functions

An Introduction

Translated from the German
by J. R. Smart and E. A. Schwandt

With 22 Figures

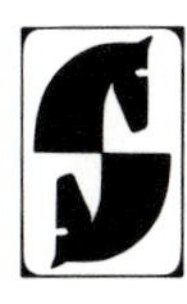

Springer-Verlag
New York Heidelberg Berlin 1974

Bruno Schoeneberg
Universität Hamburg

AMS Subject Classifications (1970):
Primary 30 A 58, 10 D 05   Secondary 20 H 10

ISBN 0-387-06382-X Springer-Verlag New York Heidelberg Berlin
ISBN 3-540-06382-X Springer-Verlag Berlin Heidelberg New York

 Library of Congress Catalog Card Number 73-8486. Printed in Germany. Monophoto typesetting and offset printing: Zechnersche Buchdruckerei, Speyer. Bookbinding: Konrad Triltsch, Würzburg.

Preface

This book is a fully detailed introduction to the theory of modular functions of a single variable. I hope that it will fill gaps which in view of the lively development of this theory have often been an obstacle to the students' progress.

The study of the book requires an elementary knowledge of algebra, number theory and topology and a deeper knowledge of the theory of functions.

An extensive discussion of the modular group SL(2, Z) is followed by the introduction to the theory of automorphic functions and automorphic forms of integral dimensions belonging to SL(2, Z). The theory is developed first via the Riemann mapping theorem and then again with the help of Eisenstein series. An investigation of the subgroups of SL(2, Z) and the introduction of automorphic functions and forms belonging to these groups follows. Special attention is given to the subgroups of finite index in SL(2, Z) and, among these, to the so-called congruence groups. The decisive role in this setting is assumed by the Riemann-Roch theorem. Since its proof may be found in the literature, only the pertinent basic concepts are outlined.

For the extension of the theory, special fields of modular functions—in particular the transformation fields of order n—are studied. Eisenstein series of higher level are introduced which, in case of the dimension -2, allow the construction of integrals of the 3rd kind. The properties of these integrals are discussed at length.

The book closes with a treatment of theta series associated with certain positive-definite quadratic forms. These series can be identified with modular forms belonging to a definite congruence group.

As a sequel to this book the reader may want to study the works of E. Hecke on elliptic modular functions. Hecke's beautiful articles still stimulate the present day research interests. The reader will now be able to appreciate them without difficulties provided he has a mature knowledge of the theory of algebraic numbers. He may find it particularly helpful to be familiar with quadratic number fields and with the theory of representations of finite groups.

The content of the first few chapters belongs almost entirely to the repertory of every scholar in the field of elliptic modular functions. Its presentation was influenced by J. Lehner's book on "Discontinuous Groups and Automorphic Functions" and by the notes of Wilhelm Maak's lectures on "Elliptische Modulfunctionen" in Goettingen, which were based on the lectures by Hecke. Chapter VII consists essentially of an article by Hecke. Chapters VIII and IX are based on the work of the author.

This book grew out of lectures which the author has given at the Universities in Hamburg, Karlsruhe and Taipei/Taiwan. The notes for the lectures in Karlsruhe proved valuable for chapters I—V, VII and VIII. They were prepared by Messrs. Helmut Bauer and Heinrich Matzat with the assistance of Mr. Sige-Nobu Kuroda. The German text was read in its entirety by Dr. Rolf-Dieter Kulle (Hamburg–Goettingen). I am indebted to him for the numerous suggestions for its improvement. The translation was undertaken by Dr. J.R. Smart (Madison) and Dr. E.A. Schwandt (Milwaukee), professors at the University of Wisconsin, who also made many valuable critical remarks. The proof sheets were corrected by these two gentlemen and by Dr. Renate Carlsson (Hamburg). Dr. Carlsson also contributed suggestions for a series of improvements. For their help I owe my gratitude to all of them.

I also owe my thanks to the editors of the "Grundlehren der mathematischen Wissenschaften" for their interest in this book and to Springer Verlag and the Zechnersche Buchdruckerei for their exemplary cooperation.

Hamburg, January 1974 Bruno Schoeneberg

Table of Contents

Chapter I. The Modular Group

§ 1. Inhomogeneous Linear Transformations

In this section are collected the necessary facts about inhomogeneous linear transformations and non-Euclidean geometry. More precise details may be found in the well-known textbooks on function theory.

1. Definition. Let $\mathbb{R}$ denote the field of real numbers, $\mathbb{C}$ the field of complex numbers, and $\hat{\mathbb{C}}$ the Riemann sphere. An *inhomogeneous* (fractional) *linear transformation* (with the usual conventions for calculating with ∞) is a map

$$L:\hat{\mathbb{C}}\to\hat{\mathbb{C}},\quad z\mapsto w=L(z):=\frac{az+b}{cz+d},\tag{1}$$

where $a,b,c,d\in\mathbb{C}$ are parameters which satisfy $ad-bc\neq 0$. The set of all inhomogeneous linear transformations shall be denoted by $\mathscr{L}$.

2. Important properties of inhomogeneous linear transformations. $L\in\mathscr{L}$ is invertible and the inverse map $L^{-1}:\hat{\mathbb{C}}\to\hat{\mathbb{C}}$ is given by

$$w\mapsto z=L^{-1}(w)=\frac{dw-b}{-cw+a}.$$

$\mathscr{L}$ is a group under the operation of composition of functions. Furthermore $L\in\mathscr{L}$ is holomorphic for $z\in\mathbb{C}$ and $z\neq -d/c$. The behavior at $z=\infty$ and $z=-d/c$ is determined in the usual way by introducing reciprocals. It follows that the $L\in\mathscr{L}$ are angle preserving. Moreover, $\mathscr{L}$ can be characterized as the set of all angle preserving maps of $\hat{\mathbb{C}}$ onto itself. Finally, $L\in\mathscr{L}$ is circle preserving, i.e. L maps the set of circles of $\hat{\mathbb{C}}$ onto itself.

3. Classification of the inhomogeneous linear transformations by their fixed points. If $L\in\mathscr{L}$ is not the identity map, then L has at most two fixed points. It follows that $L \in \mathscr{L}$ is uniquely determined by the images of three distinct points of $\hat{\mathbb{C}}$. These may be specified arbitrarily:

$$z_1 \mapsto w_1, \qquad z_2 \mapsto w_2, \qquad z_3 \mapsto w_3,$$

and the map may be realized by

$$\frac{w-w_1}{w-w_2} : \frac{w_3-w_1}{w_3-w_2} = \frac{z-z_1}{z-z_2} : \frac{z_3-z_1}{z_3-z_2}, \tag{2}$$

where suitable modifications must be made for the appearance of ∞. If we solve (2) for w, we obtain L in the form of (1). The expression

$$\frac{z-z_1}{z-z_2} : \frac{z_3-z_1}{z_3-z_2} =: D(z_1, z_2, z_3, z) \tag{3}$$

is called the *cross ratio* of the four points z_1, z_2, z_3, and z. It follows from (2) that the cross ratio (3) remains invariant if all four points are subjected to the same linear transformation $L \in \mathscr{L}$. We give the following description of $L \in \mathscr{L}$ in terms of fixed points:

Theorem 1. *$L \in \mathscr{L}$ with two distinct fixed points $z_1, z_2 \in \mathbb{C}$ has the normal form*

$$\frac{L(z)-z_1}{L(z)-z_2} = \lambda \frac{z-z_1}{z-z_2} \quad \text{with } \lambda \in \mathbb{C}, \quad \lambda \neq 0, 1.$$

$L \in \mathscr{L}$ with the fixed points $z_1 \in \mathbb{C}$ and ∞ has the normal form

$$L(z) - z_1 = \lambda(z - z_1) \quad \text{with } \lambda \in \mathbb{C}, \quad \lambda \neq 0, 1.$$

$L \in \mathscr{L}$ with one fixed point $z_0 \in \mathbb{C}$ has the normal form

$$\frac{1}{L(z)-z_0} = \frac{1}{z-z_0} + \alpha \quad \text{with } \alpha \in \mathbb{C}, \quad \alpha \neq 0.$$

$L \in \mathscr{L}$ with ∞ as the only fixed point has the normal form

$$L(z) = z + \alpha \quad \text{with } \alpha \in \mathbb{C}, \quad \alpha \neq 0.$$

The factor $\lambda \neq 0, 1$ occurring above is called the *multiplier of L*. L is called *hyperbolic* if λ is positive, *elliptic* if $|\lambda| = 1$, or *loxodromic* if λ is neither positive nor of absolute value 1. A linear transformation with only one fixed point is called *parabolic.*

4. Inhomogeneous linear transformations with real coefficients. Non-Euclidean geometry. If $a, b, c, d \in \mathbb{R}$ and $ad - bc > 0$, then L maps the real axis (including ∞) onto itself, and likewise L maps the upper half plane $\mathscr{H}$ onto itself. In particular, L maps the set of semicircles in $\mathscr{H}$ (including vertical half lines) which are orthogonal to the real axis onto itself. If one defines these as the (straight) lines of a geometry, whose points are the points of the upper half plane, then one obtains Poincaré's model of a non-Euclidean (N.E.) geometry: One measures the angle between two N.E.-lines as the Euclidean angle between their E.-tangents at the point of intersection. In order to define the N.E.-distance $\delta(z_1, z_2)$ between two points $z_1, z_2 \in \mathcal{H}$, we consider the real points ∞_1 and ∞_2 of the E.-circle through z_1 and z_2 which is orthogonal to the real axis. The points ∞_1 and ∞_2 are labeled in such a way that $\infty_1, z_1, z_2, \infty_2$ follow one another cyclically on this circle. Then we define

$$\delta(z_1, z_2) := \log D(z_1, z_2, \infty_1, \infty_2) \tag{4}$$

with real positive logarithm. This choice is seen to be possible if one maps the four-tuple $(\infty_1, z_1, z_2, \infty_2)$ by a linear transformation onto $(0, 1, \lambda, \infty)$.

The notions of ray, segment and circle will be taken over from Euclidean geometry.

If K is a N.E.-circle with center μ and radius r, then its image under the transformation

$$z \mapsto w := \frac{z-\mu}{z-\bar{\mu}}, \qquad \bar{\mu} \text{ the complex conjugate of } \mu,$$

is an E.-circle centered at $w=0$, because N.E.-lines through $z=\mu$ are transformed into E.-lines through $w=0$, the real z-axis is mapped on an E.-circle about $w=0$, and the invariance of the cross ratio under linear transformations implies that the image of K is an E.-circle. Therefore we see that the N.E.-circle K is an E.-circle, however, its E.-center and radius are different.

If z_1 and z_2 are two distinct points in $\mathcal{H}$, then the set of points of $\mathcal{H}$ with equal N.E.-distance from them is a N.E.-line which is the N.E.-perpendicular bisector of the segment $[z_1, z_2]$. One sees this by mapping z_1, z_2 onto a pair of points at the same height above the real axis.

As is well known, the transformations we have considered coincide with the set of conformal maps of $\mathcal{H}$ onto itself.

For a thorough treatment of non-Euclidean geometry see especially C. Caratheodory [1] and C.L. Siegel [2].

§ 2. Homogeneous Linear Transformations

We again confine ourselves in this section to an account of the facts which shall be needed later. For details we refer to the literature on linear algebra.

1. Definition. Let $\mathscr{A}$ be the group of invertible two-by-two matrices over $\mathbb{C}$:

$$A=\begin{pmatrix}\alpha & \beta\\ \gamma & \delta\end{pmatrix},\qquad \alpha,\beta,\gamma,\delta\in\mathbb{C},\qquad \text{with determinant, } |A|=\alpha\delta-\beta\gamma\neq 0.$$

Furthermore, let $\boldsymbol{L}$ be the group of invertible linear maps of $\mathbb{C}^2$ onto itself, which we call *homogeneous linear transformations*. For every $A\in\mathscr{A}$ the transformation

$$L_A\colon \mathbb{C}^2\to\mathbb{C}^2,\qquad \boldsymbol{z}\mapsto\boldsymbol{w}=A\cdot\boldsymbol{z}$$

belongs to L, where $A\cdot z$ is the matrix product of A with the column $\boldsymbol{z}=\begin{pmatrix}\omega_1\\ \omega_2\end{pmatrix}\in\mathbb{C}^2$, and where $\boldsymbol{w}=\begin{pmatrix}w_1\\ w_2\end{pmatrix}$.

The correspondence $\mathscr{A}\to\boldsymbol{L}$, $A\mapsto L_A$ is an isomorphism, $\mathscr{A}\cong\boldsymbol{L}$.

In the future we shall therefore use the same notation for matrices and homogeneous linear transformations. If we set

$$\boldsymbol{z}'=S\boldsymbol{z},\qquad \boldsymbol{w}'=S\boldsymbol{w},\qquad |S|\neq 0,$$

then

$$\boldsymbol{w}'=SAS^{-1}\boldsymbol{z}'.$$

2. The Jordan normal form of a homogeneous linear transformation. Two matrices $A,B\in\mathscr{A}$ are called *equivalent* if they are conjugates, i.e. if there is an $S\in\mathscr{A}$ with $B=SAS^{-1}$. The *eigenvalues* of $A=\begin{pmatrix}\alpha & \beta\\ \gamma & \delta\end{pmatrix}\in\mathscr{A}$, that is, the roots of the *characteristic polynomial*

$$\phi_A(x):=\det(xI-A)=x^2-(\alpha+\delta)x+\alpha\delta-\beta\gamma,$$

are not zero and depend only upon the equivalence class of A (I denotes the identity matrix $\begin{pmatrix}1 & 0\\ 0 & 1\end{pmatrix}$). Now we have

Theorem 2. *The classes of equivalent matrices with the distinct eigenvalues λ_1,λ_2 each contain exactly two diagonal matrices, namely,*

$$\begin{pmatrix}\lambda_1 & 0\\ 0 & \lambda_2\end{pmatrix}\quad\text{and}\quad\begin{pmatrix}\lambda_2 & 0\\ 0 & \lambda_1\end{pmatrix},$$

and are characterized by the pair λ_1,λ_2.

Proof. Suppose

$$\boldsymbol{z}^{(1)}=\begin{pmatrix}\omega_1^{(1)}\\ \omega_2^{(1)}\end{pmatrix},\qquad \boldsymbol{z}^{(2)}=\begin{pmatrix}\omega_1^{(2)}\\ \omega_2^{(2)}\end{pmatrix}$$

are linearly independent eigenvectors of a matrix A of such a class, that is,

$$A z^{(\nu)} = \lambda_\nu z^{(\nu)} \quad \text{for } \nu = 1, 2.$$

Then

$$S := \begin{pmatrix} \omega_1^{(1)} & \omega_1^{(2)} \\ \omega_2^{(1)} & \omega_2^{(2)} \end{pmatrix}^{-1} \tag{5}$$

transforms A into

$$S A S^{-1} = \begin{pmatrix} \lambda_1 & 0 \\ 0 & \lambda_2 \end{pmatrix}.$$

The characterization is now obvious. □

Theorem 3. *The classes of equivalent matrices which are different from* λI *and possess exactly one eigenvalue* λ_0, *contain the matrix* $\begin{pmatrix} \lambda_0 & 1 \\ 0 & \lambda_0 \end{pmatrix}$, *and are characterized by the number* λ_0.

Proof. If $z^{(0)} = \begin{pmatrix} \omega_1^{(0)} \\ \omega_2^{(0)} \end{pmatrix}$ is an eigenvector of a matrix $A = \begin{pmatrix} \alpha & \beta \\ \gamma & \delta \end{pmatrix}$ of such a class, then

$$S = \begin{pmatrix} \omega_1^{(0)} & 1 \\ \omega_2^{(0)} & 0 \end{pmatrix}^{-1} \quad \text{if } \omega_2^{(0)} \neq 0, \quad \text{or } S = I \quad \text{if } \omega_2^{(0)} = 0, \tag{6}$$

transforms A to $S A S^{-1} = \begin{pmatrix} \lambda_0 & \beta^* \\ 0 & \lambda_0 \end{pmatrix}$, where $\beta^* = \gamma / \omega_2^{(0)}$ or $\beta^* = \beta$, respectively.

In both cases $\beta^* \neq 0$ since $A \neq \lambda I$. One can further transform to $\begin{pmatrix} \lambda_0 & 1 \\ 0 & \lambda_0 \end{pmatrix}$ by means of $\begin{pmatrix} 1 & 0 \\ 0 & \beta^* \end{pmatrix}$. The characterization is again clear. □

The classes $\{\lambda I\}$ have the sole eigenvalue λ, the eigenspace is $\mathbb{C}^2$.

Summarizing, we have obtained: the classes of equivalent matrices distinct from $\{\lambda I\}$ are determined by their eigenvalues.

3. The connection between homogeneous and inhomogeneous linear transformations. The correspondence

$$\varphi \colon \mathscr{A} \to \mathscr{L}, \qquad A \mapsto \bar{A} \colon z \mapsto w = \bar{A}(z) = \frac{\alpha z + \beta}{\gamma z + \delta} \quad \text{for } A = \begin{pmatrix} \alpha & \beta \\ \gamma & \delta \end{pmatrix}$$

is a homomorphism with kernel $\mathbb{C}^* = \mathbb{C} \setminus \{0\}$ (after identification of $\alpha \in \mathbb{C}$ with $\alpha I \in \mathscr{A}$). If one restricts oneself to the group $\mathscr{A}^*$ of unimodular matrices

$$A=\begin{pmatrix}\alpha & \beta\\ \gamma & \delta\end{pmatrix}\in\mathscr{A}^*,\quad |A|=\alpha\delta-\beta\gamma=1\,,$$

then the kernel of φ is $\{\pm I\}$.

$$\mathscr{A}/\mathbb{C}^*\cong\mathscr{L}\,,\qquad \mathscr{A}^*/\{\pm I\}\cong\mathscr{L}\,.$$

Suppose A is a matrix with two distinct eigenvalues λ_1,λ_2, then the substitution $z':=Sz$, $\boldsymbol{w}':=S\boldsymbol{w}$, where S is the matrix occurring in the proof of Theorem **2**, transforms

$$\boldsymbol{w}=A\,z$$

into

$$\boldsymbol{w}'=\begin{pmatrix}\lambda_1 & 0\\ 0 & \lambda_2\end{pmatrix}z'\,,$$

or

$$\begin{aligned}\omega_2^{(2)}w_1-\omega_1^{(2)}w_2&=\lambda_1(\omega_2^{(2)}\omega_1-\omega_1^{(2)}\omega_2),\\ -\omega_2^{(1)}w_1+\omega_1^{(1)}w_2&=\lambda_2(-\omega_2^{(1)}\omega_1+\omega_1^{(1)}\omega_2).\end{aligned}\tag{7}$$

Set

$$\frac{\omega_1}{\omega_2}=z,\quad \frac{w_1}{w_2}=w,\quad \frac{\omega_1^{(1)}}{\omega_2^{(1)}}=z_1\quad\text{and}\quad\frac{\omega_1^{(2)}}{\omega_2^{(2)}}=z_2\,;$$

after division in (7), we obtain the equation for the corresponding inhomogeneous transformation, $w=\bar{A}z$, in its normal form:

$$\begin{aligned}\frac{w-z_1}{w-z_2}&=\lambda\frac{z-z_1}{z-z_2}\quad\text{if }\omega_2^{(1)}\text{ and }\omega_2^{(2)}\neq 0,\quad\text{or}\\ w-z_1&=\lambda(z-z_1)\quad\text{if }\omega_2^{(2)}=0.\end{aligned}\tag{8}$$

The multiplier of $\bar{A}$ is $\lambda=\dfrac{\lambda_2}{\lambda_1}$ and the fixed points are z_1,z_2. From the equations

$$A\,z^{(\nu)}=\lambda_\nu z^{(\nu)}\quad\text{or}\quad\begin{cases}\alpha\,\omega_1^{(\nu)}+\beta\,\omega_2^{(\nu)}=\lambda_\nu\omega_1^{(\nu)},\\ \gamma\,\omega_1^{(\nu)}+\delta\,\omega_2^{(\nu)}=\lambda_\nu\omega_2^{(\nu)}\end{cases}$$

we obtain the relations between fixed points and eigenvalues:

$$\lambda_\nu=\alpha+\frac{\beta}{z_\nu}\quad\text{and}\quad\lambda_\nu=\gamma\,z_\nu+\delta,\qquad \nu=1,2\,.\tag{9}$$

One of these relations must be excluded in case $\omega_1^{(\nu)}$ or $\omega_2^{(\nu)}=0$.

The matrices μA, with $\mu\neq 0$, satisfy $\overline{\mu A}=\bar{A}$. They have the eigenvalues $\mu\lambda_1$, $\mu\lambda_2$, thus the quotient of their eigenvalues is also equal to λ. Hence we have proved

Theorem 4. *If the matrix* $A=\begin{pmatrix}\alpha & \beta\\ \gamma & \delta\end{pmatrix}$ *has two distinct eigenvalues* λ_1, λ_2 *with the linearly independent eigenvectors* $\begin{pmatrix}\omega_1^{(1)}\\ \omega_2^{(1)}\end{pmatrix}$ *and* $\begin{pmatrix}\omega_1^{(2)}\\ \omega_2^{(2)}\end{pmatrix}$, *then the associated transformation* $\bar{A}$ *has two distinct fixed points*

$$z_\nu = \frac{\omega_1^{(\nu)}}{\omega_2^{(\nu)}} = \frac{\lambda_\nu - \delta}{\gamma} = \frac{\beta}{\lambda_\nu - \alpha}, \qquad \nu = 1, 2,$$

and the multiplier $\lambda = \dfrac{\lambda_2}{\lambda_1}$.

If A has a single eigenvalue λ_0, then, using the substitution S from the proof of Theorem **3**, the analogous process will yield the corresponding normal form. In this case with the fixed point z_0 we have

$$\lambda_0 = \alpha \quad \text{if } z_0 = \infty, \qquad \lambda_0 = \gamma z_0 + \delta \quad \text{if } z_0 \neq \infty. \tag{9'}$$

4. Transformations with the same fixed points. We consider the group of all $\bar{A} \in \mathscr{L}$ with the same pair of fixed points z_1, z_2, or with the same (single) fixed point z_0, respectively. By means of conjugation we see that this group is isomorphic to either the group of all $\bar{A} \in \mathscr{L}$ with the pair of fixed points 0, ∞ or to the group of $\bar{A} \in \mathscr{L}$ with the (single) fixed point ∞. Furthermore, the normal forms indicate the isomorphism of this group with either the multiplicative group of all $\lambda \in \mathbb{C}^*$ or the additive group of all $\alpha \in \mathbb{C}$, respectively. More precisely, these deliberations yield

Theorem 5. *A group of inhomogeneous linear transformations with a common pair of fixed points is isomorphic to the multiplicative group of associated multipliers; a group of inhomogeneous linear transformations with a common single fixed point is isomorphic to the additive group of associated constants.*

§ 3. The Modular Group. Fixed Points

1. Definition and classification. A *homogeneous modular transformation* is a homogeneous linear transformation whose elements are rational integers and whose determinant is 1. The homogeneous modular transformations form a group, the *homogeneous modular group;* it is isomorphic to the integral unimodular matrix group

$$\Gamma := \left\{ \begin{pmatrix} a & b\\ c & d\end{pmatrix} \,\middle|\, a, b, c, d \in \mathbb{Z},\ ad - bc = 1 \right\}.$$

The homomorphism φ given in **§ 2, 3** leads to the *inhomogeneous modular group*

$$\bar{\Gamma} := \{\bar{A} \mid A \in \Gamma\},$$

where

$$\bar{A}: z \mapsto \frac{az+b}{cz+d} \quad \text{if} \quad A = \begin{pmatrix} a & b \\ c & d \end{pmatrix} \in \Gamma.$$

The elements $\bar{A} \in \bar{\Gamma}$ are called *inhomogeneous modular transformations.* As maps they preserve the upper half plane $\mathscr{H}$, the real axis $\mathbb{R}$, and the set of rational numbers $\mathbb{Q}$. The relation between Γ and $\bar{\Gamma}$ is given by the isomorphism

$$\Gamma/\{\pm I\} \cong \bar{\Gamma}.$$

If $A = \begin{pmatrix} a & b \\ c & d \end{pmatrix} \in \Gamma$ then the characteristic polynomial

$$\phi_A(x) = \det\begin{pmatrix} a-x & b \\ c & d-x \end{pmatrix} = x^2 - \operatorname{tr}(A)\,x + 1, \text{ with trace } \operatorname{tr}(A) = a+d,$$

has the roots

$$\lambda_{1,2} = \frac{a+d}{2} \pm \frac{1}{2}\sqrt{(a+d)^2 - 4}. \tag{10}$$

Clearly $\lambda_1 \lambda_2 = 1$. Hence the eigenvalues $\lambda_{1,2}$ are algebraic integers and they are units either in the field $\mathbb{Q}$ of rational numbers, or in the quadratic number field $\mathbb{Q}(\sqrt{(a+d)^2-4})$. The units of $\mathbb{Q}$ are ± 1, and, because $\lambda_1 \lambda_2 = 1$, either $\lambda_1 = \lambda_2 = 1$, namely for $a+d=2$, or $\lambda_1 = \lambda_2 = -1$, if $a+d = -2$. It follows that $\lambda_{1,2}$ is not in $\mathbb{Q}$ for any other values of $a+d$. In particular, in all other cases the two eigenvalues must be distinct. As a result we have the following *classification*:

The parabolic case (defined in **§ 1** by having 'exactly one fixed point' and shown in **§ 2** to be equivalent to having 'exactly one eigenvalue') occurs only when $|a+d| = 2$ and consequently when $\lambda_0 = 1$ or $\lambda_0 = -1$.

The elliptic case over $\mathbb{Q}(i)$ occurs for $a+d=0$, $\lambda_{1,2} = \pm i$. By **§ 2, 3** the multiplier is $\lambda = \lambda_2/\lambda_1$, thus $\lambda = -1$. This determines an elliptic (modular) transformation.

The elliptic case over $\mathbb{Q}(\sqrt{-3})$ occurs for $|a+d| = 1$, $\lambda_{1,2} = \rho^{\pm 1}$ or $\lambda_{1,2} = (-\rho)^{\pm 1}$, where $\rho = e^{2\pi i/3}$. Here the multiplier is a cube root of unity, hence the associated modular transformation is elliptic.

The hyperbolic case is comprised of the remaining possibilities, $|a+d| > 2$, for which, by the results derived above, the eigenvalues and the multipliers are units in a real quadratic number field. The multiplier λ is positive, for $\lambda_1 \lambda_2 = 1$ implies that $\lambda = \lambda_2/\lambda_1 = \lambda_2^2$ (and thus the transformation is indeed hyperbolic).

We need the following notion of equivalence: Two points $z_1, z_2 \in \hat{\mathbb{C}}$ are called *equivalent under the modular group* $\bar{\Gamma}$, if there is an inhomogeneous modular transformation $\bar{A} \in \bar{\Gamma}$ with $\bar{A}(z_1) = z_2$. It is clear that this defines an equivalence relation.

We introduce the *discriminant* of a number ω of a quadratic number field K over $\mathbb{Q}$. For this, ω may be represented by $\omega = \alpha_1/\alpha_2$ with integral α_1, α_2 of K. Then the discriminant $\Delta(\omega)$ is defined by the expression

$$\Delta(\omega) := \left[\frac{\alpha_1 \alpha_2' - \alpha_2 \alpha_1'}{N(\alpha_1, \alpha_2)}\right]^2, \tag{11}$$

where the prime denotes conjugation and the norm in the denominator is the norm of the ideal (α_1, α_2)—the greatest common divisor of α_1 and α_2—considered as a rational integer. It is evident that $\Delta(\omega)$ is independent of the representation as a quotient of integers of K. $\Delta(\omega)$ is at the same time the discriminant of that quadratic polynomial of which ω is a root and whose coefficients are rational integers without a common divisor, namely of the polynomial

$$\frac{(\alpha_2 x - \alpha_1)(\alpha_2' x - \alpha_1')}{N(\alpha_1, \alpha_2)} = \frac{1}{N(\alpha_1, \alpha_2)} [\alpha_2 \alpha_2' x^2 - (\alpha_1 \alpha_2' + \alpha_1' \alpha_2) x + \alpha_1 \alpha_1'] . \tag{12}$$

After these preparations we shall discuss the modular transformations according to the above classification.

2. The parabolic case. The modular transformations

$$\bar{A} \in \bar{\Gamma} \quad \text{with } A = \begin{pmatrix} a & b \\ c & d \end{pmatrix} \in \Gamma \quad \text{and } |a+d| = 2$$

have the following properties:

Theorem 6. 1) *The set of fixed points consists of the rational numbers and* ∞.

2) *The* $\bar{A} \in \bar{\Gamma}$ *with the same fixed point form an infinite cyclic group which is conjugate in* $\bar{\Gamma}$ *to the group of modular transformations with the single fixed point* ∞.

3) *All rational numbers are equivalent to* ∞ *under the modular group* $\bar{\Gamma}$.

Proof. If $A = \begin{pmatrix} a & b \\ c & d \end{pmatrix} \in \Gamma$ with $|a+d| = 2$ and $c \neq 0$, then $\frac{a+d}{2}$ is the only eigenvalue. The formula (9′) implies that the rational point $\frac{a-d}{2c}$

is the only fixed point of $\bar{A}$. If, however, $c=0$, then $A=\pm\begin{pmatrix}1 & b\\ 0 & 1\end{pmatrix}$. With the (standard) notation

$$U=\begin{pmatrix}1 & 1\\ 0 & 1\end{pmatrix}, \qquad \bar{U}: z \mapsto z+1, \tag{13}$$

it follows that $\bar{A}=\bar{U}^b$. Only the transformations $\bar{U}^b$, $b\in\mathbb{Z}$, are parabolic with fixed point ∞, i.e. have ∞ as their sole fixed point; they form an infinite cyclic group generated by $\bar{U}$. For a given rational point $\frac{a'}{c'}$ with greatest common divisor $(a',c')=1$, we construct the matrix $S=\begin{pmatrix}a' & b'\\ c' & d'\end{pmatrix}$ with $b',d'\in\mathbb{Z}$ such that $a'd'-b'c'=1$. First of all, it follows that $\bar{S}(\infty)=a'/c'$ which proves *assertion* 3). Further, if

$$A'=S\,U\,S^{-1}\in\Gamma,$$

then $\bar{A}'$ has the given point $\frac{a'}{c'}$ as fixed point and is parabolic. Hence $\frac{a'}{c'}$ occurs as a fixed point which proves *assertion* 1). Moreover, the groups fixing ∞ and $\frac{a'}{c'}$ are conjugates by means of $\bar{S}\in\bar{\Gamma}$. This proves *assertion* 2). □

3. The elliptic case over the Gaussian number field. The modular transformations

$$\bar{A}\in\bar{\Gamma} \quad \text{with } A=\begin{pmatrix}a & b\\ c & d\end{pmatrix}\in\Gamma \quad \text{and } a+d=0$$

have the following properties:

Theorem 7. 1) *The set of fixed points consists of the pairs* $\frac{\pm i-d}{c}$ *with* $d^2\equiv -1 \bmod c$; *these are the numbers in* $\mathbb{Q}(i)$ *of discriminant* -4.

2) *The* $\bar{A}\in\bar{\Gamma}$ *with the same pair of such fixed points form a cyclic group of order* 2.

3) *All numbers* $\frac{i-d}{c}\in\mathbb{Q}(i)$ *with* $d^2\equiv -1 \bmod c$, $c>0$, *are equivalent under the modular group* $\bar{\Gamma}$.

Proof. Since the multiplier is $\lambda=-1$, *assertion* 2) follows immediately from the results of § **2, 4.** Now we prove *assertion* 1). The fixed points of $\bar{A}$ are computed from the eigenvalues $\lambda_{1,2}=\pm i$ by the formula $\lambda_{1,2}=c z_{1,2}+d$. Thus $z_{1,2}=\frac{\pm i-d}{c}$ and since $a=-d$, it follows from

$ad-bc=1$ that $d^2\equiv -1 \bmod c$. Conversely, if $c,d\in\mathbb{Z}$ with $d^2\equiv -1 \bmod c$, then the matrix

$$A:=\begin{pmatrix} -d & -\dfrac{d^2+1}{c} \\ c & d \end{pmatrix}$$

has integral entries, determinant 1, and trace 0; $\bar{A}$ has the fixed points $\dfrac{\pm i-d}{c}$. This proves the first part of *assertion* 1). The numbers $\dfrac{\pm i-d}{c}$ with $d^2\equiv -1 \bmod c$, are the zeros of the polynomial

$$cx^2+2dx-b \quad \text{with } b=-\frac{1+d^2}{c},$$

and, since $-bc=1+d^2$ implies that g.c.d. $(b,c,2d)=1$, they therefore have discriminant -4. Conversely, suppose $\dfrac{d'i-d}{c}\in\mathbb{Q}(i)$ is given with $c,d,d'\in\mathbb{Z}$ without common divisors and with discriminant -4. If $\gamma=(d'i-d,c)$ then $-4=(2id'c/N(\gamma))^2$ and so $d'c=\pm N(\gamma)$. If there were a prime $\pi|d'$ then, because $\pi|\gamma$ or $\pi|\gamma'$, it follows that $\pi|c$ and $\pi|d$. This contradicts g.c.d. $(c,d,d')=1$. Thus $d'=\pm 1$ and so $N(\gamma)=\pm c$. In particular, since $\gamma|\pm i-d$, it follows that $c=\pm N(\gamma)|N(\pm i-d)=1+d^2$ and so $d^2\equiv -1 \bmod c$. This proves *assertion* 1) completely. In the next section we give the proof of *assertion* 3) and of the additional remark that all $\bar{A}\in\bar{\Gamma}$ with $\mathrm{tr}(A)=0$ are conjugates (in $\bar{\Gamma}$) of $\bar{T}$ with (in standard notation)

$$T=\begin{pmatrix} 0 & -1 \\ 1 & 0 \end{pmatrix}, \qquad \bar{T}: z\mapsto -\frac{1}{z}. \tag{14}$$

4. The elliptic case over the field of cube roots of unity. The modular transformations

$$\bar{A}\in\bar{\Gamma} \quad \text{with } A=\begin{pmatrix} a & b \\ c & d \end{pmatrix}\in\Gamma \quad \text{and } |a+d|=1$$

have the following properties:

Theorem 8. 1) *The set of fixed points consists of the pairs of numbers*

$$\frac{\rho^{\pm 1}-d}{c}\in\mathbb{Q}(\rho) \quad \textit{with } d^2+d+1\equiv 0 \bmod c;$$

these are the numbers of $\mathbb{Q}(\rho)$ *of discriminant* -3.

2) *The* $\bar{A}\in\bar{\Gamma}$ *with the same pair of such fixed points form a cyclic group of order* 3.

3) *All numbers*

$$\frac{\rho-d}{c}\in\mathbb{Q}(\rho) \quad \text{with } d^2+d+1\equiv 0 \bmod c, \qquad c>0,$$

are equivalent under the modular group $\bar{\Gamma}$.

Proof. Since the multiplier is a cube root of unity, *assertion* 2) follows immediately from the results of **§ 2, 4.** Now we prove *assertion* 1). If $\bar{A}\in\bar{\Gamma}$ with $|a+d|=1$ is given, then without loss of generality one may choose $a+d=-1$. The fixed points of $\bar{A}$, $\frac{\rho^{\pm 1}-d}{c}$, are calculated from the eigenvalues $\lambda_{1,2}=\rho^{\pm 1}$. Since $a=-(1+d)$, it follows from $ad-bc=1$ that $d^2+d+1\equiv 0 \bmod c$. Conversely, for $c,d\in\mathbb{Z}$ with $d^2+d+1\equiv 0 \bmod c$, the matrix

$$A:=\begin{pmatrix} -(1+d) & -\dfrac{d^2+d+1}{c} \\ c & d \end{pmatrix}$$

has integral entries, determinant 1, and trace -1; and $\bar{A}$ has the fixed points $\frac{\rho^{\pm 1}-d}{c}$. This proves the first part of *assertion* 1). The numbers

$$\frac{\rho^{\pm 1}-d}{c} \quad \text{with } d^2+d+1\equiv 0 \bmod c$$

are roots of the polynomial

$$cx^2+(1+2d)x-b \quad \text{with } b=-\frac{d^2+d+1}{c},$$

which has discriminant -3. It follows from this that g.c.d. $(c,1+2d,b)=1$ and that therefore the numbers $\frac{\rho^{\pm 1}-d}{c}$ also have discriminant -3. Conversely, let $\frac{d'\rho-d}{c}\in\mathbb{Q}(\rho)$ be given with $c,d,d'\in\mathbb{Z}$ without common divisors and with discriminant -3 (the case $\frac{d'\bar{\rho}-d}{c}$ is included here). If $\gamma=(d'\rho-d,c)$ then $-3=\left(\frac{d'c(\rho-\bar{\rho})}{N(\gamma)}\right)^2$, so $d'c=\pm N(\gamma)$. If there were a prime $\pi|d'$, then, because $\pi|\gamma$ or $\pi|\gamma'$, it follows that $\pi|c$ and $\pi|d$. This contradicts the fact that g.c.d. $(c,d,d')=1$. Thus $d'=\pm 1$, and by simultaneously changing the signs of c,d and d', if need be, we may suppose $d'=1$. Since $\gamma|\rho-d$, it follows that $c=\pm N(\gamma)|N(\rho-d)=d^2+d+1$, and thus $d^2+d+1\equiv 0 \bmod c$. This proves *assertion* 1).

In the next section we will give the proof of *Assertion* 3) and of the additional remark that all $\bar{A}\in\bar{\Gamma}$ with $|\mathrm{tr}(A)|=1$ are conjugate (in $\bar{\Gamma}$) to $\bar{R}$ or $\bar{R}^2$ with (in standard notation)

$$R=TU=\begin{pmatrix}0 & -1\\ 1 & 1\end{pmatrix}, \qquad \bar{R}: z\mapsto \frac{-1}{z+1}. \tag{15}$$

5. The hyperbolic case. The modular transformations

$$\bar{A}\in\bar{\Gamma} \quad\text{with}\quad A=\begin{pmatrix}a & b\\ c & d\end{pmatrix}\in\Gamma \quad\text{and}\quad |a+d|>2$$

have the following properties:

Theorem 9. 1) *The set of fixed points consists of the pairs of distinct algebraic conjugate numbers of an arbitrary real quadratic number field.*

2) *The $\bar{A}\in\bar{\Gamma}$ with the same pair of such fixed points form an infinite cyclic group.*

3) *Two real quadratic irrationalities $z=\dfrac{\alpha_1}{\alpha_2}$ (α_1,α_2 integral) and $w=\dfrac{\beta_1}{\beta_2}$ or $w=\dfrac{\beta_2}{\beta_1}$ (β_1,β_2 integral) are equivalent under the modular group if and only if the $\mathbb{Z}$-modules (α_1,α_2) and (β_1,β_2) are similar.*

The $\mathbb{Z}$-module (α,β) is defined by

$$(\alpha,\beta):=\{a\alpha+b\beta\,|\,a,b\in\mathbb{Z}\}.$$

Two $\mathbb{Z}$-modules (α_1,α_2) and (β_1,β_2) are similar if the elements α_1,α_2, β_1,β_2 belong to the same field and if there exists a number σ in this field such that $(\alpha_1,\alpha_2)=\sigma(\beta_1,\beta_2)$.

Proof. We already know that the multiplier $\lambda=\lambda_2^2$ is the square of a unit of the real quadratic number field $\mathbb{Q}(\sqrt{(a+d)^2-4})$. We conclude from the existence of a fundamental unit in a real quadratic number field (see H. Hasse [1], p. 288ff. or W.J. Le Vegue [1] p. 74ff.) that every proper subgroup of the group of units not containing -1 is a cyclic subgroup of infinite order. In particular this is true for the subgroup of multipliers; by § **2, 4** this proves *assertion* 2).

Proof of *assertion* 1): Because the eigenvalues are algebraic conjugates, so are the fixed points, since $\lambda_\nu=cz_\nu+d$; furthermore, the fixed points are not rational. Suppose now that $z=\omega_1/\omega_2\in\mathbb{Q}(\sqrt{D})$, $D>0$ not a rational square, and ω_1,ω_2 are integral. We further suppose that $z\notin\mathbb{Q}$, so that ω_1,ω_2 is a basis of the $\mathbb{Z}$-module (ω_1,ω_2). Choose a unit $\varepsilon\equiv 1 \bmod(\omega_1\omega_2'-\omega_2\omega_1')$ with $\varepsilon\neq\pm 1$ and $\varepsilon\varepsilon'=1$; such a unit exists because the ring of residues $\bmod(\omega_1\omega_2'-\omega_2\omega_1')$ is finite and the order of the group of units with positive norm is infinite. From the equations

$$\begin{aligned} \varepsilon\omega_1 &= a\omega_1 + b\omega_2, \\ \varepsilon\omega_2 &= c\omega_1 + d\omega_2 \end{aligned} \tag{16}$$

where $a, b, c, d \in \mathbb{Q}$ and their conjugates, we calculate the coefficients

$$a = \frac{\varepsilon\omega_1\omega_2' - \varepsilon'\omega_2\omega_1'}{\omega_1\omega_2' - \omega_2\omega_1'} \in \mathbb{Z}, \qquad b = \frac{(\varepsilon' - \varepsilon)\omega_1\omega_1'}{\omega_1\omega_2' - \omega_2\omega_1'} \in \mathbb{Z},$$

$$d = \frac{\varepsilon'\omega_1\omega_2' - \varepsilon\omega_2\omega_1'}{\omega_1\omega_2' - \omega_2\omega_1'} \in \mathbb{Z}, \qquad c = \frac{(\varepsilon - \varepsilon')\omega_2\omega_2'}{\omega_1\omega_2' - \omega_2\omega_1'} \in \mathbb{Z}.$$

The fact that $a, b, c, d \in \mathbb{Z}$ follows from $\varepsilon \equiv \varepsilon' \equiv 1 \bmod (\omega_1\omega_2' - \omega_2\omega_1')$. We see from

$$\det\begin{pmatrix}\omega_1 & \omega_1' \\ \omega_2 & \omega_2'\end{pmatrix} = \det\begin{pmatrix}\varepsilon\omega_1 & \varepsilon'\omega_1' \\ \varepsilon\omega_2 & \varepsilon'\omega_2'\end{pmatrix} = \det\begin{pmatrix}a & b \\ c & d\end{pmatrix}\det\begin{pmatrix}\omega_1 & \omega_1' \\ \omega_2 & \omega_2'\end{pmatrix}$$

that $ad - bc = 1$. We see from (16) that the inhomogeneous transformation corresponding to $\begin{pmatrix}a & b \\ c & d\end{pmatrix}$ is not the identity and has the fixed point $z = \omega_1/\omega_2$. This proves *assertion* 1).

Now we prove *assertion* 3): In the case of similarity there exists, perhaps with an interchange of β_1 and β_2, an integral unimodular transformation $\begin{pmatrix}a & b \\ c & d\end{pmatrix}$ with

$$\begin{aligned} \sigma\beta_1 &= a\alpha_1 + b\alpha_2, \\ \sigma\beta_2 &= c\alpha_1 + d\alpha_2, \end{aligned} \tag{17}$$

whose corresponding inhomogeneous transformation takes $z_1 = \dfrac{\alpha_1}{\alpha_2}$ into $w = \dfrac{\beta_1}{\beta_2}$. Conversely, if z and w are equivalent, we have the equations (17) with $ad - bc = 1$, and so

$$\sigma(\beta_1, \beta_2) = (a\alpha_1 + b\alpha_2, c\alpha_1 + d\alpha_2) = (\alpha_1, \alpha_2). \quad \square$$

§ 4. Generators and Relations

1. Generation. First we prove

Theorem 10. *The homogeneous modular group is generated by the elements* $U = \begin{pmatrix}1 & 1 \\ 0 & 1\end{pmatrix}$ *of infinite order and* $T = \begin{pmatrix}0 & -1 \\ 1 & 0\end{pmatrix}$ *of order* 4.

The inhomogeneous modular group is generated by the transformations $\bar{U}: z \mapsto z+1$ *of infinite order and* $\bar{T}: z \mapsto \frac{-1}{z}$ *of order* 2.

Proof. If $A = \begin{pmatrix} a & b \\ c & d \end{pmatrix} \in \Gamma$ and $k \in \mathbb{Z}$, then

$$TA = \begin{pmatrix} -c & -d \\ a & b \end{pmatrix}, \qquad U^k A = \begin{pmatrix} a+kc & b+kd \\ c & d \end{pmatrix}.$$

We assume that $|c| \leqq |a|$ (for if this is not true for A, then we start with TA). If $c=0$, then $A = \pm U^{q_0}$. If, however, $c \neq 0$, we apply the Euclidean algorithm to a and c (in modified form):

$$a = q_0 c + r_1, \qquad -c = q_1 r_1 + r_2, \qquad r_1 = q_2 r_2 + r_3, \dots, (-1)^n r_{n-1} = q_n r_n + 0,$$

which ends with $r_n = \pm 1$, since $(a, c) = 1$. Premultiplying

$$T U^{-q_n} T \dots T U^{-q_0}$$

by A, we obtain $\pm U^{q_{n+1}}$ with $q_{n+1} \in \mathbb{Z}$. Thus we have—the case $|c| > |a|$ included—:

$$A = T^m U^{q_0} T U^{q_1} \dots T U^{q_n} T U^{q_{n+1}} \tag{18}$$

with $m = 0, 1, 2$ or 3; $q_0, q_1, \dots, q_{n+1} \in \mathbb{Z}$ and $q_0, \dots, q_n \neq 0$. The assertion concerning $\bar{\Gamma}$ follows immediately. □

In **§ 5, 2** we shall give another proof of Theorem **10** depending upon the existence of the fundamental region.

2. Defining relations. Since U and T generate the group Γ, so do T and $R_1 = -R$, for by (15) $U = -TR$. We now prove

Theorem 11. *The generators* T *and* R_1 *of the group* Γ *satisfy the relations*

$$T^4 = R_1^3 = I, \qquad R_1 T^2 = T^2 R_1, \tag{19}$$

and these are defining relations for Γ.

Proof. One easily verifies that T and R_1 satisfy the relations (19). Suppose now that an arbitrary relation is given which we assume without restriction to have the form

$$T^{\varepsilon_1} R_1^{\alpha_1} T^{\varepsilon_2} R_1^{\alpha_2} \dots T^{\varepsilon_n} R_1^{\alpha_n} T^{\varepsilon_{n+1}} = I, \qquad \varepsilon_\nu, \alpha_\nu \in \mathbb{Z}.$$

By applying (19), we transform this to

$$T^{\varepsilon_1} R_1^{\alpha_1} T R_1^{\alpha_2} \dots T R_1^{\alpha_n} T^{\varepsilon_{n+1}} = I,$$

where $\varepsilon_1, \varepsilon_{n+1}=0,1,2$ or 3, $\alpha_\nu=1$ or 2, and $n\geqq 0$. After a suitable multiplication by T we obtain

$$S:=R_1^{\alpha_1}T\ldots R_1^{\alpha_n}T=T^\alpha,\quad \alpha_\nu=1 \text{ or } 2,\quad \alpha=0,1,2 \text{ or } 3,\quad n\geqq 0. \tag{20}$$

However, such a relation cannot exist for $n\geqq 1$. Indeed, if we set $S=\begin{pmatrix} a & -b\\ -c & d\end{pmatrix}$ in (20), then we show by induction that for $n\geqq 1$ the entries of S satisfy the conditions

$$a,b,c,d\geqq 0,\quad b+c>0 \quad\text{or}\quad a,b,c,d\leqq 0,\quad b+c<0,$$

whereas the entries of T^α, $\alpha=0,1,2,3$ do not. The relation (20) is only possible for $n=0$, i.e. it is trivial. Our claim for the entries of S is correct for S equal to

$$R_1T=\begin{pmatrix} 1 & 0\\ -1 & 1\end{pmatrix} \quad\text{and}\quad R_1^2T=-\begin{pmatrix} 1 & -1\\ 0 & 1\end{pmatrix}.$$

If it is correct for the matrix $S=\begin{pmatrix} a & -b\\ -c & d\end{pmatrix}$, then

$$R_1TS=\begin{pmatrix} a & -b\\ -(a+c) & b+d\end{pmatrix},\quad R_1^2TS=-\begin{pmatrix} a+c & -(b+d)\\ -c & d\end{pmatrix}$$

implies the claim for all S with $n\geqq 1$. □

In place of the defining relations (19) one has

$$\bar{T}^2=\bar{R}^3=\bar{I} \tag{21}$$

as defining relations for the inhomogeneous group $\bar{\Gamma}$. The last relation of (19) becomes trivial since $\bar{T}^2=\bar{I}$. Thus we have

Theorem 12. *The group $\bar{\Gamma}$ is isomorphic to the free product of a cyclic group of order 2 and a cyclic group of order 3.*

For an essentially different method of proof see F. Klein-R. Fricke [1], vol. 1, p. 452ff.

§ 5. Fundamental Region

1. Fundamental region. Hereafter the rational points of the real axis and the point ∞, for which we now prefer the symbol $i\infty$, shall be adjoined to the upper half plane. This extended upper half plane shall be denoted by $\mathscr{H}^*$, and its variable by τ.

Definition. A set $\mathscr{F}$ is called a *fundamental set* of the modular group $\bar{\Gamma}$ for $\mathscr{H}^*$ if $\mathscr{F}$ contains exactly one point from each class of points of $\mathscr{H}^*$ equivalent under $\bar{\Gamma}$.

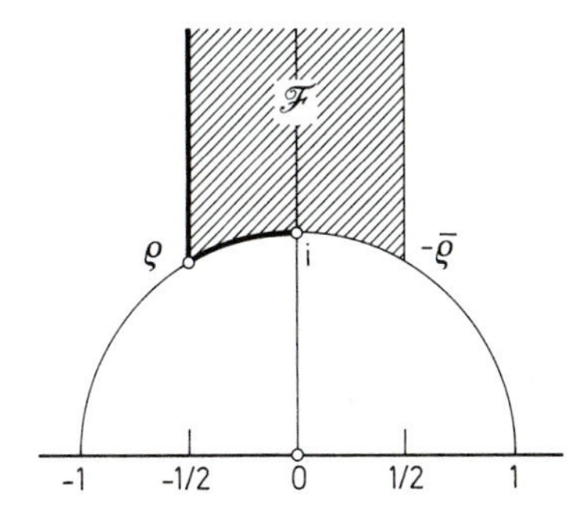

Fig. 1

A set $\mathscr{F}$ is called a *fundamental region* if it contains a fundamental set, and if

$$\tau \in \mathscr{F}, \qquad \bar{S}(\tau) \in \mathscr{F}, \qquad \bar{I} \neq \bar{S} \in \bar{\Gamma}$$

imply that τ is a boundary point of $\mathscr{F}$.

Now we prove the following important

Theorem 13. *The set*

$$\mathscr{F} := \{\tau \,|\, \tau \in \mathscr{H}, |\mathrm{Re}\,\tau| < \tfrac{1}{2}, |\tau| > 1\} \cup \{i\infty\} \cup \{\tau \,|\, \mathrm{Re}\,\tau = -\tfrac{1}{2}, |\tau| \geqq 1\}$$
$$\cup \{\tau \,|\, |\tau| = 1, -\tfrac{1}{2} \leqq \mathrm{Re}\,\tau \leqq 0\}$$

is a fundamental region of $\bar{\Gamma}$ *for* $\mathscr{H}^*$.

Proof. First we prove that each equivalence class under $\bar{\Gamma}$ has a representative in

$$\mathscr{F}^- := \{\tau \,|\, \tau \in \mathscr{H}, |\mathrm{Re}(\tau)| \leqq \tfrac{1}{2}, |\tau| \geqq 1\} \cup \{i\infty\}.$$

For this purpose we apply for $A = \begin{pmatrix} a & b \\ c & d \end{pmatrix} \in \Gamma$ and $\tau \in \mathscr{H}$ the formula

$$\mathrm{Im}\,\bar{A}(\tau) = \frac{\mathrm{Im}(\tau)}{|c\tau + d|^2}.$$

For a fixed τ with $\mathrm{Im}\,\tau > 0$, the points $c\tau + d, c, d \in \mathbb{Z}$, form a lattice in $\mathbb{C}$, thus $|c\tau + d|$ has a positive minimum. Now suppose that $A = \begin{pmatrix} a & b \\ c & d \end{pmatrix} \in \Gamma$ has been chosen so that $|c\tau + d|$ is minimal. Then $\tau_0 = \bar{A}(\tau)$ has the maximum imaginary part of elements of its class. This implies that

$|\tau_0| \geqq 1$ holds for every τ_0 with maximum imaginary part, because if $T = \begin{pmatrix} 0 & -1 \\ 1 & 0 \end{pmatrix}$ and $|\tau_0| < 1$, then

$$\operatorname{Im} \bar{T}(\tau_0) = \operatorname{Im}\left(\frac{\tau_0}{|\tau_0|^2}\right) > \operatorname{Im}(\tau_0).$$

For such a τ_0 we determine $k \in \mathbb{Z}$ so that $|\operatorname{Re} \bar{U}^k(\tau_0)| \leqq \frac{1}{2}$. Then $\bar{U}^k(\tau_0)$ also has maximal imaginary part (namely, the same as τ_0). It follows from the previous argument that $|\bar{U}^k(\tau_0)| \geqq 1$. Finally, recall that every rational point is equivalent to $i\infty$.

Next, we investigate which finite points in $\mathscr{F}^-$ are equivalent under the modular group. To this end, suppose $\tau = x + iy$, $\tau' = x' + iy'$ are given in $\mathscr{F}^-$, where $\tau' = \bar{A}(\tau)$, $\bar{A} \in \bar{\Gamma}$, and where $y \leqq y'$, say. Since $y' = y/|c\tau + d|^2$, it follows that

$$|c\tau + d|^2 \leqq 1. \tag{22}$$

We distinguish a number of cases. First suppose that $c = 0$; then $d = \pm 1$, say $d = 1$, then

$$A = \begin{pmatrix} 1 & b \\ 0 & 1 \end{pmatrix} = U^b.$$

The b's with $|b| > 1$ are omitted from consideration; the boundary arcs $(\rho, i\infty)$ and $(-\bar{\rho}, i\infty)$ are equivalent under $\bar{U}$:

$$\bar{U}(-\tfrac{1}{2} + iy) = \tfrac{1}{2} + iy. \tag{23}$$

Next assume $c \neq 0$. Then $|\tau + d/c| \leqq 1/|c|$, thus $|c| = 1$, say $c = 1$, so $|\tau + d| \leqq 1$. Hence $|d| \leqq 1$. Here we have used the fact that $\tau \in \mathscr{F}^-$. If $c = 1$ and $d = 0$, then

$$A = \begin{pmatrix} a & -1 \\ 1 & 0 \end{pmatrix} = U^a T \quad \text{with } |a| \leqq 1,$$

for if $|a| > 1$, $\bar{A}$ maps $\mathscr{F}^-$ outside of $\mathscr{F}^-$. If $a = 0$ then $A = T$, and

$$\bar{T}(-x + iy) = x + iy \quad \text{for } |\tau| = 1 \tag{24}$$

shows the equivalence of the boundary arcs (ρ, i) and $(-\bar{\rho}, i)$. If $a = 1$, then $\bar{A}(-\bar{\rho}) = -\bar{\rho}$, and if $a = -1$, then $\bar{A}(\rho) = \rho$.

If $c = 1$ and $d = \pm 1$, then

$$A = \begin{pmatrix} a & \pm a - 1 \\ 1 & \pm 1 \end{pmatrix} = U^a T U^{\pm 1} \quad \text{with } a = 0, \quad \text{or } a = d = \pm 1,$$

for in the other cases $\bar{A}$ again maps $\mathscr{F}^-$ outside of $\mathscr{F}^-$. Now,

$$\bar{A}(\rho)=\rho \quad \text{if } a=0, d=1; \qquad \bar{A}(-\bar{\rho})=-\bar{\rho} \quad \text{if } a=0, d=-1;$$
$$\bar{A}(\rho)=-\bar{\rho} \quad \text{if } a=d=1; \quad \text{and } \bar{A}(-\bar{\rho})=\rho \quad \text{if } a=d=-1.$$

The interior points of $\mathscr{F}$ are never equivalent to one another. Lastly, $\bar{A}(i\infty)=\tau_0\in\mathscr{F}^-$ if and only if $\tau_0=i\infty$ and $A=\pm U^k, k\in\mathbb{Z}$. Thus $\mathscr{F}$ is a fundamental region in addition to being a fundamental set. □

The boundary of $\mathscr{F}$ consists of pairs of '*conjugate*' sides, namely

$$(\rho,i\infty)\sim(-\bar{\rho},i\infty) \quad \text{and} \quad (\rho,i)\sim(-\bar{\rho},i),$$

which are mapped one onto the other by the generators $\bar{U}$ or $\bar{U}^{-1}$ and $\bar{T}$ *(boundary transformations)*, respectively.

$\mathscr{F}$ is a very special form of the fundamental region of $\bar{\Gamma}$; we shall have occasion to use yet other forms.

2. Consequences of the existence of a fundamental region. Let $\Gamma^*=[\bar{U},\bar{T}]\subset\bar{\Gamma}$ be the group generated by $\bar{U}$ and $\bar{T}$, then for every $\tau\in\mathscr{H}$ there is a point equivalent to τ under Γ^* with maximal imaginary part. This follows because the points $c\tau+d$ where c and d occur as coefficients in the transformations of Γ^* belong to the lattice mentioned above. This implies that each equivalence class under Γ^* has a representative in $\mathscr{F}^-$, since the transformations $\bar{U}$ and $\bar{T}$ already lie in Γ^*. Now suppose $\bar{A}\in\bar{\Gamma}$ is given and τ is an interior point of $\mathscr{F}$, then corresponding to $\bar{A}(\tau)$ there is a $\bar{B}\in\Gamma^*$ with $\overline{BA}(\tau)\in\mathscr{F}$. Now $\overline{BA}=\bar{I}$ so $\bar{A}=\bar{B}^{-1}\in\Gamma^*$ and thus $\Gamma^*=\bar{\Gamma}$. We have given a second proof for the assertion that $\bar{U}$ and $\bar{T}$ generate $\bar{\Gamma}$ which again leads to Theorem **10**.

Now we want to present the proofs which were deferred in § **3, 3** and **3, 4**.

All numbers

$$\frac{i-d}{c}\in\mathbb{Q}(i) \quad \textit{with } d^2\equiv-1 \bmod c, \quad c>0,$$

are equivalent under $\bar{\Gamma}$. *All* $\bar{A}\in\bar{\Gamma}$ *with* $\operatorname{tr}(A)=0$ *are conjugate in* $\bar{\Gamma}$ *to* $\bar{T}$.

Proof. Let τ be such a number. The representative of its class in $\mathscr{F}$ is likewise the fixed point of a transformation of $\bar{\Gamma}$ and it also is in $\mathbb{Q}(i)$. Clearly this must be i. If τ and $\bar{\tau}$ are the fixed points of $\bar{A}$, where $\operatorname{tr}(A)=0$, then there is an $\bar{S}\in\bar{\Gamma}$ with $\bar{S}(\tau)=i$ and so $\overline{SAS^{-1}}(i)=i$. Hence $\overline{SAS^{-1}}=\bar{T}$, for this is the only modular transformation fixing i. □

In an analogous way one proves:

All numbers

$$\frac{\rho-d}{c}\in\mathbb{Q}(\rho) \quad \textit{with } d^2+d+1\equiv 0 \bmod c, \quad c>0,$$

are equivalent under the modular group. All $\bar{A} \in \bar{\Gamma}$ *with* $|\mathrm{tr}(A)| = 1$ *are conjugate in* $\bar{\Gamma}$ *to* $\bar{R}$ *or* $\bar{R}^2$.

3. Tessellation of the upper half plane. An inhomogeneous modular transformation is uniquely determined by the image of a point $\tau \in \mathscr{F}$ distinct from ρ, i and $i\infty$. This result may also be formulated as follows: every point τ of the upper half plane $\mathscr{H}^*$, not equivalent to i, ρ or $i\infty$, occurs exactly once as the image under $\bar{\Gamma}$ of a point in $\mathscr{F}$. Clearly the points equivalent to i are covered twice, the points equivalent to ρ are covered three times, and the points equivalent to $i\infty$ are covered infinitely often. By a *modular triangle* $\mathscr{F}_S$ we understand the image of $\mathscr{F}$ under $\bar{S}$ (for arbitrary $S \in \Gamma$):

$$\mathscr{F}_S = \bar{S}(\mathscr{F}).$$

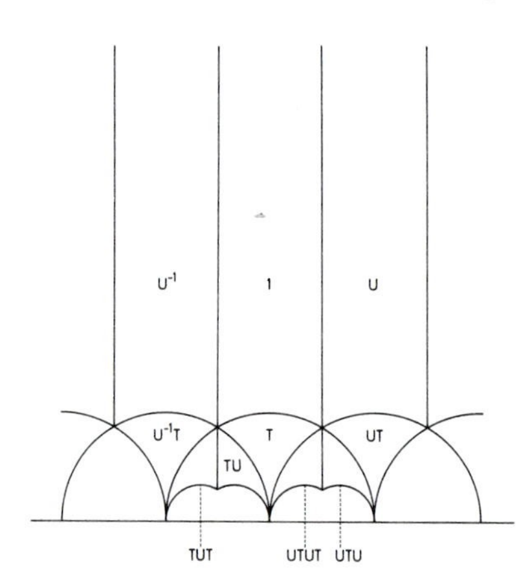

Fig. 2

Of course, $\mathscr{F}_S$ is a fundamental region. The set of modular triangles covers the upper half plane $\mathscr{H}^*$. Every modular triangle $\mathscr{F}_S$ is adjacent to three modular triangles, namely

$$\mathscr{F}_{SU} = \overline{SUS^{-1}}(\mathscr{F}_S), \quad \mathscr{F}_{SU^{-1}} = \overline{SU^{-1}S^{-1}}(\mathscr{F}_S), \quad \mathscr{F}_{ST} = \overline{STS^{-1}}(\mathscr{F}_S). \tag{25}$$

It is clear that six modular triangles abut at the point $\bar{S}(\rho)$. There the angle of each triangle is $\pi/3$. The transformations

$$\bar{S}\bar{R}\bar{S}^{-1}$$

are N.E.-rotations about $\bar{S}(\rho)$ through the angle $2\pi/3$.

Two triangles abut at the point $\bar{S}(i)$. The transformations

$$\bar{S}\bar{T}\bar{S}^{-1}$$

are N.E.-reflections in $\bar{S}(i)$.

At the rational point $\bar{S}(i\infty)$ countably many modular triangles

$$F_{SU^n}, \quad n \in \mathbb{Z},$$

come together. The adjacent sides of such a triangle form the angle 0 at $\bar{S}(i\infty)$ and make an angle of $\pi/2$ with the real axis. We say that $\bar{S}(i\infty)$ is a *cusp* of the modular triangle $\mathscr{F}_{SU^n}$ and say more generally that every rational point is a rational cusp. The transformations

$$\bar{S}\,\bar{U}^{\pm 1}\bar{S}^{-1}$$

have $\bar{S}(i\infty)$ as fixed point and map each modular triangle touching $\bar{S}(i\infty)$ on a neighboring triangle touching $\bar{S}(i\infty)$. The covering of $\mathscr{H}^*$ by the modular triangles $\mathscr{F}_S$ is called the *modular tessellation.* We infer

Theorem 14. *For a fixed $\tau_0 \in \mathscr{H}$, the set*

$$\{\bar{S}(\tau_0) \mid S \in \bar{\Gamma}\}$$

has no accumulation point in $\mathscr{H}$.

For otherwise, it has already an accumulation point in $\mathscr{F} \setminus \{i\infty\}$, which is obviously not the case.

Therefore we say: *$\bar{\Gamma}$ is discontinuous on $\mathscr{H}$.*

4. Neighborhoods. Canonical form of the fundamental region. For every $\tau \in \mathscr{H}^*$ different from ρ, i and $i\infty$ and their equivalent points, there is a neighborhood in $\mathbb{C}$ that does not contain a pair of equivalent points.

If τ_0 is equivalent to i under $\bar{\Gamma}$, then every neighborhood (in $\mathbb{C}$) contains pairs of equivalent points. If $\tau_0 = \bar{S}(i)$ say, and if the neighborhood $\mathscr{U}$ of τ_0 is the interior of a sufficiently small N.E.-circle about $S(i)$, then $\mathscr{U} \cap \mathscr{F}_S$ contains exactly one representative of each such pair of equivalent points of $\mathscr{U}$.

If τ_0 is equivalent to ρ under $\bar{\Gamma}$, then every neighborhood contains triples of equivalent points. If $\tau_0 = \bar{S}(\rho)$ say, and the neighborhood $\mathscr{U}$ of τ_0 is the interior of a sufficiently small N.E.-circle about $\bar{S}(\rho)$, then

$$\mathscr{U} \cap \{\mathscr{F}_S \cup \mathscr{F}_{ST}\}$$

contains exactly one representative of each triple of equivalent points in $\mathscr{U}$.

If τ_0 is equivalent to $i\infty$ under $\bar{\Gamma}$, say $\tau_0 = \bar{S}(i\infty)$, then each $\hat{\mathbb{C}}$-neighborhood of τ_0 contains infinitely many points equivalent to τ_0 as well as points of the lower half plane. We therefore introduce the sets

$$\mathscr{U} := \{\tau \mid \operatorname{Im} \bar{S}^{-1}(\tau) > 1\} \cup \{\tau_0\},$$

thus for $\tau_0 = i\infty$

$$\mathscr{U} = \{\tau \mid \operatorname{Im} \tau > 1\} \cup \{i\infty\}.$$

Every set

$$\mathcal{U} \cap \mathcal{F}_{SU^n}$$

contains exactly one of the points equivalent to $\tau \in \mathcal{U}$. These considerations will be continued in Chapter **IV**, **§6.** If one wants to obtain connected sets equivalent to the sets of points for i, ρ and $i\infty$ just described, then one makes use of the so-called *canonical form* of the fundamental region. For arbitrary $\alpha \in \mathcal{H}$ with $\operatorname{Re}\alpha = \frac{1}{2}$, $|\alpha| > 1$, let

$$\Delta(i, \rho, \overline{T}(\alpha)) \quad \text{and} \quad \Delta(i, -\overline{\rho}, \alpha)$$

denote the N.E.-triangles with vertices i, ρ and $\overline{T}(\alpha)$, and $i, -\overline{\rho}$ and α, respectively. Then

$$\Delta(i, \rho, \overline{T}(\alpha)) = \overline{T}(\Delta(i, -\overline{\rho}, \alpha)),$$

and hence

$$\sphericalangle(\rho, i, \overline{T}(\alpha)) = \sphericalangle(-\overline{\rho}, i, \alpha).$$

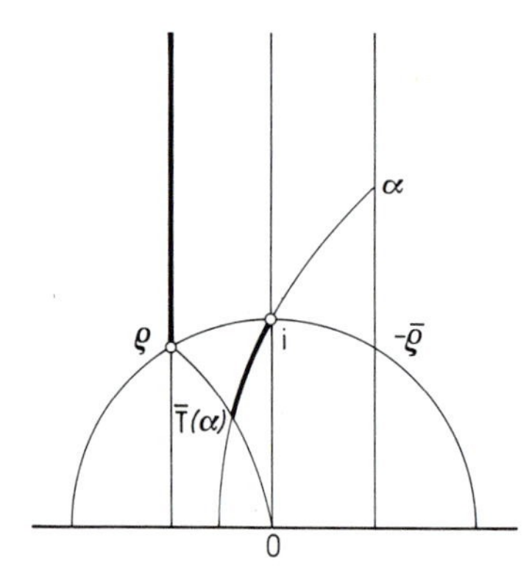

Fig. 3

Consequently,

$$\sphericalangle(\overline{T}(\alpha), i, \alpha) = \pi,$$

and the quadrilateral

$$(i\infty, \rho, \overline{T}(\alpha), \alpha)$$

is a fundamental region with the required properties.

For the concept of a canonical fundamental region see J. Lehner [1], p. 241 ff.

We have used a special case of a general method. From the classical fundamental region $\mathcal{F}$ of Theorem **13** we have obtained a new fundamental region by cutting off a finite number of pieces by means of N.E.-lines and circles and by replacing them with equivalent (under $\overline{\Gamma}$) pieces. A particularly interesting case is discussed in the next section.

5. The normal polygon. The geometric concepts which occur in what follows shall always be understood in the sense of N.E.-geometry. In particular the distance between two points of $\mathcal{H}$ is defined by equation (4). Let $\tau_0 \in \mathcal{H}$ be given such that τ_0 is not equivalent to i or ρ. It is our goal to obtain a fundamental region whose interior points τ have a smaller distance $\delta(\tau, \tau_0)$ from τ_0 than from any of the equivalent points $\bar{S}(\tau_0)$, $\pm I \neq S \in \Gamma$.

Without restriction assume that τ_0 belongs to the closure $\mathcal{F}^-$ of $\mathcal{F}$. First we treat the case

$$|\tau_0| > 1, \quad \operatorname{Re} \tau_0 > 0.$$

Note the symmetry in $\operatorname{Re} \tau = 0$ of the case

$$|\tau_0| > 1, \quad \operatorname{Re} \tau_0 < 0.$$

The two cases

$$|\tau_0| = 1 \quad \text{and} \quad \operatorname{Re} \tau_0 = 0$$

are special.

Let

$$\tau_0 \in \mathcal{F}^-, \quad |\tau_0| > 1, \quad \text{and } \operatorname{Re} \tau_0 > 0.$$

We set

$$R_1 := U\,T, \quad \text{so } \bar{R}_1(-\bar{\rho}) = -\bar{\rho}, \quad \bar{R}_1^3 = \bar{I}.$$

The bisector ω_1 of the angle $\sphericalangle(\tau_0, -\bar{\rho}, \bar{R}_1(\tau_0))$ is the set of points equidistant from τ_0 and $\bar{R}_1(\tau_0)$. Hence ω_1 divides $\mathcal{H}$ into two half planes, one of which contains all the points closer to τ_0 than to $\bar{R}_1(\tau_0)$. The corresponding result is true for the angle bisector ω_2 of $\sphericalangle(\tau_0, -\bar{\rho}, \bar{R}_1^2(\tau_0))$ and for the perpendicular bisector ℓ of the segment $[\tau_0, \bar{T}(\tau_0)]$. In this case the lines ℓ and ω_2 intersect in a point $\tau_1 \in \mathcal{H}$. Now we set

$$\tau_2 := \bar{T}(\tau_1) \in \ell, \quad \tau_3 := \bar{R}_1(\tau_1) \in \omega_1.$$

Since

$$-\bar{\rho} = \bar{U}(\rho) \quad \text{and} \quad \tau_3 = \bar{U}(\tau_2),$$

it follows that

$$\Delta(-\bar{\rho}_1\, \tau_3, i\infty) = \bar{U}(\Delta(\rho_1\, \tau_0, i\infty)).$$

In addition,

$$\Delta(i, \tau_1, -\bar{\rho}) = \bar{T}(\Delta(i, \tau_2, \rho)),$$

which implies that *the closed pentagon*

$$\mathcal{F}_0 := (i\infty, \tau_2, \tau_1, -\bar{\rho}, \tau_3) \tag{27}$$

is a fundamental region. Its interior contains no equivalent points. The pairs of conjugate sides and the associated boundary substitutions are indicated by

$$\begin{aligned}\bar{U}(i\infty,\tau_2) &= (i\infty,\tau_3),\\ \bar{T}(i,\tau_2) &= (i,\tau_1), \quad \text{and}\\ \bar{R}_1(-\bar{\rho},\tau_1) &= (-\bar{\rho},\tau_3).\end{aligned}$$

It follows from

$$\delta(\tau_0,\tau_1) = \delta(\tau_0,\tau_2) = \delta(\tau_0,\tau_3)$$

and

$$\operatorname{Im}(\tau_2) = \operatorname{Im}(\tau_3),$$

that

$$\operatorname{Re}(\tau_0) = \operatorname{Re}\left(\frac{\tau_2+\tau_3}{2}\right),$$

and consequently that

$$\mathscr{F}_0 = \{\tau \mid \delta(\tau,\tau_0) \leqq \delta(\tau,\bar{S}(\tau_0)), \quad S = U^{\pm 1}, R_1^{\pm 1}\} \cup \{i\infty\}.$$

If

$$|\tau_0| = 1, \qquad 0 < \operatorname{Re}\tau_0 < \tfrac{1}{2},$$

then

$$\tau_1 = 0, \qquad \tau_2 = \tau_3 = i\infty,$$

and in this case the fundamental region is the closed triangle

$$\mathscr{F}_0 := \Delta(i\infty, 0, -\bar{\rho})$$

with the two equivalent cusps $i\infty$ and 0. The conjugate sides and the associated boundary substitutions are indicated by

$$\bar{T}(i,i\infty) = (i,0), \qquad \bar{R}_1(-\bar{\rho},0) = (-\bar{\rho},i\infty).$$

Finally,

$$\mathscr{F}_0 = \{\tau \mid \delta(\tau,\tau_0) \leqq \delta(\tau,\bar{S}(\tau_0)), S = R_1^{\pm 1}, T\} \cup \{i\infty\}.$$

The case

$$|\tau_0| = 1, \qquad -\tfrac{1}{2} < \operatorname{Re}(\tau_0) < 0$$

is symmetric to this one.

For the case

$$\operatorname{Re}(\tau_0) = 0, \qquad \operatorname{Im}(\tau_0) > 1,$$

there is no point of intersection τ_1. We consider $\bar{U}^{\pm 1}(\tau_0)$ and $\bar{T}(\tau_0)$ and obtain as fundamental region

$$\mathscr{F}_0 := \Delta(i\infty, \rho, -\bar{\rho})$$

with the two equivalent vertices ρ and $-\bar{\rho}$. Moreover,

$$\mathscr{F}_0 = \{\tau \mid \delta(\tau, \tau_0) \leqq \delta(\tau, \bar{S}(\tau_0)), S = U^{\pm 1}, T\} \cup \{i\infty\}$$

is the closure of the classical fundamental region $\mathscr{F}$.

For each admissible τ_0 the fundamental region $\mathscr{F}_0$ is defined by certain inequalities. Moreover:

$$\delta(\tau, \tau_0) > \delta(\tau, \bar{S}(\tau_0)) \tag{28}$$

for no point $\tau \in \mathscr{F}_0$ and no $S \in \Gamma$.

If this inequality were to hold for a pair $\tau \in \mathscr{F}_0$ and $S_0 \in \Gamma$, then it would be satisfied by an interior point τ of $\mathscr{F}_0$ with the same S_0. Then, with this τ, $\mathscr{F}_0 \setminus \{\tau\}$ does not contain a fundamental set, while the subset

$$\mathscr{M} := \{\tau \mid \delta(\tau, \tau_0) \leqq \delta(\tau, \bar{S}(\tau_0)), \text{ for all } S \in \Gamma\} \cup \{i\infty\} \tag{29}$$

contains a fundamental set.

We prove this last claim as follows: for every $\tau \in \mathscr{H}$ there is an S_1 so that

$$\delta(\tau, \bar{S}_1(\tau_0)) \leqq \delta(\tau, \bar{S}(\tau_0)) \quad \text{for all } S \in \Gamma.$$

Therefore,

$$\delta(\bar{S}_1^{-1}(\tau), \tau_0) \leqq \delta(\bar{S}_1^{-1}(\tau), \bar{S}_1^{-1}\bar{S}(\tau_0)) \quad \text{for all } S \in \Gamma.$$

Consequently,

$$\bar{S}_1^{-1}(\tau) \in \mathscr{M}.$$

If τ_0 is an arbitrary point of $\mathscr{H}$, equivalent neither to i nor to ρ under $\bar{\Gamma}$, then by transformation we obtain

Theorem 15. *The set*

$$\mathscr{F}_0 := \{\tau \mid \delta(\tau, \tau_0) \leqq \delta(\tau, \bar{S}(\tau_0)) \text{ for all } S \in \Gamma\} \cup \{(i\infty)'\}, \tag{30}$$

where τ_0 and $(i\infty)' \in \mathbb{Q}$ lie in the same modular triangle, is a fundamental region for $\bar{\Gamma}$.

$\mathscr{F}_0$ is called a *normal polygon.*

For the concept of a normal polygon see J. Lehner [1], p. 146ff.

Chapter II. The Modular Functions of Level One

§ 1. Definition and Properties of Modular Functions

We begin by agreeing that henceforth we shall drop the bar when we denote the image of z by $\bar{A} \in \bar{\Gamma}$, thus we shall write $A(z)$ in place of $\bar{A}(z)$. Furthermore, in this section we define and discuss modular functions without regard to the question of their existence—this question will be treated in **§ 3**.

1. Definition. Suppose, as before, that

$$\mathscr{H} = \{\tau \in \mathbb{C} \mid \operatorname{Im} \tau > 0\}$$

denotes the upper half plane,

$$\mathscr{H}^* = \mathscr{H} \cup \mathbb{Q} \cup \{i\infty\}$$

the extended upper half plane, and $\hat{\mathbb{C}}$ the compactified complex plane.

A map $f: \mathscr{H}^* \to \hat{\mathbb{C}}$ is called a *modular function* (of level one) if

a) f is meromorphic in $\mathscr{H}$, i.e. holomorphic except for poles,

b) $f(A(\tau)) = f(\tau)$ for all $A \in \Gamma$ and all $\tau \in \mathscr{H}^*$,

c) there is an $a > 0$, such that for $\operatorname{Im} \tau > a$ $f(\tau)$ has an expansion of the form

$$f(\tau) = \sum_{\nu \geq h} b_\nu e^{2\pi i \nu \tau}, \quad h \in \mathbb{Z}, \quad b_h \neq 0;$$

and so $f(i\infty) = 0$ if $h > 0$, $f(i\infty) = b_0$ if $h = 0$ and $f(i\infty) = \infty$ if $h < 0$.

2. Power series expansions. From the expansion c) at $i\infty$ we can obtain the expansion of the modular function f about any rational point r. We set $r = -d/c$ with $c, d \in \mathbb{Z}$, $(c, d) = 1$ and determine $a, b \in \mathbb{Z}$ so that $ad - bc = 1$. Then $S(r) = i\infty$ for the modular transformation $S = \begin{pmatrix} a & b \\ c & d \end{pmatrix}$. Due to the invariance of f under $S \in \Gamma$, we obtain for $\tau \in \mathscr{H}$

and $\operatorname{Im}\frac{a\tau+b}{c\tau+d}$ sufficiently large, the expansion

$$f(\tau)=f(S(\tau)) = \sum_{\nu \geqq h} b_\nu \left(e^{2\pi i \frac{a\tau+b}{c\tau+d}}\right)^\nu, \qquad h\in\mathbb{Z}. \tag{1}$$

One easily verifies that $e^{2\pi i \frac{a\tau+b}{c\tau+d}}$ is independent of a, b. Since modular functions assume the same value at the rational points and since $\mathbb{Q}$ is dense in $\mathbb{R}$, it is not possible to continue analytically a non-constant modular function to the lower half plane.

We take the expansion of the modular function f about the point $\rho=e^{2\pi i/3}$ in the form

$$f(\tau) = \sum_{\nu \geqq m} c_\nu \left(\frac{\tau-\rho}{\tau-\bar{\rho}}\right)^\nu.$$

The transformation $\bar{R}$ with $R=TU$ has the fixed points $\rho, \bar{\rho}$ and the normal form

$$\frac{R(\tau)-\rho}{R(\tau)-\bar{\rho}} = \bar{\rho}\,\frac{\tau-\rho}{\tau-\bar{\rho}}.$$

Therefore it follows that

$$f(R(\tau)) = \sum_{\nu \geqq m} c_\nu \bar{\rho}^\nu \left(\frac{\tau-\rho}{\tau-\bar{\rho}}\right)^\nu,$$

and the condition $f(R(\tau))=f(\tau)$ implies that $c_\nu=0$ for $\nu \not\equiv 0 \bmod 3$. Consequently, the expansion at ρ assumes the form

$$f(\tau) = \sum_{\nu \geqq m} c_\nu \left(\frac{\tau-\rho}{\tau-\bar{\rho}}\right)^{3\nu}, \qquad m\in\mathbb{Z}, \tag{2}$$

and the expansion about points equivalent to ρ has an analogous form. Correspondingly, the invariance of a modular function f under $\bar{T}\in\bar{\Gamma}$ leads to the expansion about the point i:

$$f(\tau) = \sum_{\nu \geqq n} d_\nu \left(\frac{\tau-i}{\tau+i}\right)^{2\nu}, \qquad n\in\mathbb{Z}. \tag{3}$$

Also, for points equivalent to i there is an analogous expansion. For points $\tau_0\in\mathscr{H}$, which are not equivalent to ρ or i, we have an expansion of the form

$$f(\tau) = \sum_{\nu \geqq k} a_\nu \left(\frac{\tau-\tau_0}{\tau-\bar{\tau}_0}\right)^\nu, \qquad k\in\mathbb{Z}. \tag{4}$$

The functions

$$t_{i\infty}:=e^{2\pi i\tau}, \quad t_\rho:=\left(\frac{\tau-\rho}{\tau-\bar{\rho}}\right)^3, \quad t_i:=\left(\frac{\tau-i}{\tau+i}\right)^2, \quad \text{and } t:=\frac{\tau-\tau_0}{\tau-\bar{\tau}_0},$$

which occur in these expansions, are called the *local uniformizing variables* at $i\infty$, ρ, i and τ_0, respectively. They are defined analogously at equivalent points. These functions map the sets introduced in **I, § 4, 6,** which have no equivalent points, onto neighborhoods (in $\mathbb{C}$) of 0.

3. The order at c-points. We write the above expansions uniformly:

$$\sum_{\nu \geqq \tilde{k}} a_\nu t_{\tilde{\tau}}^\nu, \qquad a_{\tilde{k}} \neq 0,$$

where either $\tilde{\tau}$ is one of the points $i\infty, \rho, i, \tau_0$ or is equivalent to one of them. For $c \in \hat{\mathbb{C}}$ the *c-point order* $n(c, \tilde{\tau})$ of f at $\tau \in \mathscr{H}^*$ is defined as follows:

$$n(\infty, \tilde{\tau}) = \max(0, -\tilde{k}); \qquad n(0, \tilde{\tau}) = \max(\tilde{k}, 0). \tag{5}$$

If $c \neq 0, \infty$ one considers the above expansion for $f(\tau) - c$. Then $n(c, \tilde{\tau})$ of f at $\tilde{\tau}$ is defined to be the order $n(0, \tilde{\tau})$ of the 0 of $f(\tau) - c$ at $\tilde{\tau}$.

The expansion (1) of a modular function f at a rational point r shows that the c-point order is the same at all points equivalent to $i\infty$.

If $\tau_0 \in \mathscr{H}$ is not equivalent to ρ or i, then our concept of c-point order coincides with the usual idea of order as measured in the variable $\tau - \tau_0$; at the points equivalent to ρ it differs by a factor of 3, and at points equivalent to i it differs by a factor of 2. To prove the invariance of the c-point order under the transition to equivalent points $\tau \in \mathscr{H}$, it is therefore sufficient to consider the order at the c-point as measured in the variable $\tau - \tau_0$ for arbitrary $\tau_0 \in \mathscr{H}$.

These orders agree, for let

$$f(\tau) = \sum_{\nu \geqq k} a_\nu \left(\frac{\tau - \tau_0}{\tau - \bar{\tau}_0} \right)^\nu, \qquad f(\tau') = \sum_{\nu \geqq k'} a'_\nu \left(\frac{\tau' - \tau'_0}{\tau' - \bar{\tau}'_0} \right)^\nu \tag{6}$$

be the expansions of f at τ_0 and τ'_0. If we set

$$S = \begin{pmatrix} a & b \\ c & d \end{pmatrix} \in \Gamma, \qquad \tau' = S(\tau), \qquad \tau'_0 = S(\tau_0),$$

then

$$\tau' - \tau'_0 = \frac{\tau - \tau_0}{(c\tau + d)(c\tau_0 + d)}, \quad \text{and} \quad \frac{\tau' - \tau'_0}{\tau' - \bar{\tau}'_0} = \frac{c\bar{\tau}_0 + d}{c\tau_0 + d} \cdot \frac{\tau - \tau_0}{\tau - \bar{\tau}_0}. \tag{7}$$

The invariance of f implies the equations

$$a_\nu = a'_\nu \left(\frac{c\bar{\tau}_0 + d}{c\tau_0 + d} \right)^\nu, \quad \text{for } \nu \geqq k = k'. \tag{8}$$

This proves

Theorem 1. *A modular function has the same c-point order at equivalent points.*

4. The number of c-points. In what follows f is a non-constant modular function. For such f we define the *c-order* by

$$N(c) := \sum_{\tau \in \mathscr{F}} n(c, \tau), \tag{9}$$

and we remark that this sum is finite since the c-points in $\mathscr{F}$ of f do not have a limit point. Our next goal is the proof of

Theorem 2. *$N(c)$ is the same for all $c \in \hat{\mathbb{C}}$, i.e. a modular function takes each value c equally often in $\mathscr{F}$ provided one measures the multiplicity in the local variable.*

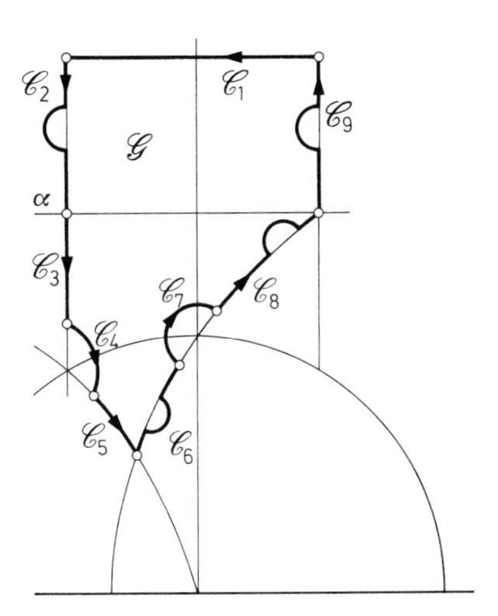

Fig. 4

Proof. We consider a fixed $c \neq \infty$ and prove that $N(c) = N(\infty)$. We cut the canonical form of the fundamental region along a line parallel to the real axis in such a way that the finite portion contains all c-points and poles except, perhaps, $i\infty$. Likewise, we remove by cuts along N.E.-circular arcs about i and ρ such regions which contain no c-points or poles with the possible exception of those at i and ρ. Furthermore, c-points or poles on the boundary are avoided as indicated in the figure. We obtain a region $\mathscr{G}$ whose boundary $\mathscr{C}$ contains all c-points and poles in its interior except those at ρ, i or $i\infty$. We decompose $\mathscr{C}$ as the sum of the subpaths $\mathscr{C}_1$ through $\mathscr{C}_9$; the pairs of subpaths $\mathscr{C}_2$ and $\mathscr{C}_9$, $\mathscr{C}_3$ and $\mathscr{C}_5$, and $\mathscr{C}_6$ and $\mathscr{C}_8$ are equivalent under Γ. Defining

$$N_0(c) = \sum_{\tau \in \mathscr{G}} n(c, \tau), \qquad N_0(\infty) = \sum_{\tau \in \mathscr{G}} n(\infty, \tau),$$

we obtain

$$N_0(c)-N_0(\infty)=\frac{1}{2\pi i}\oint_{\mathscr{C}}\frac{f'(\zeta)}{f(\zeta)-c}\,d\zeta\,.$$

We evaluate this integral by writing it as a sum of integrals over the subpaths $\mathscr{C}_1$ through $\mathscr{C}_9$. Since the pairs of equivalent subpaths given above are traversed in opposite directions, the sum of a pair of corresponding integrals is 0. Indeed, if $S\in\Gamma$ and $\mathscr{W}\subset\mathscr{H}$ is an arbitrary path, we have

$$\int_{\mathscr{W}}\frac{f'(\zeta)}{f(\zeta)-c}\,d\zeta=\int_{S(\mathscr{W})}\frac{1}{f(S^{-1}(\tilde\zeta))-c}\left(\frac{df}{d\tilde\zeta}\cdot\frac{d\tilde\zeta}{d\zeta}\right)\left(\frac{d\zeta}{d\tilde\zeta}\right)d\tilde\zeta=\int_{S(\mathscr{W})}\frac{f'(\tilde\zeta)}{f(\tilde\zeta)-c}\,d\tilde\zeta. \tag{10}$$

Hence only the integrals along $\mathscr{C}_1, \mathscr{C}_4$ and $\mathscr{C}_7$ need to be calculated. To compute the integral along $\mathscr{C}_1$ we introduce $t_{i\infty}=e^{2\pi i\tau}$ and for sufficiently large a we obtain

$$\begin{aligned}\frac{1}{2\pi i}\int_{\mathscr{C}_1}\frac{f'(\zeta)}{f(\zeta)-c}\,d\zeta&=-\frac{1}{2\pi i}\oint_{|t_{i\infty}|=e^{-2\pi a}}\frac{f'(t_{i\infty})}{f(t_{i\infty})-c}\,dt_{i\infty}=-h\\&=n(\infty,i\infty)-n(c,i\infty).\end{aligned}$$

For the definition of h, and m and n below see (1), (2), (3). Because of (10), the integral along $\mathscr{C}_4$ coincides with the integral along $R(\mathscr{C}_4)$ and likewise with the integral along $R^2(\mathscr{C}_4)$. Moreover,

$$\frac{1}{2\pi i}\int_{\mathscr{C}_4+R(\mathscr{C}_4)+R^2(\mathscr{C}_4)}\frac{f'(\zeta)}{f(\zeta)-c}\,d\zeta=-3m,$$

hence

$$\frac{1}{2\pi i}\int_{\mathscr{C}_4}\frac{f'(\zeta)}{f(\zeta)-c}\,d\zeta=-m=n(\infty,\rho)-n(c,\rho)\,.$$

Similarly,

$$\frac{1}{2\pi i}\int_{\mathscr{C}_7}\frac{f'(\zeta)}{f(\zeta)-c}\,d\zeta=-n=n(\infty,i)-n(c,i)\,.$$

Since

$$N(c)=N_0(c)+n(c,i\infty)+n(c,\rho)+n(c,i)$$

is also valid with $c=\infty$, it therefore follows that $N(c)=N(\infty)$. □

Of course, equation (10) implies the invariance of the c-point order at equivalent points. As a consequence of this theorem we define the *order* (or *valence*) *of* the *modular function* f to be the constant $N(c)$, $c\in\hat{\mathbb{C}}$.

For complex $\alpha, \beta, \gamma, \delta$ with $\alpha\delta - \beta\gamma \neq 0$ it is easy to show that

$$\frac{\alpha f + \beta}{\gamma f + \delta}$$

is a modular function of the same order as f. Two modular functions of order 1 are related by such a linear fractional transformation. Obviously, sums, differences, products, and quotients of modular functions with non-0 denominator are likewise invariant under the $\bar{A} \in \bar{\Gamma}$; their power series can be immediately written down. Thus follows

Theorem 3. *The modular functions form a field.*

5. Normalization. To aid in the construction of modular functions we prove the following useful theorem.

Theorem 4. *If modular functions of order 1 exist, then there is a modular function of order 1 which maps the interior of*

$$\mathscr{F}^* := \{\tau \mid \tau \in \mathscr{H}, -\tfrac{1}{2} \leqq \operatorname{Re}\tau \leqq 0, |\tau| \geqq 1\} \cup \{i\infty\}$$

one-to-one onto the upper half plane and maps the boundary of $\mathscr{F}^*$ onto the real axis.

Proof. First of all, we show that if $f(\tau)$ is a modular function, then so is $\overline{f(-\bar{\tau})}$, and that both have the same order. If, namely,

$$f(\tau) = \sum_{\nu \geqq k} a_\nu (\tau - \tau_0)^\nu$$

is the expansion of f at $\tau_0 \in \mathscr{H}$, then

$$\overline{f(-\bar{\tau})} = \sum_{\nu \geqq k} \bar{a}_\nu (-\tau - \bar{\tau}_0)^\nu = \sum_{\nu \geqq k} (-1)^\nu \bar{a}_\nu (\tau - (-\bar{\tau}_0))^\nu$$

is the expansion at $-\bar{\tau}_0$. Thus $\overline{f(-\bar{\tau})}$ is meromorphic in $\mathscr{H}$ and its order at $-\bar{\tau}_0$ is the same as the order of f at τ_0. The analogous result holds for the expansions at $i\infty$. The invariance under $\begin{pmatrix} a & b \\ c & d \end{pmatrix} \in \Gamma$ may be shown directly:

$$\overline{f\left(-\frac{a\bar{\tau}+b}{c\bar{\tau}+d}\right)} = \overline{f\left(\frac{a(-\bar{\tau})-b}{-c(-\bar{\tau})+d}\right)} = \overline{f(-\bar{\tau})},$$

since $\begin{pmatrix} a & -b \\ -c & d \end{pmatrix}$ is also in Γ.

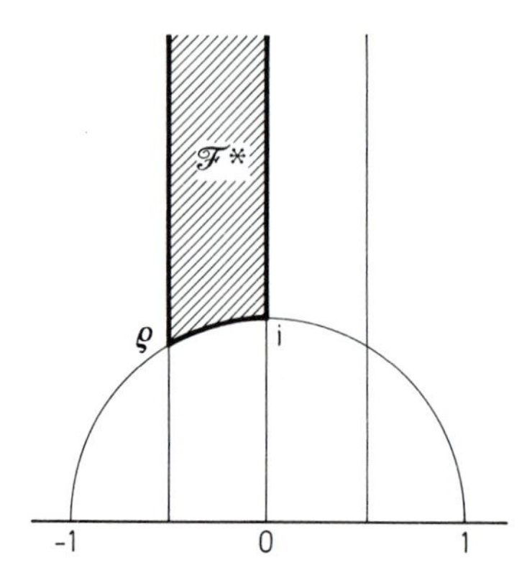

Fig. 5

Now suppose f is a modular function of order 1. Since it takes each value exactly once in $\mathscr{F}$, it is a bijection of $\mathscr{F}$ onto $\hat{\mathbb{C}}$ and has exactly one pole of order 1. We suppose that this pole is at $i\infty$ (otherwise we apply a suitable linear transformation). If necessary we multiply by a constant to obtain a real residue at $i\infty$ in order that the modular function

$$f(\tau)-\overline{f(-\bar{\tau})}$$

be holomorphic at $i\infty$ and thus be holomorphic everywhere. Therefore

$$f(\tau)-\overline{f(-\bar{\tau})}=c.$$

Setting $\tau=iy$ we see that $c=ir$, $r\in\mathbb{R}$. If we replace f by $f-ir/2$ and again call this function f, we have

$$f(\tau)=\overline{f(-\bar{\tau})}.$$

$$\text{If}\quad \tau=iy,\quad \text{then } \bar{\tau}=-\tau,\quad \text{and so } f(\tau)=\overline{f(\tau)}\in\mathbb{R};$$

$$\text{if } |\tau|=1,\quad \text{then } \bar{\tau}=\frac{1}{\tau}\quad \text{and } f(\tau)=\overline{f\left(\frac{-1}{\tau}\right)}=\overline{f(\tau)}\in\mathbb{R};$$

$$\text{if } \tau=-\tfrac{1}{2}+iy,\quad \text{then } \bar{\tau}=-\tau-1,\quad \text{and so } f(\tau)=\overline{f(\tau+1)}=\overline{f(\tau)}\in\mathbb{R}.$$

Thus f maps the boundary of $\mathscr{F}^*$ into the real axis. Suppose now that τ is an interior point of $\mathscr{F}^*$ and assume that $f(\tau)\in\mathbb{R}$, so $f(\tau)=\overline{f(\tau)}$. It then follows that $f(\tau)=f(-\bar{\tau})$ and, since f is bijective, that τ is equivalent to $-\bar{\tau}$. Furthermore, $-\bar{\tau}$ is also an inner point of $\mathscr{F}$. This implies $\tau=-\bar{\tau}$ or $\operatorname{Re}\tau=0$. This contradicts the fact that τ was in the interior of $\mathscr{F}^*$. Thus the assumption $f(\tau)\in\mathbb{R}$ is false. Since every point of the real line is the image of a point of $\mathscr{F}$, it follows that $\mathbb{R}$ is the image of the boundary of $\mathscr{F}^*$. □

By composition with a linear transformation we now obtain an f taking $\rho \mapsto 0$, $i \mapsto 1$ and $i\infty \mapsto \infty$. Obviously, f is uniquely determined by this normalization and the interior of $\mathscr{F}^*$ is mapped onto $\mathscr{H}$.

§ 2. Extension of the Modular Group by Reflections

In this section we denote the linear transformations $\bar{U}, \bar{T}, \bar{A}, \ldots$ by $U, T, A, \ldots$ omitting the bar.

1. Reflections in $\mathbb{C}$. Let

$$O_1: \hat{\mathbb{C}} \to \hat{\mathbb{C}}, \quad z \mapsto -\bar{z}.$$

Definition. A map

$$S: \hat{\mathbb{C}} \to \hat{\mathbb{C}}, \quad z \mapsto \frac{\alpha \bar{z} + \beta}{\gamma \bar{z} + \delta}$$

is called a *reflection* if there exists an inhomogeneous linear transformation L such that S can be written in the form

$$S = L O_1 L^{-1}.$$

Since $O_1^2 = I$ we also have $S^2 = I$. Furthermore

$$S L(iy) = L O_1(iy) = L(iy), \quad y = \operatorname{Im}(z),$$

i.e. the circular image of the imaginary axis under L is left pointwise fixed by S. We call S a reflection in this circle. In particular O_1 is the reflection in the imaginary axis σ_1.

2. Extension of $\bar{\Gamma}$ by reflections. For the transformations O_1, T, U one has

$$U O_1 = O_1 U^{-1}, \quad U^{-1} O_1 = O_1 U, \quad T O_1 = O_1 T. \tag{11}$$

From these we obtain

Theorem 5. *The set $\Gamma^* = \bar{\Gamma} \cup O_1 \bar{\Gamma}$ of transformations is a group under the usual definition of product of transformations. The group Γ^* is generated by the transformations O_1, T and U.*

If one chooses O_1, T and R as generators, then

$$O_1^2 = T^2 = R^3 = I, \quad T O_1 = O_1 T, \quad R O_1 = O_1 R^{-1}$$

are defining relations for Γ^*. These are obtained in a fashion similar to the derivation for $\bar{\Gamma}$. For what follows it is important to have Γ^*

generated by three reflections:

$$O_1; \quad O_2 := O_1 U, \quad z \mapsto -\bar{z} - 1; \quad O_3 := O_1 T, \quad z \mapsto \frac{1}{\bar{z}}.$$

Here

$$O_2 = L O_1 L^{-1} \quad \text{with } L: z \mapsto z - \tfrac{1}{2},$$

is a reflection in $\sigma_2 := \{z \mid \operatorname{Re} z = -\frac{1}{2}\}$, and

$$O_3 = L O_1 L^{-1} \quad \text{with } L: z \mapsto \frac{1+z}{1-z},$$

is the reflection in the unit circle σ_3.

3. Fundamental region for Γ^*. We carry over the definition of fundamental region from I **§4, 3** and prove

Theorem 6. *The set*

$$\mathscr{F}^* := \{\tau \mid \tau \in \mathscr{H}, -\tfrac{1}{2} \leqq \operatorname{Re} \tau \leqq 0, |\tau| \geqq 1\} \cup \{i\infty\}$$

is a fundamental region for Γ^*

Proof. Every point $\tau \in \mathscr{H}^*$ can be mapped into $\mathscr{F}$ by $\bar{\Gamma}$. If the real part of this image is positive, we apply O_1. Now assume that two points $\tau_1, \tau_2 \in \mathscr{F}^*$ are equivalent under Γ^*. Then $\tau_2 = A(\tau_1)$ or $\tau_2 = O_1 A(\tau_1)$ with $A \in \bar{\Gamma}$. In the first case, if $A \neq I$, it follows from earlier results that either: $\tau_1 = \tau_2 = i$ and $A = T$; $\tau_1 = \tau_2 = \rho$ and $A = R$ or R^2; or $\tau_1 = \tau_2 = i\infty$ and $A = U^n$. In the second case, $-\bar{\tau}_2 = A(\tau)$ and so as earlier either: $\operatorname{Re}(\tau_1) = 0$ and $A = I$, thus $O_1 A = O_1$; $\operatorname{Re}(\tau_1) = -\frac{1}{2}$ and $A = U$, thus $O_1 A = O_2$; $|\tau| = 1$ and $A = T$, thus $O_1 A = O_3$; or finally, $\tau_1 = \rho$ and $A = UR$ or UR^2. In any case $\tau_1 = \tau_2$. Hence, all boundary points are non-trivially self-equivalent. Other equivalences do not exist. $\mathscr{F}^*$ is a fundamental set and is obviously a fundamental region. □

4. Tessellation of the upper half plane.

Theorem 7. *The upper half plane $\mathscr{H}^*$ can be covered by repeated reflections as follows: reflect the fundamental region $\mathscr{F}^*$ in the three circles σ_1, σ_2, and σ_3; reflect the images so obtained again in these circles, and so on.*

Proof. We see that

$$\mathscr{H}^* = \bigcup_{A \in \bar{\Gamma}} A(\mathscr{F}^* \cup O_1 \mathscr{F}^*) = \bigcup_{M^* \in \Gamma^*} M^*(\mathscr{F}^*).$$

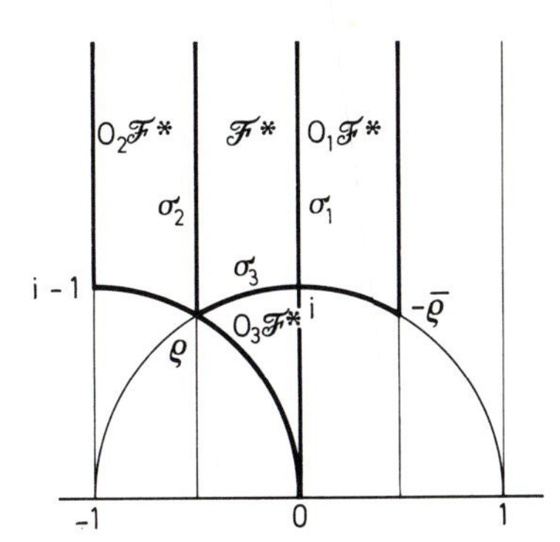

Fig. 6

Thus $\mathcal{H}^*$ is covered by images of $\mathcal{F}^*$ under Γ^*. One obtains $M^*(\mathcal{F}^*)$ through a series of reflections since M^* is a product of the reflections O_1, O_2 and O_3. The inner points of all $M^*(\mathcal{F}^*)$ are covered once. The points of $M^*(\sigma_\nu)$, $\nu=1,2,3$, with the exception of their end points are covered twice; the points equivalent to i under Γ^* are covered four times; the points equivalent to ρ are covered six times; finally, the points equivalent to $i\infty$ are covered infinitely often. □

Incidentally, one can obtain the adjacent images $M^* O_\nu(\mathcal{F}^*)$, $\nu=1,2,3$ of $M^*(\mathcal{F}^*)$ by means of the reflections $M^* O_\nu M^{*-1}$ in their common sides. We remark that $M^* \in \Gamma^*$ lies in $\bar{\Gamma}$ if and only if M^* can be written as the product of an even number of the reflections O_1, O_2 and O_3.

§ 3. Existence of Modular Functions. The Absolute Modular Invariant J

1. Construction of the modular function J. By the Riemann mapping theorem there is a function f holomorphic in $\mathcal{F}^{*0}$, the interior or $\mathcal{F}^*$, which maps $\mathcal{F}^{*0}$ $1-1$ onto $\mathcal{H}$. This map extends to a homeomorphism of $\mathcal{F}^*$ onto $\mathcal{H}^-$ in $\hat{\mathbb{C}}$ (cf. W. Rudin [1] p. 281 or H. Behnke and F. Sommer [1], p. 357ff.).

We normalize the extension by $\rho\mapsto 0$, $i\mapsto 1$, and $i\infty\mapsto\infty$. We continue it by reflections:

$$\begin{aligned} \overline{f(-\bar{\tau})} &= f(\tau) \quad \text{in } \sigma_1 \text{ into } O_1\mathcal{F}^*, \\ \overline{f(-\bar{\tau}-1)} &= f(\tau) \quad \text{in } \sigma_2 \text{ into } O_2\mathcal{F}^*, \\ \overline{f(1/\bar{\tau})} &= f(\tau) \quad \text{in } \sigma_3 \text{ into } O_3\mathcal{F}^*. \end{aligned} \tag{12}$$

It follows from **§ 2, 4** that the unlimited repetition of such reflections yields a single valued function J—J is independent of the various modes of the reflections. The Schwarz reflection principle implies that J is holomorphic in $\mathcal{H}$ with the possible exception of those points equivalent to i and ρ. However, J is also holomorphic at these points, for the points equivalent to i and ρ are isolated and are points of continuity for J. By construction

$$J(\tau+1)=J(U(\tau))=J(O_1\,O_2(\tau))=\overline{J(O_2(\tau))}=\overline{\overline{J(\tau)}}=J(\tau),$$

and as a periodic function it has an expansion

$$J(\tau)=\sum_{\nu\in\mathbb{Z}} b_\nu\, e^{2\pi i\nu\tau}.$$

This converges for $\mathrm{Im}(\tau)>0$, because J is holomorphic for $\mathrm{Im}(\tau)>0$. J takes each value in $\mathcal{F}$ exactly once and therefore has a pole of order 1 in $\mathcal{F}$. Thus $b_\nu=0$ for $\nu<-1$ and $b_{-1}\neq 0$. The invariance of J under modular transformations is implied by an argument similar to that which proved periodicity: J is invariant for products of pairs of reflections and only the products of an even number of O_1, O_2 and O_3 lie in $\bar{\Gamma}$. This implies the form (1) of the expansion at the cusps. In conclusion we have:

J is a modular function of order one. We call J the *absolute modular invariant.*

Since $\overline{J(-\bar{\tau})}=J(\tau)$, the coefficients b_ν are real.

2. Main theorem. The special importance of the function J is shown in

Theorem 8. *The field of modular functions is the rational function field obtained by adjoining J to $\mathbb{C}$: $\mathbb{C}(J)$.*

Proof. Suppose h is a modular function with $\bar{\Gamma}$-inequivalent zeros α_μ of order r_μ, $(\mu=1,\dots,m)$ and poles β_ν of order s_ν, $(\nu=1,\dots,n)$. Then $r_1+\cdots+r_m=s_1+\cdots+s_m$. Set

$$Q(\tau)=\frac{\prod_{\mu=1}^{m}(J(\tau)-J(\alpha_\mu))^{r_\mu}}{\prod_{\nu=1}^{n}(J(\tau)-J(\beta_\nu))^{s_\nu}} \tag{13}$$

where $J(\tau)-J(\alpha_\mu)$ or $J(\tau)-J(\beta_\nu)$ is to be replaced by 1 in case α_μ or β_ν equals $i\infty$. Q is a modular function which has the same zeros and poles with the same multiplicities as h. Consequently $\frac{Q}{h}$ is everywhere regular and is therefore a constant. □

3. The Riemann surface of the inverse of J. The function J maps the fundamental region $\mathcal{F}^*$ of the group Γ^* one-to-one onto $\mathcal{H}\cup\mathbb{R}\cup\{\infty\}$ in the closed w-plane. Likewise J maps every image $A(\mathcal{F}^*)$ with $A\in\bar{\Gamma}$

one-to-one onto $\mathscr{H} \cup \mathbb{R} \cup \{\infty\}$. However, J maps the images $O_1 A(\mathscr{F}^*)$, $A \in \bar{\Gamma}$, one-to-one onto $-\mathscr{H} \cup \mathbb{R} \cup \{\infty\}$, where $-\mathscr{H}$ denotes the lower half-plane. The Riemann surface of the inverse of J—the images of the point $i\infty$ and of the rational points adjoined—therefore consists of in-

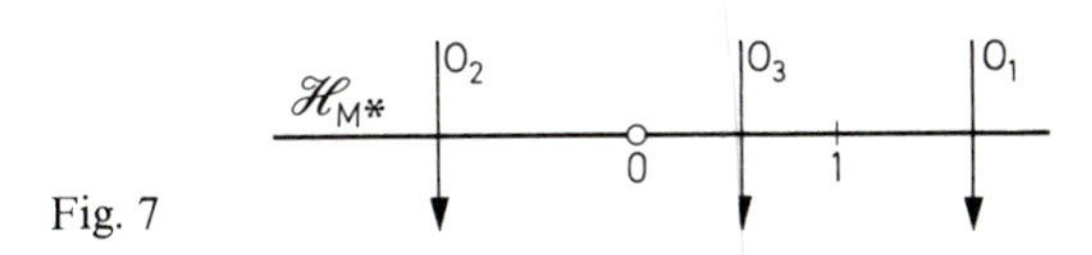

Fig. 7

finitely many copies of the upper w-half-plane and of infinitely many copies of the lower w-half-plane together with their boundaries. The copies of the half-planes may be indexed by M^* in Γ^*, where $M^*(\mathscr{F}^*)$ denotes the corresponding pre-image under J. Thus the segment $[0,1]$ of the real axis in $\mathscr{H}_{M^*}$, the image of σ_3 under M^*, will be identified with the segment $[0,1]$ in $\mathscr{H}_{M^*O_3}$, which is the image of σ_3 under $M^* O_3$. Similarly, we identify $M^*(\sigma_1)$ with $M^* O_1(\sigma_1)$ and $M^*(\sigma_2)$ with $M^* O_2(\sigma_2)$. The image of $\hat{i}$, a point equivalent to i, is a branch point of order 1 over $w=1$. This follows from the inversion of the expansion

$$w = J(\tau) = 1 + c_1 \left(\frac{\tau - \hat{i}}{\tau + \hat{i}} \right)^2 + \cdots \quad \text{with } c_1 \neq 0 .$$

Analogously, the image of a point equivalent to ρ is a branch point of order 2 over $w=0$. For the point $i\infty$ we consider the inversion of the expansion

$$J(\tau) = w = \frac{b_{-1}}{q} + b_0 + b_1 q + \cdots, \qquad q = e^{2\pi i \tau}, \quad b_1 \neq 0,$$

which for sufficiently large $|w|$ has the form

$$q = \frac{1}{w} \left(d_1 + \frac{d_2}{w} + \cdots \right), \qquad d_1 \neq 0.$$

Using the principal branch of the logarithm, we obtain

$$\tau = -\frac{1}{2\pi i} \log w + \text{power series in } \frac{1}{w} + k, \qquad k \in \mathbb{Z},$$

where k denotes the respective distinct pre-images under J (translation by U^k).

We have just described the behavior of J^{-1} in a neighborhood of ∞ considered as a pre-image of a neighborhood of $i\infty$. J^{-1} behaves analogously in a neighborhood of ∞ considered as a pre-image of a neighborhood of the rational point $-\frac{d}{c}$, i.e. J^{-1} behaves analogously in the approach to ∞ on those sheets of the Riemann surface whose pre-images have a rational cusp $-\frac{d}{c}$. J^{-1} has infinite logarithmic branch points over $w=\infty$. This completes the description of the behavior of J^{-1} at $w=0, 1$ and ∞. Circuits about these points subject the branches to a modular transformation.

We can present a second proof of the main theorem. Let f be a modular function; as such it is a single-valued function of w, since w determines the equivalence class of τ under $\bar{\Gamma}$. Moreover, it displays the same analytic behavior in w and τ at all points distinct from $w=0, 1, \infty$. At these points the function has a limit and is thus rational in $w=J(\tau)$.

§ 4. Modular Forms

Modular forms represent an important class of functions in the development of the theory of elliptic modular functions.

1. Definition. A complex valued function h of two complex variables ω_1, ω_2, defined for $\frac{\omega_1}{\omega_2} \in \mathscr{H}$, is called a *homogeneous modular form of dimension* $-k \in \mathbb{Z}$ if

a) $h(\lambda\omega_1, \lambda\omega_2) = \lambda^{-k} h(\omega_1, \omega_2)$ for all complex $\lambda \neq 0$,

b) $h(a\omega_1 + b\omega_2, c\omega_1 + d\omega_2) = h(\omega_1, \omega_2)$ for all $\begin{pmatrix} a & b \\ c & d \end{pmatrix} \in \Gamma$,

c) $h(\tau, 1)$ defines a holomorphic function except for poles in $\operatorname{Im}\tau > 0$,

d) in a neighborhood $\{\tau \mid \operatorname{Im}\tau > \alpha\}$ of $i\infty$

$$h(\tau, 1) = \sum_{\nu \geqq \nu_0} b_\nu e^{2\pi i \tau \nu}.$$

The condition b) says that h is a function of the oriented lattice spanned by ω_1, ω_2.

A function f of one complex variable τ is called an *inhomogeneous modular form* if $\omega_2^{-k} f\left(\frac{\omega_1}{\omega_2}\right)$ defines a homogeneous modular form.

In particular, $h(\tau, 1)$ is an inhomogeneous modular form. The behavior of the inhomogeneous modular form f under the modular transformation

$$\tau \mapsto A(\tau) = \frac{a\tau + b}{c\tau + d}$$

follows from conditions a) and b):

$$f\left(\frac{a\tau+b}{c\tau+d}\right)=(c\tau+d)^k f(\tau), \qquad \begin{pmatrix} a & b \\ c & d \end{pmatrix}=A\in\Gamma. \tag{14}$$

Conversely, if $h(\omega_1,\omega_2)=\omega_2^{-k} f\left(\frac{\omega_1}{\omega_2}\right)$, then b) is a consequence of (14) and a) is trivial.

For $k\in\mathbb{Z}$, arbitrary functions f on $\mathscr{H}$, and for arbitrary real $A=\begin{pmatrix} a & b \\ c & d \end{pmatrix}$ with $|A|=ad-bc>0$, we define the functions $f|_k A$ by

$$(f|_k A)(\tau)=|A|^{\frac{1}{2}}(c\tau+d)^{-k} f\left(\frac{a\tau+b}{c\tau+d}\right).$$

We obtain equation (14) in the form

$$f|_k A=f, \qquad A\in\Gamma.$$

For two real matrices A, B with positive determinant we have

$$(f|_k A)|_k B=f|_k(AB),$$

and so (14) holds for all $A\in\Gamma$ if it is true for the generators. As an immediate consequence of the defining properties for homogeneous and inhomogeneous modular forms we have

Theorem 9. *The modular forms of the same dimension form a vector space over the field of complex numbers $\mathbb{C}$. The dimension of a product of modular forms is the sum of their dimensions. The quotient of two modular forms of the same dimension is a modular form of dimension 0 if the denominator is not the 0-function, and it is a modular function if the values at the rational points are suitably chosen.*

As yet we have said nothing about the existence of modular forms. We prove

Theorem 10. *The only modular form of odd dimension is the 0-function.*

Proof. Since $\begin{pmatrix} -1 & 0 \\ 0 & -1 \end{pmatrix}\in\Gamma$, $h(-\omega_1,-\omega_2)=h(\omega_1,\omega_2)$, but also $h(-\omega_1,-\omega_2)=(-1)^k h(\omega_1,\omega_2)$. Either $h=0$ or $(-1)^k=1$, which implies that k is even. □

2. Power series expansions. Let $r=-\frac{d}{c}$, $c\neq 0$, $(c,d)=1$, be a rational point, and let a and b be chosen so that $ad-bc=1$. Then, as a consequence of the property d) of the definition and of (14), we have the expansion

$$c^k(\tau-r)^k f(\tau) = \sum_{\nu \geqq \nu_0} b_\nu e^{2\pi i \frac{a\tau+b}{c\tau+d}\nu} \tag{15}$$

at r, which is valid for $\operatorname{Im}\dfrac{a\tau+b}{c\tau+d} > \alpha$, with some $\alpha>0$. If $\tau_0 \in \mathscr{H}$ is an arbitrary point, then we start with the expansion

$$(\tau-\bar{\tau}_0)^k f(\tau) = \sum_{\nu \geqq \nu_0} a_\nu \left(\frac{\tau-\tau_0}{\tau-\bar{\tau}_0}\right)^\nu,$$

which is valid in a neighborhood of τ_0. In particular, suppose that $\tau_0=\tau_\rho$ is a fixed point of

$$\bar{R}_1: \tau \mapsto \tau' = R_1(\tau) = \frac{a\tau+b}{c\tau+d} \quad \text{of order } 3,$$

which we apply to both sides of the above equation. Then from (7) and (14) we obtain

$$(c\bar{\tau}_\rho+d)^{-k}(\tau-\bar{\tau}_\rho)^k f(\tau) = \sum_{\nu \geqq \nu_0} a_\nu \left(\frac{c\bar{\tau}_\rho+d}{c\tau_\rho+d}\right)^\nu \left(\frac{\tau-\tau_\rho}{\tau-\bar{\tau}_\rho}\right)^\nu.$$

Furthermore, because of **I, § 2, (9)** with $\lambda_1 = \pm\rho^{\pm 1}$, $\lambda = \dfrac{\lambda_2}{\lambda_1}$, we obtain

$$(\tau-\bar{\tau}_\rho)^k f(\tau) = \sum_{\nu \geqq \nu_0} a_\nu \lambda^{\nu-k} \left(\frac{\tau-\tau_\rho}{\tau-\bar{\tau}_\rho}\right)^\nu.$$

Consequently,

$$a_\nu = a_\nu \lambda^{\nu-k}, \qquad a_\nu = 0 \quad \text{for } \nu \not\equiv k \bmod 3.$$

In terms of the local variable

$$t_{\tau_\rho} = \left(\frac{\tau-\tau_\rho}{\tau-\bar{\tau}_\rho}\right)^3,$$

we finally obtain the expansion

$$(\tau-\bar{\tau}_\rho)^k f(\tau) = \sum_{\nu \geqq \nu_3} c_\nu t_{\tau_\rho}^{\nu+\frac{k_3}{3}}, \qquad k_3 \equiv k \bmod 3, \qquad 0 \leqq k_3 < 3.$$

Similarly, we obtain the expansion

$$(\tau-\bar{\tau}_i)^k f(\tau) = \sum_{\nu \geqq \nu_2} d_\nu t_{\tau_\rho}^{\nu+\frac{k_2}{2}}, \qquad k_2 \equiv \frac{k}{2} \bmod 2, \qquad 0 \leqq k_2 < 2$$

at τ_i equivalent to i under Γ with

$$t_{\tau_i} = \left(\frac{\tau-\tau_i}{\tau-\bar{\tau}_i}\right)^2, \qquad \lambda = -1.$$

We obtain a uniform representation for all $\tilde{\tau} \in \mathscr{H}^*$ if we denote the local variable used above by $t_{\tilde{\tau}}$, as was previously done in § **1**, **3**. We formulate

Theorem 11. *If f is a modular form of dimension $-k$ and if $\tilde{\tau}$ is in $\mathscr{H}^*$, then in a neighborhood of $\tilde{\tau}$ we have*

$$(\tau-\bar{\tilde{\tau}})^k f(\tau) = \sum_{\nu \geqq \tilde{\nu}} a_\nu t_{\tilde{\tau}}^{\nu+\tilde{\kappa}}. \tag{16}$$

$\tilde{\kappa}$ is the smallest non-negative residue modulo 1 of $k/4$ or $k/3$ when $\tilde{\tau}$ is equivalent to i or ρ, otherwise $\tilde{\kappa}$ is 0. The factor $(\tau-\tilde{\tau})^k$ is omitted if $\tilde{\tau}=i\infty$.

3. Zeros and poles. If in (16) the coefficient $a_{\tilde{\nu}} \neq 0$ and $\tilde{\nu} \geqq 0$, then we say that f has a *zero of order* $\tilde{\nu}+\tilde{\kappa}$ at $\tilde{\tau}$; if $\tilde{\nu}<0$, we say that f has a *pole of order* $-(\tilde{\nu}+\tilde{\kappa})$ at $\tilde{\tau}$. For these—not necessarily integral—orders we have

Theorem 12. *A modular form has the same order of zeros or of poles at equivalent points.*

The proof for rational points follows from equation (15). For points $\tau_0 \in \mathscr{H}$ replace $f(\tau)$ and $f(\tau')$ in (6) by $(\tau-\bar{\tilde{\tau}})^k f(\tau)$ and $(\tau'-\bar{\tilde{\tau}}')^k f(\tau')$ and continue as in the proof of Theorem **1**.

It is not useful to consider c-points $c \neq 0, \infty$ of a modular form of dimension $k \neq 0$, because they are not invariant under $\bar{\Gamma}$.

We want to prove a theorem for modular forms analogous to Theorem **2**. We define the *total zero order* $N(0)$ as the (naturally finite) sum of the orders of zeros over a maximal system of inequivalent zeros; for the *total pole order* $N(\infty)$ we use the corresponding sum for poles. We prove

Theorem 13. *For a modular form of dimension $-k$ the difference of the total zero order and the total pole order is*

$$N(0)-N(\infty) = \frac{k}{12}. \tag{17}$$

In order to carry over the proof of Theorem **2** we first make the following remarks. If we set $\tau' = S(\tau)$ in the transformation formula

$$f\left(\frac{a\tau+b}{c\tau+d}\right) = (c\tau+d)^k f(\tau), \qquad \begin{pmatrix} a & b \\ c & d \end{pmatrix} = S \in \Gamma$$

of a modular form of dimension $-k$, we obtain upon differentiating with respect to τ

$$\frac{df(\tau')}{d\tau'}\,\frac{d\tau'}{d\tau} = kc(c\tau+d)^{k-1} f(\tau) + (c\tau+d)^k f'(\tau). \tag{18}$$

This implies that

$$\frac{f'(\tau')}{f(\tau')} = kc(c\tau+d)+(c\tau+d)^2\frac{f'(\tau)}{f(\tau)}.$$

Hence for an arbitrary path $\mathscr{W}\subset\mathscr{H}$ not passing through zeros or poles of f, we have

$$\int\limits_{S(\mathscr{W})}\frac{f'(\tau)}{f(\tau)}d\tau = \int\limits_{\mathscr{W}}\frac{kc}{(c\tau+d)}d\tau + \int\limits_{\mathscr{W}}\frac{f'(\tau)}{f(\tau)}d\tau. \tag{19}$$

Proof of Theorem **13**. We refer to Fig. 4 used in the proof of Theorem **2** (and choose the path $\mathscr{C}$ in a similar manner). Suppose that $f(\tau)$ is a modular form of dimension $-k$ and that $N_0(0)$ and $N_0(\infty)$ denote respectively the numbers of zeros and poles of f within the curve $\mathscr{C}$. Then

$$N_0(0)-N_0(\infty)=\frac{1}{2\pi i}\int\limits_{\mathscr{C}}\frac{f'(\tau)}{f(\tau)}d\tau.$$

Now we write this integral as the sum of the integrals along the paths $\mathscr{C}_1$ through $\mathscr{C}_9$. The radii of the circular arcs $\mathscr{C}_4$ about ρ and $\mathscr{C}_7$ about i will tend to 0. Then, using (19) and the fact that $\mathscr{C}_5=-R(\mathscr{C}_3)$, $R=\begin{pmatrix}0 & -1\\ 1 & 1\end{pmatrix}$, we see that the sum of the integrals

$$\int\limits_{\mathscr{C}_3}+\int\limits_{\mathscr{C}_5}=-\int\limits_{\mathscr{C}_3}\frac{k\,d\tau}{\tau+1}\quad\text{converges to}\quad -k\log\frac{\rho+1}{\alpha+1}.$$

Similarly,

$$\int\limits_{\mathscr{C}_6}+\int\limits_{\mathscr{C}_8}=-\int\limits_{\mathscr{C}_6}\frac{k\,d\tau}{\tau}\quad\text{converges to}\quad -k\log\frac{\alpha+1}{i}.$$

The sum of the four integrals thus converges to

$$-k\log\frac{\rho+1}{i}=\frac{2\pi i k}{12}.$$

The integral over $\mathscr{C}_4$ converges to $\dfrac{2\pi i n_\rho}{3}$, where n_ρ is the order of the zero at ρ measured in terms of τ, or $-n_\rho$ is the pole order at ρ. Indeed, if one sets $t=\dfrac{\tau-\rho}{\tau-\bar{\rho}}$, then

$$\frac{1}{f}\frac{df}{dt}=\frac{n_\rho}{t}+c_0+c_1t+\cdots,$$

and so

$$\int_{\mathscr{C}_4} \frac{f'(\tau)}{f(\tau)} d\tau = \int_{\mathscr{C}_4'} \frac{1}{f} \frac{df}{dt} dt,$$

where $\mathscr{C}_4'$ is a circular arc centered at $t=0$, of radius r, subtending an angle of $\frac{2\pi}{3}$. Then

$$\lim_{r\to 0} \int_{\mathscr{C}_4'} \frac{1}{f} \frac{df}{dt} dt = \frac{2\pi i n_\rho}{3}.$$

Similarly, the integral along $\mathscr{C}_7$ converges to $\frac{2\pi i n_i}{2}$.

The integrals along $\mathscr{C}_2$ and $\mathscr{C}_9$ cancel; the integral over $\mathscr{C}_1$ is $-2\pi i n_\infty$, where n_∞ is the order of the zero, or $-n_\infty$ is the order of the pole of f at $i\infty$ as measured in $t=e^{2\pi i\tau}$. We obtain

$$N_0(0) - N_0(\infty) = \frac{k}{12} - \frac{n_\rho}{3} - \frac{n_i}{2} - n_\infty,$$

where $\frac{n_\rho}{3}$ and $\frac{n_i}{2}$ are the order of the zeros or the negative of the order of the poles as defined at the beginning of this subsection. Thus Theorem **12** is proved:

$$N(0) - N(\infty) = N_0(0) - N_0(\infty) + \frac{n_\rho}{3} \quad \frac{n_i}{2} + n_\infty = \frac{k}{12}. \quad \square \tag{20}$$

We note that two modular forms which coincide in the location and order of their zeros and poles differ by a constant factor. For their quotient is everywhere regular and has no zeros.

§ 5. Entire Modular Forms

1. Definition. The *inhomogeneous* modular form f is called *entire* if it is holomorphic in $\mathscr{H}$ and if in its expansion at $i\infty$,

$$f(\tau) = \sum_{\nu \geqq \nu_0} b_\nu e^{2\pi i \nu\tau}, \qquad b_{\nu_0} \neq 0,$$

with the index $\nu_0 \geqq 0$. The entire modular form is called a *cusp form* if $\nu_0 \geqq 1$. The *homogeneous* modular form h is called *entire* if the inhomogeneous modular form $\omega_2^{-k} h(\omega_1, \omega_2) = h(\tau, 1)$ is entire.

Clearly, the entire (homogeneous or inhomogeneous) modular forms of the same dimension—including the function $f=0$—form a vector space over $\mathbb{C}$. The product of two entire modular forms of dimensions $-k_1$ and $-k_2$, respectively, is an entire modular form of dimension $-(k_1+k_2)$.

Equation (20), when applied to entire modular forms, says: the number of zeros of an entire modular form of dimension $-k$ is

$$N(0)=\frac{k}{12}=\frac{n_\rho}{3}+\frac{n_i}{2}+n_\infty+N_0(0)\,, \tag{21}$$

where all terms are non-negative. This equation is not solvable for

$$k\equiv 1 \bmod 2, \qquad k=2 \quad \text{or } k<0\,.$$

Thus we have

Theorem 14. *There are no modular forms of odd dimension (cf. Theorem* **10***); there are no entire modular forms of dimension* -2 *or of positive dimension.*

2. The derivative of modular functions. The existence of modular functions implies the existence of modular forms of arbitrary even integral dimension.

Theorem 15. *If f is a modular function, then* $\dfrac{df}{d\tau}$ *is an inhomogeneous modular form of dimension* -2.

Proof. If f is a modular function, $\begin{pmatrix} a & b \\ c & d \end{pmatrix}=S\in\Gamma$, and $\tau'=S(\tau)$, then

$$\frac{df(\tau')}{d\tau'}=\frac{df(S\tau)}{d\tau}\,\frac{d\tau}{d\tau'}=\frac{df(\tau)}{d\tau}(c\tau+d)^2.$$

Incidentally, this is the special case $k=0$ of equation (18). The remaining defining properties of modular forms are easily verified. □

In particular J', the derivative of the absolute modular invariant, is a modular form of dimension -2. We consider its expansion about the point $\tau_0\in\mathscr{H}^*$. If, first of all, $\tau_0\in\mathscr{H}$, then

$$J(\tau)=\sum_{\nu\geqq 0} c_\nu(\tau_0)\left(\frac{\tau-\tau_0}{\tau-\bar{\tau}_0}\right)^{e_{\tau_0}\nu}, \qquad c_1(\tau_0)\neq 0\,,$$

where

$e_{\tau_0}=2$ if $\tau_0\sim i$, $\quad e_{\tau_0}=3$ if $\tau_0\sim\rho$ under $\bar{\Gamma}$, and

$e_{\tau_0}=1$ in the remaining cases.

The expansion of $J'(\tau)$ shows: as measured in the local variable $\left(\dfrac{\tau-\tau_0}{\tau-\bar{\tau}_0}\right)^{e_{\tau_0}}$, $J'(\tau)$ has a zero of order $\frac{1}{2}$ at $\tau_0\sim i$ under $\bar{\Gamma}$, a zero of order $\frac{2}{3}$ at $\tau_0\sim\rho$ under $\bar{\Gamma}$, and no zeros at the remaining points.

At $i\infty$

$$J(\tau)=\sum_{\nu\geqq -1} c_\nu e^{2\pi i\nu\tau}, \quad c_{-1}\neq 0,$$

and $J'(\tau)$ has therefore a pole of order 1 at $i\infty$ and at all rational points. Thus

$$N(0)-N(\infty)=\tfrac{1}{6} \quad \text{for} \quad J'(\tau).$$

From the expansions of $J'(\tau)$ we obtain those for $(J'(\tau))^k$, $k\in\mathbb{Z}$, which is a modular form of dimension $-2k$. From this follows Theorem **11** again, since the quotient of two modular forms of the same dimension is a modular function. Analogously, it follows that

$$N(0)-N(\infty)=\frac{k}{6} \quad \text{for} \quad (J'(\tau))^k,$$

and so for each modular form of dimension $-2k$. This again proves Theorem **13**.

3. Construction of entire modular forms. We shall now construct entire modular forms with the help of $J'(\tau)$. We consider the scheme

	ρ	i	$i\infty$
J	1	0	-1
$J-1$	0	1	-1
J'	$\frac{2}{3}$	$\frac{1}{2}$	-1

of orders of zeros and negative pole orders for the functions $J, J-1, J'$. From it we conclude

Theorem 16. *The function*

$$f_{a,b,c}(\tau)=\frac{J'^a}{J^b(J-1)^c}, \qquad a,b,c\in\mathbb{Z}$$

is an entire modular form of dimension $-2a$ *if*

$$a\geqq 2, \quad 2c\leqq a, \quad 3b\leqq 2a, \quad b+c\geqq a, \quad \textit{hence also } b,c>0.$$

For $a=2, b=c=1$ we obtain the entire form

$$G_4^*=\frac{J'^2}{J(J-1)} \tag{22}$$

of dimension -4 with a zero of order $\frac{1}{3}$ at ρ; and for $a=3, b=2, c=1$ we obtain

$$G_6^*=\frac{J'^3}{J^2(J-1)}, \tag{23}$$

a form of dimension -6 with its zero at i of order $\frac{1}{2}$. We have entire forms of dimensions -8, -10, and -14 in

$$G_8^* = G_4^{*^2}, \quad G_{10}^* = G_4^* G_6^*, \quad \text{and} \quad G_{14}^* = G_4^{*^2} G_6^*,$$

respectively.

Now, however, equation (21) for the number of zeros of an entire form of dimension $-k$, i.e.

$$\frac{k}{12} = \frac{n_\rho}{3} + \frac{n_i}{2} + n_\infty + N_0(0),$$

has exactly one solution in the cases $k=4, 6, 8, 10$ and 14. That is, apart from a multiplicative constant, there is exactly one entire modular form of these dimensions, since the quotient of two such modular forms with the same zeros of the same orders is a constant. The functions

$$G_4^{*^3} \quad \text{and} \quad G_6^{*^2}$$

are entire modular forms of dimension -12 with zeros of order 1 at ρ and i, respectively, and which are therefore linearly independent. The function

$$\Delta^* := G_4^{*^3} - G_6^{*^2} = \frac{J'^6}{J^4(J-1)^3} \tag{24}$$

is an entire modular form of dimension -12. One can see from the expansions in $t_{i\infty}$ that it has a zero of order 1 at $i\infty$; and because $N(0)=1$, it therefore has no zeros in $\mathscr{H}$. Δ^* differs by a constant factor from the discriminant of the theory of elliptic functions and shall here also be called the *discriminant*. We formulate

Theorem 17. *The discriminant Δ^* is an entire modular form of dimension -12 that vanishes at the point $i\infty$. It is thereby characterized up to a multiplicative constant. It is never zero in the upper half plane.*

We need not require that Δ^* have a zero of order one because an entire modular form of dimension -12 has exactly one zero.

If f_{12} is an arbitrary entire modular form of dimension -12, then there are constants c and c' such that

$$f_{12} - c\, G_4^{*^3} - c'\, \Delta^*$$

has a zero of order at least two at $i\infty$. Since $N(0)=1$, it must therefore vanish identically. This implies: the $\mathbb{C}$-dimension of the vector space over $\mathbb{C}$ of entire modular forms of dimension -12 is 2.

4. Basis for the entire modular forms. We shall give a basis for the vector space over $\mathbb{C}$ of entire modular forms of dimension $-k \equiv 0 \bmod 2$. This has been done for $4 \leqq k \leqq 14$ in subsection **3**.

Suppose first, that $0<k\equiv 0 \bmod 12$, so $k=12k'$. Then the functions

$$G_4^{*3(k'-\nu)}\Delta^{*\nu}, \qquad \nu=0,1,\ldots,k'=\frac{k}{12}, \tag{25}$$

are entire modular forms of dimension $-12k'$, which have a zero at $i\infty$ of order ν, respectively, and which are therefore linearly independent. If $f_{12k'}$ is an entire modular form of dimension $-12k'$, then for suitable c_ν, $\nu=0,1,\ldots,k'$, the function

$$f_{12k'}-c_0 G_4^{*3k'}-c_1 G_4^{*3k'-3}\Delta^*-\cdots-c_{k'}\Delta^{*k'}$$

has a zero at $i\infty$ of order at least $(k'+1)$, but since $N(0)=k'$, it must vanish identically. Hence the functions (25) form a basis when $k\equiv 0 \bmod 12$.

Next, let $k\geqq 4$ be even. We determine three non-negative integers k', α, β so that the system

$$k=12k'+4\alpha+6\beta, \qquad 12 \neq 4\alpha+6\beta\leqq 14 \tag{26}$$

is satisfied. This is uniquely possible. If f_k is an entire modular form of dimension $-k$, then it has a zero at ρ of order $\frac{\alpha}{3}+\gamma$ by Theorem **11**; likewise at i it has a zero of order $\frac{\beta}{2}+\delta$ with nonnegative $\gamma,\delta\in\mathbb{Z}$. Since $\frac{\alpha}{3}$ and $\frac{\beta}{2}$ are exactly the orders of the zeros of $G_4^{*\alpha}$ at ρ and $G_6^{*\beta}$ at i, respectively, the function

$$f_k G_4^{*-\alpha} G_6^{*-\beta}$$

is an entire modular form of dimension $-12k'$ with

$$k'=\frac{k-4\alpha-6\beta}{12}.$$

Thus this function can be uniquely represented as a homogeneous linear combination of the functions in (25). Using the constants α,β,k' as defined by (26) we obtain

Theorem 18. *The vector space over $\mathbb{C}$ of entire modular forms of dimension $-k \equiv 0 \bmod 2$ has the $\mathbb{C}$-dimension*

$$D=\begin{cases}\left[\frac{k}{12}\right] & \text{if } k\equiv 2 \bmod 12,\\ \left[\frac{k}{12}\right]+1 & \textit{in the remaining cases,}\end{cases}$$

and the basis

$$G_4^{*\alpha}G_6^{*\beta}G_4^{*3(k'-\nu)}\Delta^{*\nu} \qquad \nu=0,1,\ldots,k'. \tag{27}$$

We now prove

Theorem 19. *Every entire modular form f_k of dimension $-k$ can be represented in the form*

$$f_k = \sum_{\substack{\mu,\nu \geqq 0 \\ 4\mu+6\nu=k}} c_{\mu,\nu} G_4^{*\mu} G_6^{*\nu}$$

with uniquely determined coefficients $c_{\mu,\nu} \in \mathbb{C}$.

Proof. Since $\Delta^* = G_4^{*3} - G_6^{*2}$, f_k has such a representation by Theorem **18**. The uniqueness is already known for $k \leqq 14$. Suppose there is a representation of the 0-function of the form

$$\sum_{\substack{4\mu+6\nu=k \\ \mu,\nu \geq 0}} c_{\mu,\nu} G_4^{*\mu} G_6^{*\nu} = 0, \qquad k \geqq 16 .$$

As is easily seen by setting $\tau = i$, if $k \equiv 0 \bmod 6$, then $c_{0,\frac{k}{6}} = 0$. Analogously, if $k \equiv 0 \bmod 4$, $c_{\frac{k}{4},0} = 0$. For the remaining values of k the exponents $\mu = 0$ or $\nu = 0$ do not occur. In every case one can factor either G_4^* or G_6^* from the left hand side to obtain a representation of the 0-function in the above form, but with $k' < k$. Complete induction gives the proof. □

5. Representations of modular functions as quotients of entire modular forms. Let f be a modular function with the $\bar{\Gamma}$-inequivalent poles $\tau_1, \ldots, \tau_m$ of orders $n_1, \ldots, n_m$ measured in $\tau - \tau_\mu$ for *finite* τ_μ and in $e^{2\pi i\tau}$ for $\tau_\mu = i\infty$. Now choose k so large that the dimension R of the space M_k of entire modular forms of dimension $-k$ is larger than the sum P of the orders of poles. Let $f_1, \ldots, f_R$ be a basis of M_k, and let $f_\nu^{(\lambda)}$ denote the λth derivative of f_ν with respect to τ or with respect to $e^{2\pi i\tau}$ if $\tau_\mu = i\infty$. Then the system of equations

$$\sum_{\nu=1}^{R} x_\nu f_\nu^{(\lambda)}(\tau_\mu) = 0, \qquad \mu = 1, 2, \ldots, m; \; \lambda = 0, 1, \ldots, n_\mu - 1 ,$$

has a non-trivial solution $c_1, \ldots, c_R$, and so the function

$$f(\tau) \sum_{\nu=1}^{R} c_\nu f_\nu(\tau)$$

is an entire modular form. We have

Theorem 20. *Every modular function is a quotient of entire modular forms.*

One may reason as follows: Suppose G'_{12}, G''_{12} is a basis for M_{12}. Then for each $\tau_0 \in \mathscr{H}^*$ there exists a form

$$G_{12}(\tau;\tau_0)=c'\,G'_{12}(\tau)+c''\,G''_{12}(\tau),$$

which has a single zero at τ_0 of order 1. Thus we obtain f in the form

$$f(\tau)=c\,\frac{\prod\limits_{\nu} G_{12}(\tau;\tau_\nu)}{\prod\limits_{\nu} G_{12}(\tau;\tau'_\nu)},\quad c \text{ constant}, \tag{29}$$

where the products extend over all zeros τ_ν and poles τ'_ν of $f(\tau)$ with proper multiplicities.

In particular, (24) implies for $J(\tau)$ that

$$\Delta^* J = \frac{J'^6}{J^3(J-1)^3} = G_4^{*3}, \qquad J = \frac{G_4^{*3}}{G_4^{*3}-G_6^{*2}}, \tag{30}$$

from which Theorem **20** follows again since by Theorem **8** each modular function is a rational function of $J(\tau)$.

Theorems **9**, **10**, **14**, **18**, **19**, **20** are true independent of whether a modular form is understood to be homogeneous or inhomogeneous. This is implied by the connection fixed by the definition between the two kinds of modular forms.

Chapter III. Eisenstein Series

§ 1. The Eisenstein Series in the Case of Absolute Convergence

We shall now represent modular forms by certain analytic expressions, the so-called *Eisenstein series*. This provides another approach to the theory of modular functions and modular forms that is independent of the Riemann mapping theorem.

1. Definition of Eisenstein series. Let k be a rational integer, $k \equiv 0 \bmod 2$, and $k > 2$; let ω_1, ω_2 be complex variables with $\operatorname{Im}\left(\frac{\omega_1}{\omega_2}\right) > 0$. Series of the form

$$G_k(\omega_1, \omega_2) = \sum_{m_1, m_2 \in \mathbb{Z}}' (m_1 \omega_1 + m_2 \omega_2)^{-k} \tag{1}$$

are called *homogeneous Eisenstein series*. The prime on the summation sign indicates that $(m_1, m_2) \neq (0, 0)$. If $\tau = \frac{\omega_1}{\omega_2}$, then

$$G_k(\tau) = \omega_2^k \, G_k(\omega_1, \omega_2) = \sum_{m_1, m_2 \in \mathbb{Z}}' (m_1 \tau + m_2)^{-k} \tag{2}$$

is the corresponding *inhomogeneous Eisenstein series*. We claim

Theorem 1. *For* $k \equiv 0 \bmod 2$, $k \geqq 4$, $G_k(\omega_1, \omega_2)$ *is a homogeneous entire modular form of dimension* $-k$.

Proof. To show that $G_k(\tau)$ is holomorphic in $\mathscr{H}$ we prove that the series for $G_k(\tau)$ converges absolutely and uniformly on each compact set $K \subset \mathscr{H}$. We consider $|x\tau + y|$ for fixed $\tau \in K$ and all real x, y with $x^2 + y^2 = 1$. The minimum $\mu(\tau)$ of $|x\tau + y|$ is attained and is positive. Also $\mu(\tau)$ is a continuous positive function of τ, and so $\mu = \min_{\tau \in K} \mu(\tau)$ is attained and is likewise positive. We therefore have the estimate

$$|m_1 \tau + m_2|^2 \geqq \mu^2(\tau)(m_1^2 + m_2^2) \geqq \mu^2 (m_1^2 + m_2^2) \quad \text{for} \quad \tau \in K,$$

Consequently the series for $G_k(\tau)$ is majorized by

$$M=\mu^{-k}\sum_{m_1,m_2\in\mathbb{Z}}{}'(m_1^2+m_2^2)^{-k/2}.$$

There are at most $8N$ pairs (m_1,m_2) with $|m_1|=N$ or $|m_2|=N$, and for these $m_1^2+m_2^2\geqq N^2$. Thus we obtain

$$\mu^{-k}\sum_{N\geqq 1}\frac{8N}{N^k}=8\mu^{-k}\sum_{N\geqq 1}N^{1-k}$$

as majorant for M. This converges since $k>2$. We have proved that $G_k(\tau)$ is holomorphic in the upper half plane. The condition

$$G_k(\lambda\omega_1,\lambda\omega_2)=\lambda^{-k}G_k(\omega_1,\omega_2)\qquad\text{for }\lambda\neq 0$$

is obviously fulfilled. In order to prove the invariance under $A\in\Gamma$, we note that $\frac{\omega_1}{\omega_2}\in\mathscr{H}$ implies that ω_1,ω_2 span an oriented lattice in the complex plane. The sum extends over all non-zero lattice points and, because of absolute convergence, is independent of the order of summation. A modular transformation maps this lattice onto itself preserving the zero point. Thus it changes only the order of summation but not the value of $G_k(\omega_1,\omega_2)$. This invariance under modular transformations which we just proved says that G_k is a function on the lattices in the complex plane.

We remark that the invariance under $A\in\Gamma$ would still be valid if the sum were only over the pairs (m_1,m_2) with g.c.d. $(m_1,m_2)=1$, because the lattice points, whose coordinates are relatively prime, are also transformed onto themselves by such A.

2. Fourier coefficients of the Eisenstein series. For the proof of Theorem **1** we must investigate the behavior of $G_k(\tau)$ at $\tau=i\infty$. Since $G_k(\tau)$ is holomorphic in $\mathscr{H}$ and periodic with period 1, it has a Fourier expansion

$$G_k(\tau)=\sum_{n\in\mathbb{Z}}a_n e^{2\pi i\tau n},$$

valid for $\tau\in\mathscr{H}$. We sum the defining series for G_k in the following way:

$$G_k(\tau)=2\zeta(k)+2\sum_{m_1\geqq 1}\sum_{m_2\in\mathbb{Z}}(m_1\tau+m_2)^{-k},\tag{3}$$

where $\zeta(s)$ denotes the Riemann zeta function. First, we expand the inner sum with $m_1=1$ in a Fourier series (we include the case $k=2$ for later reference):

$$\sum_{n=-\infty}^{\infty}(\tau+n)^{-k}=\sum_{\nu\in\mathbb{Z}}\alpha_\nu e^{2\pi i\tau\nu}\qquad(\operatorname{Im}\tau>0).$$

We show that $\alpha_\nu=0$ for $\nu\leqq 0$. For all ν we have

$$\alpha_\nu=\int_{\tau_0}^{\tau_0+1}\Big(\sum_{n\in\mathbb{Z}}(\tau+n)^{-k}\Big)e^{-2\pi i\nu\tau}d\tau \quad \text{for arbitrary } \tau_0 \text{ with } \operatorname{Im}\tau_0>0.$$

If $k\geqq 2$ this series converges uniformly on the path of integration so that

$$\alpha_\nu=\sum_{n\in\mathbb{Z}}\int_{\tau_0}^{\tau_0+1}(\tau+n)^{-k}e^{-2\pi i\nu\tau}d\tau=\int_{-\infty+iy_0}^{\infty+iy_0}\tau^{-k}e^{-2\pi i\nu\tau}d\tau \quad \text{for}$$

$$y_0=\operatorname{Im}\tau_0>0 \quad \text{and} \quad k\geqq 2.$$

If $\tau=x+iy_0$, then

$$|\alpha_\nu|\leqq e^{2\pi\nu y_0}\int_{-\infty}^{\infty}|x+iy_0|^{-k}dx=e^{2\pi\nu y_0}\int_{-\infty}^{\infty}(x^2+y_0^2)^{-k/2}dx$$

$$=e^{2\pi\nu y_0}y_0^{1-k}\int_{-\infty}^{\infty}(x^2+1)^{-k/2}dx.$$

This improper integral, which we denote by c_k, converges since $k\geqq 2$. Thus

$$|\alpha_\nu|\leqq\frac{e^{2\pi\nu y_0}}{y_0^{k-1}}c_k.$$

However, y_0 may be chosen arbitrarily large. This implies that

$$\alpha_\nu=0 \quad \text{for } \nu\leqq 0.$$

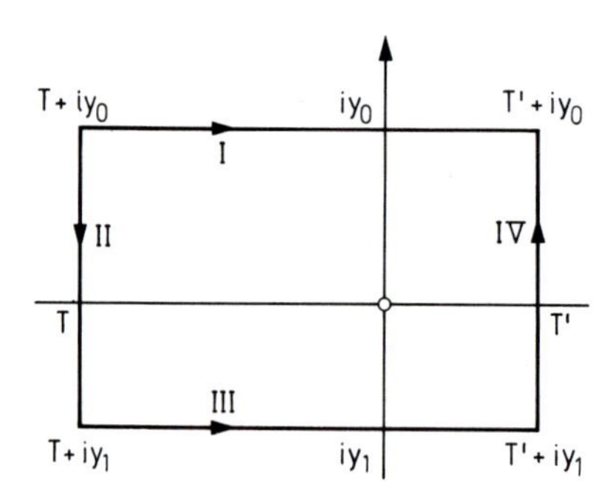

Fig. 8

We now compute the coefficients

$$\alpha_\nu=\int_{-\infty+iy_0}^{\infty+iy_0}\tau^{-k}e^{-2\pi i\nu\tau}d\tau, \qquad y_0>0,$$

for $v>0$. We integrate along the paths I through IV as illustrated in the figure and then apply the residue theorem:

$$\int_{\mathrm{I}} \tau^{-k} e^{-2\pi i v\tau} d\tau = \int_{\mathrm{II}+\mathrm{III}+\mathrm{IV}} \tau^{-k} e^{-2\pi i v\tau} d\tau - 2\pi i \operatorname*{Res}_{\tau=0} \tau^{-k} e^{-2\pi i v\tau}.$$

To estimate the integral along path II we note that

$$|e^{-2\pi i v(T+iy)}| = e^{2\pi v y}$$

is bounded on it and $(T+iy)^{-k}$ has the limit 0 as $T\to-\infty$. Thus

$$\lim_{T\to-\infty} \int_{\mathrm{II}} \tau^{-k} e^{-2\pi i v\tau} d\tau = 0,$$

and likewise

$$\lim_{T'\to\infty} \int_{\mathrm{IV}} \tau^{-k} e^{-2\pi i v\tau} d\tau = 0.$$

Since the integral $\int_{-\infty+iy_0}^{+\infty+iy_0} \tau^{-k} e^{-2\pi i v\tau} d\tau$ converges, it now follows that

$$\alpha_v = \int_{-\infty+iy_1}^{\infty+iy_1} \tau^{-k} e^{-2\pi i v\tau} d\tau - 2\pi i \operatorname*{Res}_{\tau=0} (\tau^{-k} e^{-2\pi i v\tau}),$$

We conclude as above that this integral is 0 since $v>0$ and $y_1<0$. The residue on the right is calculated from the exponential series. For $\tau\in\mathscr{H}$ it follows that

$$\sum_{n\in\mathbb{Z}} (\tau+n)^{-k} = \sum_{v\geqq 1} \alpha_v e^{2\pi i v\tau}, \quad \alpha_v = (-2\pi i)^k \frac{v^{k-1}}{(k-1)!}, \quad \text{for } v>0 \text{ and } k\geqq 2. \tag{4}$$

With this result we obtain

Theorem 2. *If $\tau\in\mathscr{H}, k>2$ and $k\equiv 0 \bmod 2$, then G_k has the expansion*

$$G_k(\tau) = 2\zeta(k) + 2\sum_{m_1\geqq 1} \frac{(2\pi i)^k}{(k-1)!} \sum_{v\geqq 1} v^{k-1} e^{2\pi i m_1 v\tau}$$

$$= 2\zeta(k) + \frac{2(2\pi i)^k}{(k-1)!} \sum_{n\geqq 1} \sigma_{k-1}(n) e^{2\pi i n\tau},$$

where

$$\sigma_m(n) = \sum_{\substack{d|m \\ d>0}} d^m$$

denotes the sum of the mth powers of the divisors of n.

Because of the absolute convergence of the series the order of summation is immaterial. It follows in particular from our representation of G_k that it is not identically 0. We will treat the case $k=2$ in **§ 7.** Theorem **1** is now proved. □

3. Computation of the zeta-function for positive even integral arguments. If $k>0$ and $k\equiv 0 \bmod 2$ there is an interesting formula for the value $\zeta_{(k)}$ of the Riemann zeta function. This value already appeared as a constant in Theorem **2.** We now derive this formula. First, we define the *Bernoulli numbers* B_n by the equation

$$\frac{x}{e^x-1}=\sum_{n=0}^{\infty}\frac{B_n}{n!}x^n \quad \text{or} \quad x=\sum_{n=0}^{\infty}\frac{B_n}{n!}x^n\sum_{n=1}^{\infty}\frac{x^n}{n!}. \tag{5}$$

Equating coefficients of like powers we obtain the recursion

$$B_0=1; \quad \binom{n}{0}B_0+\binom{n}{1}B_1+\cdots+\binom{n}{n-1}B_{n-1}=0, \qquad n>1.$$

The values of the first few Bernoulli numbers are:

$$B_0=1, \quad B_1=-\tfrac{1}{2}, \quad B_2=\tfrac{1}{6}, \quad B_4=-\tfrac{1}{30}, \quad B_6=\tfrac{1}{42},$$
$$B_8=-\tfrac{1}{30}, \quad B_{10}=\tfrac{5}{66}, \quad B_{12}=-\tfrac{691}{2730}, \quad B_{14}=\tfrac{7}{6}, \quad B_{16}=-\tfrac{3617}{510}.$$

If $n>1$ and $n\equiv 1 \bmod 2$, then $B_n=0$.

We derive a relation between B_k and $\zeta(k)$ for $k\equiv 0 \bmod 2$ and $k>0$ from equation (4). We have

$$\sum_{n=-\infty}^{\infty}(\tau+n)^{-2}=-4\pi^2\sum_{n=1}^{\infty}nq^n \quad \text{with } q=e^{2\pi i\tau}$$

or

$$\sum_{\substack{n=-\infty\\ n\neq 0}}^{\infty}\frac{1}{(\tau+n)^2}=\frac{-4\pi^2 q}{(1-q)^2}-\frac{1}{\tau^2}.$$

Indefinite integration yields

$$-\sum_{\substack{n=-\infty\\ n\neq 0}}^{\infty}\left(\frac{1}{\tau+n}-\frac{1}{n}\right)=\frac{2\pi i}{1-q}+\frac{1}{\tau}+C.$$

Now on the left side we put

$$\frac{1}{\tau+n}=-\sum_{\nu=0}^{\infty}\tau^{\nu}(-n)^{-(\nu+1)}, \qquad \left|\frac{\tau}{n}\right|<1,$$

and on the right hand side we use (5) to get

$$2\sum_{n=1}^{\infty}\zeta(2n)\tau^{2n-1}=-2\pi i\sum_{n=1}^{\infty}\frac{B_n(2\pi i\tau)^{n-1}}{n!}+C, \qquad |\tau|<1.$$

Therefore it follows that

$$\zeta(k)=-\frac{(2\pi i)^k}{2(k!)}B_k \quad \text{for } k\equiv 0 \bmod 2,\ k>0, \tag{6}$$

and $C=-\pi i$, since $B_1=-\frac{1}{2}$.

With these values for $\zeta(k)$ the expansion for $G_k(\tau)$ assumes the form

$$G_k(\tau)=2\frac{(2\pi i)^k}{(k-1)!}\left(-\frac{B_k}{2k}+\sum_{n\geqq 1}\sigma_{k-1}(n)\,e^{2\pi i\tau n}\right). \tag{7}$$

4. Discriminant and absolute invariant. Using

$$\zeta(4)=\frac{\pi^4}{2\cdot 3^2\cdot 5} \quad\text{and}\quad \zeta(6)=\frac{\pi^6}{3^3\cdot 5\cdot 7},$$

we obtain in particular the expansions

$$G_4(\tau)=\frac{\pi^4}{3^2\cdot 5}\left(1+2^4\cdot 3\cdot 5\sum_{n\geqq 1}\sigma_3(n)\,e^{2\pi i\tau n}\right),$$

$$G_6(\tau)=\frac{2\pi^6}{3^3\cdot 5\cdot 7}\left(1-2^3\cdot 3^2\cdot 7\sum_{n\geqq 1}\sigma_5(n)\,e^{2\pi i\tau n}\right).$$

By **II § 5, 3** these functions agree with the modular forms G_4^* and G_6^* of equations (22) and (23) to within a constant factor. Obviously, $20\,G_4^3-49\,G_6^2$ is an entire modular form of dimension -12 which has a zero at $i\infty$, and as such differs by at most a constant factor from the modular form $\Delta^*(\tau)$ defined in **II**, (24). The following multiple of this function

$$\Delta=2^4\cdot 3^3\cdot 5^2(20\,G_4^3-49\,G_6^2), \tag{8}$$

which in the future we shall call the *discriminant*, has $(2\pi)^{12}$ as coefficient of $e^{2\pi i\tau}$ in its Fourier expansion. Moreover, we have

Theorem 3. *The coefficients in the Fourier expansion of $(2\pi)^{-12}\Delta(\tau)$ are rational integers.*

Proof. If we set $q=e^{2\pi i\tau}$, then

$$\begin{aligned}(2\pi)^{12}\Delta(\tau)&=\frac{1}{2^6\cdot 3^3}\left(1+2^4\cdot 3\cdot 5\sum_{n\geqq 1}\sigma_3(n)q^n\right)^3\\&\quad-\frac{1}{2^6\cdot 3^3}\left(1-2^3\cdot 3^2\cdot 7\sum_{n\geqq 1}\sigma_5(n)q^n\right)^2\\&=\frac{1}{2^6\cdot 3^3}\left(1+2^4\cdot 3\cdot 5\sum_{n\geqq 1}\sigma_3(n)q^n\right)^3\\&\quad-\frac{1}{2^6\cdot 3^3}\left(1-2^4\cdot 3^2\cdot 7\sum_{n\geqq 1}\sigma_5(n)q^n\right)\\&\quad+\text{power series in } q \text{ with coefficients in } \mathbb{Z}\\&=\frac{1}{12}\sum_{n\geqq 1}(5\,\sigma_3(n)+7\,\sigma_5(n))\,q^n\\&\quad+\text{power series in } q \text{ with coefficients in } \mathbb{Z}.\end{aligned}$$

Further,

$$5\sigma_3(n)+7\sigma_5(n)=\sum_{d\mid n}(5d^3+7d^5)\equiv 5\sum_{d\mid n}d^3(1-d^2)\bmod 12\,.$$

However,

$$d^3(1-d^2)\equiv 0 \bmod 12$$

is always valid. Therefore our assertion is proved. □

The first few coefficients in the expansion

$$(2\pi)^{-12}\Delta(\tau)=\sum_{n\geqq 1}\tau(n)e^{2\pi i\tau n}, \tag{9}$$

valid in $\mathscr{H}$, can be calculated from (8):

$$\tau(1)=1,\quad \tau(2)=-24,\quad \tau(3)=252,\quad \tau(4)=-1472,$$
$$\tau(5)=4830,\quad \tau(6)=-6048,\quad \tau(7)=-1674.$$

S. Ramanujan [1] calculated the values of $\tau(n)$ for n up to 30, and G. N. Watson [1] calculated them to $n=1{,}000$.

The function

$$2^6\cdot 3^3\cdot 5^3\,\frac{G_4^3}{\Delta}$$

is a modular function of order 1 as is easily seen from the expansions of numerator and denominator. Moreover, this function has a zero at $\tau=\rho$, and since $\Delta(i)=2^6\,3^3\,5^3\,G_4^3(i)\neq 0$ it takes the value 1 at $\tau=i$. Thus this quotient is the absolute modular invariant

$$J(\tau)=2^6\cdot 3^3\cdot 5^3\,\frac{G_4^3(\tau)}{\Delta(\tau)}=\frac{(2\pi)^{12}}{2^6\,3^3\,\Delta(\tau)}\left(1+2^4\cdot 3\cdot 5\sum_{n\geqq 1}\sigma_3(n)q^n\right)^3. \tag{10}$$

It follows from Theorem **3** and $\tau(1)=1$ that $(2\pi)^{12}\Delta^{-1}(\tau)$ has an expansion in powers of q with rational integral coefficients. This implies

Theorem 4. *The modular function*

$$j:=1728\,J$$

has an expansion

$$j(\tau)=\frac{1}{q}+\sum_{\nu\geqq 0}a_\nu q^\nu,\qquad q=e^{2\pi i\tau},$$

with rational integers as coefficients.

The expansion begins with

$$j(\tau)=\frac{1}{q}+744+196884\,q+21493760\,q^2+864299970\,q^3+\cdots.$$

H. S. Zuckerman [1] calculated the list of the coefficients a_ν for $\nu\leqq 24$.

The construction (10) of the absolute invariant J makes it possible to avoid the application of the Riemann mapping theorem. The modular forms G_4^* and G_6^* which were constructed in **II** (22) and (23) and which played a decisive role in the theory of entire modular forms, are no longer needed. They will be replaced by the Eisenstein series G_4 and G_6. Theorem **II 13**, regarding the difference of the number of zeros and the number of poles, retains its importance.

The representation (10) for J in terms of G_4 and Δ implies that J is a modular function of order 1 with the property

$$J(-\bar{\tau})=\overline{J(\tau)}.$$

As was shown in **II, §1, 5**, it further follows that the boundary of the region $\mathscr{F}^*$ is mapped on the real axis and the interior of $\mathscr{F}^*$ is mapped onto one of the half planes $\pm\mathscr{H}$. In fact it is mapped on $+\mathscr{H}$, since by (10) J maps the ordered triple $(\rho, i, i\infty)$ onto $(0, 1, \infty)$.

We recognize from the notation of Weierstrass in the theory of elliptic functions that

$$g_2(\omega_1, \omega_2)=60\, G_4(\omega_1, \omega_2), \qquad g_3(\omega_1, \omega_2)=140\, G_6(\omega_1, \omega_2),$$

in terms of which

$$\Delta(\omega_1, \omega_2)=g_2^3(\omega_1, \omega_2)-27\, g_3^2(\omega_1, \omega_2),$$

and

$$J(\tau)=J(\omega_1, \omega_2)=\frac{g_2^3(\omega_1, \omega_2)}{\Delta(\omega_1, \omega_2)}.$$

5. Applications. For our first application we pursue the remark of subsection **1** that $G_k(\omega_1, \omega_2)$ is a function on the lattices of the complex plane. We claim

Theorem 5. *For each pair of complex numbers*

$$(a_4, a_6) \quad \text{with} \quad 20\, a_4^3-49\, a_6^2 \neq 0$$

there is a unique lattice generated by the pair

$$(\omega_1, \omega_2), \qquad \omega_1 \neq 0, \qquad \omega_2 \neq 0, \qquad \operatorname{Im}\left(\frac{\omega_1}{\omega_2}\right)>0$$

with

$$G_4(\omega_1, \omega_2)=a_4 \quad \textit{and} \quad G_6(\omega_1, \omega_2)=a_6.$$

Proof. First we treat the case $a_4 \neq 0, a_6 \neq 0$. Obviously, it is necessary and sufficient to solve

$$\frac{G_4(\omega_1, \omega_2)}{G_6(\omega_1, \omega_2)}=\frac{a_4}{a_6}, \qquad \frac{G_4^3(\omega_1, \omega_2)}{G_6^2(\omega_1, \omega_2)}=\frac{a_4^3}{a_6^2}.$$

Since the quotient

$$\frac{G_4^3(\omega_1,\omega_2)}{G_6^2(\omega_1,\omega_2)}=\frac{G_4^3(\tau)}{G_6^2(\tau)},\qquad \tau=\frac{\omega_1}{\omega_2},$$

is a modular function of order 1, the equation

$$\frac{G_4^3(\tau)}{G_6^2(\tau)}=\frac{a_4^3}{a_6^2},$$

has a solution τ_0, which is determined up to equivalence under $\bar{\Gamma}$ and which lies in $\mathscr{H}$ since $20a_4^3-49a_6^2\neq 0$ (note that $\Delta=C(20G_4^3-49G_6^2)$, thus $\dfrac{G_4^3(i\infty)}{G_6^2(i\infty)}=\dfrac{49}{20}$).

With this τ_0

$$\frac{G_4^3(\tau_0\omega_2,\omega_2)}{G_6^2(\tau_0\omega_2,\omega_2)}=\frac{a_4^3}{a_6^2}\quad\text{for all }\omega_2\neq 0.$$

If we now choose ω_2 so that

$$\frac{G_4(\omega_2\tau_0,\omega_2)}{G_6(\omega_2\tau_0,\omega_2)}=\frac{G_4(\tau_0)}{G_6(\tau_0)}\cdot\omega_2^2=\frac{a_4}{a_6},$$

then

$$G_4(\tau_0\omega_2,\omega_2)=a_4,\qquad G_6(\tau_0\omega_2,\omega_2)=a_6.$$

Now we turn to the case $a_4=0$, $a_6\neq 0$. If one takes τ_0 equivalent to ρ under $\bar{\Gamma}$, then $G_4(\tau_0)=0$. It only remains to solve $G_6(\rho\omega_2,\omega_2)=\omega_2^{-6}G_6(\rho)=a_6$, which is possible since $G_6(\rho)\neq 0$. The case $a_4\neq 0$, $a_6=0$ is handled similarly.

We still have to prove the uniqueness. As above

$$G_4(\omega_1,\omega_2)=G_4(\omega_1^*,\omega_2^*),\qquad G_6(\omega_1,\omega_2)=G_6(\omega_1^*,\omega_2^*),$$

imply that

$$\frac{\omega_1^*}{\omega_2^*}=\frac{a\omega_1+b\omega_2}{c\omega_1+d\omega_2}\quad\text{with}\begin{pmatrix}a&b\\c&d\end{pmatrix}\in\Gamma.$$

Thus for suitable $\lambda\in\mathbb{C}$

$$\omega_1^*=\lambda(a\omega_1+b\omega_2),\qquad \omega_2^*=\lambda(c\omega_1+d\omega_2).$$

If $a_4\cdot a_6\neq 0$, then it follows from the invariance and homogeneity properties of G_4 and G_6 that

$$\lambda^4=\lambda^6=1,\quad\text{so }\lambda=\pm 1,$$

that is, (ω_1, ω_2) and (ω_1^*, ω_2^*) determine the same lattice. If $a_4 = 0$, $a_6 \neq 0$, we know that $\dfrac{\omega_1}{\omega_2} \sim \rho$ and $\lambda^6 = 1$. In this case the lattice (ω_1, ω_2) coincides with the lattice $\lambda(\omega_1, \omega_2)$ and thus coincides with the lattice (ω_1^*, ω_2^*), since ω_1, ω_2 determine an equilateral triangle. If $a_4 \neq 0$, $a_6 = 0$, then $\dfrac{\omega_1}{\omega_2} \sim i$ and $\lambda^4 = 1$, so an analogous proof gives the desired result. □

The next application is to sums of divisors. The functions

$$G_4(\tau) = \frac{\pi^4}{3^2 \cdot 5}\left(1 + 2^4 \cdot 3 \cdot 5 \sum_{n \geqq 1} \sigma_3(n) q^n\right),$$

$$G_8(\tau) = \frac{\pi^8}{3^3 \cdot 5^2 \cdot 7}\left(1 + 2^5 \cdot 3 \cdot 5 \sum_{n \geqq 1} \sigma_7(n) q^n\right)$$

(from (7) with $B_4 = B_8 = -\frac{1}{30}$) satisfy

$$G_4^2(\tau) = c\, G_8(\tau)$$

on account of Theorem **II 18**. Obviously $c = \frac{7}{3}$. Thus,

$$3\, G_4^2 = \frac{\pi^8}{3^3 \cdot 5^2}\left[1 + 2^5 \cdot 3 \cdot 5 \sum_{n \geqq 1} \sigma_3(n) q^n + 2^8 \cdot 3^2 \cdot 5^2 \sum_{n \geqq 2} \sum_{n_1 + n_2 = n} \sigma_3(n_1) \sigma_3(n_2) q^n\right],$$

and, upon equating coefficients of like powers,

$$\sigma_7(n) = \sigma_3(n) + 120 \sum_{k=1}^{n-1} \sigma_3(k) \sigma_3(n-k). \tag{11}$$

In particular, for primes $p \neq 2$,

$$p^3(p^4 - 1) = 240 \sum_{k=1}^{\frac{p-1}{2}} \sigma_3(k) \sigma_3(p-k).$$

By similar calculations one obtains from $G_4 G_6 = c' G_{10}$ and $G_4^2 G_6 = c'' G_{14}$ relations between $\sigma_3, \sigma_5, \sigma_9$ and σ_{13}.

We know that any two of the four entire modular forms

$$G_4^3,\ G_6^2,\ G_{12}, \quad \text{and } \Delta,$$

of dimension -12 form a basis. Of the many possible relations between these functions there is one of particular interest:

Multiplying G_6 and G_{12} of (7) by suitable constants we obtain

$$\tilde{G}_6(\tau) = 1 - 2^3 \cdot 3^3 \cdot 7 \sum_{n \geqq 1} \sigma_5(n) e^{2\pi i \tau n},$$

and

$$\tilde{G}_{12}(\tau) = 691 + 2^4 \cdot 3^3 \cdot 5 \cdot 7 \cdot 13 \sum_{n \geqq 1} \sigma_{11}(n) e^{2\pi i \tau n},$$

which are modular forms of dimension -6 and -12, respectively. Between these functions we have the relation

$$\tilde{G}_{12} - 691\,\tilde{G}_6^2 = C\,\Delta, \quad \text{for some constant } C,$$

since the left side has a zero at $i\infty$. Thus

$$C\,\tau(n) \equiv 2^4 \cdot 3^3 \cdot 5 \cdot 7 \cdot 13\,\sigma_{11}(n) \bmod 691\,.$$

Since $\tau(1) = \sigma_{11}(1) = 1$, it follows that $C \not\equiv 0 \bmod 691$ and, moreover, that

$$\tau(n) \equiv \sigma_{11}(n) \bmod 691\,. \tag{12}$$

This is one of the many known congruence properties for the function $\tau(n)$. Cf. F. van der Blij [1].

6. The zeros of Eisenstein series. We know the zeros in $\mathscr{F}$ of G_k for $k = 4, 6, 8, 10$ and 14; they are either at $\tau = \rho$ or $\tau = i$. We can also comment on the general case. We apply Theorem **II 18** with G_4^*, G_6^* and Δ^* replaced by G_4, G_6 and Δ. It implies the representation

$$G_k = G_4^\alpha G_6^\beta \sum_{\nu=0}^{k'} c_\nu\, G_4^{3(k'-\nu)}\, \Delta^\nu \tag{13}$$

with suitable constants c_ν and integers α, β and k' that satisfy the conditions

$$\alpha, \beta \geqq 0, \qquad 4\alpha + 6\beta + 12k' = k, \qquad 12 \neq 4\alpha + 6\beta \leqq 14\,.$$

The representations (7) for the Eisenstein series and (8) for the discriminant imply that the coefficients c_ν are rational. Dividing both sides of (13) by $G_4^\alpha\, G_6^\beta\, \Delta^{k'}$ and recalling the representation (10) for $J(\tau)$ and the equation $j = 1728\,J$, we obtain

$$\frac{G_k}{G_4^\alpha\, G_6^\beta\, \Delta^{k'}} = \sum_{\nu=0}^{k'} c'_\nu j^{k'-\nu} \tag{14}$$

with rational

$$c'_\nu = (2^{12}\, 3^6\, 5^3)^{\nu - k'}\, c_\nu\,.$$

G_k has a zero of order α at $\tau = \rho$ and a zero of order β at $\tau = i$. All other zeros, possibly including i and ρ, are zeros of the right side of (14). From $j(\rho) = 0$ and $j(i) = 1728$ we deduce

Theorem 6. *The value of j at a zero of G_k is an algebraic number.*

The case $k'=0$ is known. For $k'\geqq 1$ we must solve the equation

$$\sum_{\nu=0}^{k'} c'_\nu j^{k'-\nu}=0 .$$

The constants c_ν and hence c'_ν can be calculated by means of (13). If $1\leqq k'\leqq 3$, the solutions prove to be positive reals. Thus the zeros of G_k in the fundamental region $\mathscr{F}$ lie on the boundary of $\mathscr{F}^*$. In particular, they lie on the unit circle between ρ and i. Indeed, $j(\tau)<0$ if $\tau\in\{\tau|\operatorname{Re}\tau=-\frac{1}{2},\ \operatorname{Im}\tau>\operatorname{Im}\rho\}$; and for $\tau\in\{\tau|\operatorname{Re}\tau=0,\ \operatorname{Im}\tau>1\}$, G_k is positive if $k\equiv 0 \bmod 4$, and is different from 0 if $k\equiv 2(4)$. This is seen from (7) and the distribution of signs of the B_k. If G_k has a zero for $\operatorname{Re}\tau=0$, it must be at $\tau=i$, and this occurs when $k\equiv 2 \bmod 4$. Thus for $4\leqq k\leqq 46$ and $k=50$ the zeros of G_k in $\mathscr{F}$ lie on the unit circle between the points ρ and i.

Following F. K. C. Rankin and H. P. F. Swinnerton-Dyer [1] we prove in an elementary way the general

Theorem 7. *For even* $k>2$ *the zeros in* $\mathscr{F}$ *of the Eisenstein series* G_k *lie on the unit circle.*

Proof. We set

$$k=12k'+s \quad \text{with } s=4,6,8,10,0 \text{ or } 14 .$$

Since these G_k have zeros at $\tau=i$ and $\tau=\rho$ of order together at least $\frac{s}{12}$, and since their total number of zeros is $\frac{s}{12}+k'$, it suffices to prove that G_k has at least k' zeros on the open segment (ρ, i) of the unit circle.

We replace G_k by E_k:

$$E_k(\tau):=\frac{1}{2}\sum_{\substack{m_1,m_2\in\mathbb{Z}\\(m_1,m_2)=1}}(m_1\tau+m_2)^{-k}=\frac{1}{2\zeta(k)}G_k(\tau).$$

If $\tau=e^{i\vartheta}$, then

$$F_k(\vartheta):=e^{ik\vartheta/2}E_k(e^{i\vartheta})=\tfrac{1}{2}\sum_{\substack{m_1,m_2\in\mathbb{Z}\\(m_1,m_2)=1}}(m_1 e^{i\vartheta/2}+m_2 e^{-i\vartheta/2})^{-k}$$

is a real function of the real variable ϑ. From this sum we split off those terms for which $m_1^2+m_2^2=1$ and we obtain

$$F_k(\tau)=2\cos k\vartheta/2+R_1 .$$

Since

$$2\cos k\vartheta/2=\begin{cases}2\\-2\end{cases} \text{for } \vartheta=\frac{2\pi m}{k} \quad \text{with} \quad \begin{matrix}\text{even } m,\\ \text{odd } m,\end{matrix}$$

and since

$$|R_1|<2 \quad \text{for } \frac{\pi}{2} \leqq \vartheta \leqq \frac{2\pi}{3},$$

as we will presently prove, it follows that the number of zeros of F_k in the open interval $\left(\frac{\pi}{2}, \frac{2\pi}{3}\right)$ is at least as large as one less than the number of m's satisfying the condition

$$\frac{\pi}{2} \leqq \frac{2\pi m}{k} \leqq \frac{2\pi}{3}.$$

Hence, as is easily verified, it is at least k'.

We now show that

$$|R_1|<2 \quad \text{for } \vartheta \in \left[\frac{\pi}{2}, \frac{2\pi}{3}\right].$$

For these ϑ

$$|m_1 e^{i\vartheta/2} + m_2 e^{-i\vartheta/2}|^2 = m_1^2 + m_2^2 + 2 m_1 m_2 \cos\vartheta \geqq \tfrac{1}{2}(m_1^2 + m_2^2).$$

The number of solutions of $m_1^2 + m_2^2 = N$ is at most $2(2N^{\frac{1}{2}}+1)$, thus is smaller than $5N^{\frac{1}{2}}$ for $N \geqq 5$. The sum of the terms of the remainder R_1 with $m_1^2 + m_2^2 = 2$ has the value

$$(2\cos\vartheta/2)^{-k} + (2\sin\vartheta/2)^{-k} \leqq 1 + 2^{-k},$$

and the sum of the terms of the remainder with $m_1^2 + m_2^2 = 5$ has a value bounded by $4(\frac{5}{2})^{-k/2}$. Altogether we obtain

$$\begin{aligned}|R_1| &\leqq 1 + 2^{-k/2} + 4(\tfrac{5}{2})^{-k/2} + \sum_{N=10}^{\infty} 5N^{\frac{1}{2}}(\tfrac{1}{2}N)^{-k/2} \\ &\leqq 1 + 2^{-k/2} + 4(\tfrac{5}{2})^{-k/2} + \frac{20\sqrt{2}}{k-3}(\tfrac{9}{2})^{(3-k)/2}.\end{aligned}$$

The last inequality uses

$$\sum_{N=10}^{\infty} N^{(1-k)/2} \leqq \int_9^{\infty} x^{(1-k)/2}\, dx = 9^{(3-k)/2}.$$

Hence $|R_1|$ is majorized by a monotone decreasing function of k that is smaller than 1.1 for $k=12$. Since the validity of Theorem **7** has been verified for $k<12$, the general result follows. □

§ 2. The Eisenstein Series in the Case of Conditional Convergence

The series

$$\sum_{m_1, m_2 \in \mathbb{Z}}' (m_1 \tau + m_2)^{-2}, \qquad \tau \in \mathscr{H}$$

is not absolutely convergent. However, if we specify the order of summation as follows:

$$G_2^*(\tau) := 2\zeta(2) + 2 \sum_{m \geqq 1} \sum_{n \in \mathbb{Z}} (m\tau + n)^{-2}, \tag{15}$$

then the series defines a holomorphic function on $\mathscr{H}$. For, as we showed in **§ 1, 2**, where we included the case $k=2$,

$$\sum_{n \in \mathbb{Z}} (\tau + n)^{-2} = \sum_{\nu \geqq 1} (-2\pi i)^2 \nu e^{2\pi i \nu \tau}$$

if τ is in $\mathscr{H}$. This implies the representation

$$G_2^*(\tau) = \frac{\pi^2}{3} + 2 \sum_{m \geqq 1} (-4\pi^2) \sum_{\nu \geqq 1} \nu e^{2\pi i \tau \nu m} = \frac{\pi^2}{3} - 8\pi^2 \sum_{n \geqq 1} \sigma_1(n) e^{2\pi i \tau n} \tag{16}$$

in $\mathscr{H}$.

We now study the behavior of this function under modular transformations. Following the lead of E. Hecke [3], or [4] Nr. 24 we investigate the series

$$\Phi(\tau, s) = \sum_{m_1, m_2 \in \mathbb{Z}}' (m_1 \tau + m_2)^{-2} |m_1 \tau + m_2|^{-s}, \quad \operatorname{Re} s > 0. \tag{17}$$

By the methods of **§ 1, 1**, applied to the series for $G_k(\tau)$, $k \geqq 4$, one proves that this series converges absolutely and uniformly in s for $\operatorname{Re} s \geqq \varepsilon > 0$ for fixed τ with $\operatorname{Im} \tau > 0$. Hence for fixed τ, $\Phi(\tau, s)$ is holomorphic in s for $\operatorname{Re} s > 0$. We write

$$\Phi(\tau, s) = 2\zeta(2+s) + 2 \sum_{m \geqq 1} \sum_{n \in \mathbb{Z}} (m\tau + n)^{-2} |m\tau + n|^{-s}.$$

In order to apply to this series a method of summation originated by Poisson, we define for $\operatorname{Im} \tau > 0$, $m > 0$, and $\operatorname{Re} s > 0$ the following function of u:

$$\psi(m, u, \tau, s) := \sum_{n \in \mathbb{Z}} (m\tau + n + u)^{-2} \{(m\tau + n + u)(m\bar{\tau} + n + u)\}^{-\frac{s}{2}}.$$

Later we shall set $u=0$. As a function of u, ψ is holomorphic in the strip

$$\mathscr{S} = \{u \,|\, |\operatorname{Im} u| < |\operatorname{Im} \tau|\}$$

provided for the definition of

$$\{(m\tau + u)(m\bar{\tau} + u)\}^{-\frac{s}{2}}$$

we choose that branch of the holomorphic function

$$-\frac{s}{2}\log\{(m\tau+u)(m\bar{\tau}+u)\}$$

in $\mathscr{S}$ that is real for real u. Obviously ψ is periodic in u with period 1 and, as such, has a Fourier expansion

$$\psi(m,u,\tau,s)=\sum_{k\in\mathbb{Z}} c_{k,m}(\tau,s)\,e^{2\pi iuk} \quad \text{for } u\in\mathscr{S}\,.$$

Hence there is also an expansion

$$\Phi(\tau,s)=2\zeta(2+s)+2\sum_{m\geqq 1}\left(\sum_{k\in\mathbb{Z}} c_{k,m}(\tau,s)\right) \tag{18}$$

with the coefficients

$$c_{k,m}(\tau,s)=\int_0^1 \sum_{n\in\mathbb{Z}}(m\tau+n+u)^{-2}\{(m\tau+n+u)(m\bar{\tau}+n+u)\}^{-\frac{s}{2}}e^{-2\pi iku}\,du$$

which, because of the uniform convergence of the series on the path of integration, we may write as

$$c_{k,m}(\tau,s)=m^{-1-s}\int_{-\infty}^{\infty}(\tau+u)^{-2}\{(\tau+u)(\bar{\tau}+u)\}^{-\frac{s}{2}}e^{-2\pi ikmu}\,du\,. \tag{19}$$

In order to obtain an estimate of this integral for $k<0$, we look at the analytic continuation of the integrand to the upper u-half-plane cut from $-\bar{\tau}$ to $i\infty$. Along the path I of Figure 9, $|e^{-2\pi ikmu}|$ is bounded,

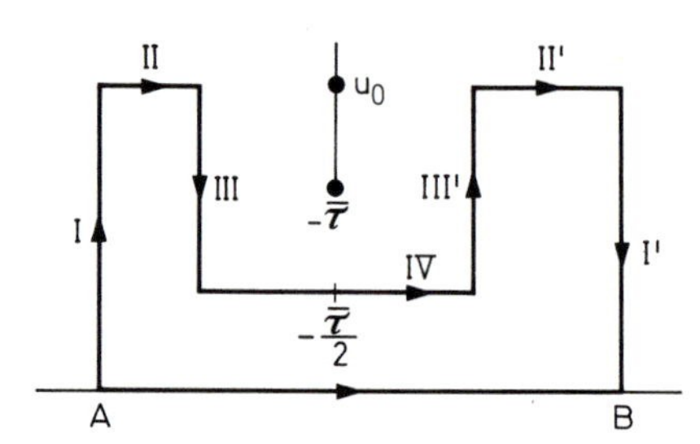

Fig. 9

indeed, uniformly bounded with respect to A. Since $\operatorname{Re}s>0$, the integral along I converges to 0 as $A\to-\infty$ because of the factor $(\tau+u)^{-2}$. The same is true of the integral along I′ as $B\to+\infty$. Hence using the notation of Figure 10 we obtain

$$m^{1+s}c_{k,m}(\tau,s)=\int_{\mathrm{II}^*}+\int_{\mathrm{III}}+\int_{\mathrm{IV}}+\int_{\mathrm{III}'}+\int_{\mathrm{II}'^*}\,.$$

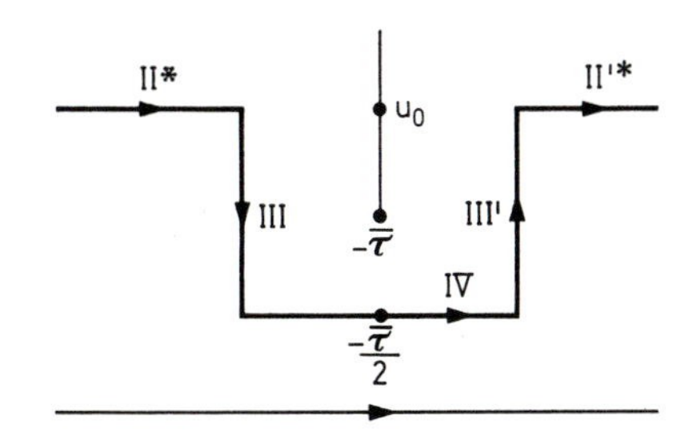

Fig. 10

Since $|e^{-2\pi ikmu_0}|=e^{2\pi km\,\mathrm{Im}(u_0)}$, the integrals along II* and II'* tend to 0 for $k<0$ as $\mathrm{Im}\,u_0\to\infty$. We obtain

$$m^{1+s}c_{k,m}(\tau,s)=\int\limits_{\mathrm{III}^*}+\int\limits_{\mathrm{IV}}+\int\limits_{\mathrm{III}'^*},$$

referring to the notation of Figure 11. This representation continues $c_{k,m}(\tau,s)$ analytically to the entire s-plane. In all three integrals

$$|e^{-2\pi iumk}|\leqq e^{\pi ymk}, \quad \text{where } y=\mathrm{Im}\left(-\frac{\bar{\tau}}{2}\right).$$

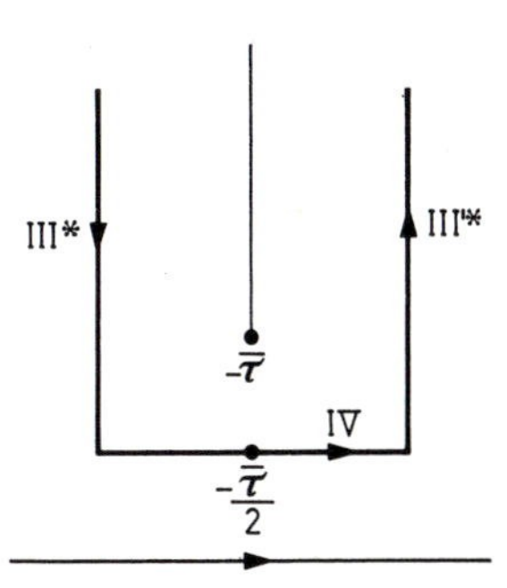

Fig. 11

Hence there is a constant $C(\tau,s)$ depending solely on τ and s such that

$$|c_{k,m}(\tau,s)|\leqq|m^{-1-s}|\,e^{\pi ymk}\cdot c(\tau,s), \qquad k<0. \tag{20}$$

If s is restricted to lie in a compact set K, then $C(\tau,s)$ may be replaced by a constant $C(\tau,K)$ depending only on τ and K. If $k>0$, there is an analogous approximation with $y=\mathrm{Im}\left(\frac{-\tau}{2}\right)$ as is easily seen by bending the path of integration into the lower half plane. Together these two approximations imply:

The series $\sum_{\substack{k\in\mathbb{Z}\\ k\neq 0\\ m\geqq 1}} c_{k,m}(\tau,s)$ *defines an entire function of s for* $\operatorname{Im}\tau>0$.

Now we investigate $c_{0,m}(\tau,s)$. For $\operatorname{Im}\tau=y>0$ and $\operatorname{Re}s>0$ we calculate

$$\int_{-\infty}^{\infty}\frac{du}{(\tau+u)^2|\tau+u|^s}=\int_{-\infty}^{\infty}\frac{du}{(iy+u)^2(y^2+u^2)^{\frac{s}{2}}}$$

$$=\int_0^{\infty}\frac{du}{(iy-u)^2(y^2+u^2)^{\frac{s}{2}}}+\int_0^{\infty}\frac{du}{(iy+u)^2(y^2+u^2)^{\frac{s}{2}}}$$

$$=2\int_0^{\infty}\frac{u^2-y^2}{(u^2+y^2)^{\frac{s}{2}+2}}\,du\,.$$

We make the substitutions $u\to uy$, $u^2\to u$, $u\to u-1$ and $u\to\frac{1}{u}$, and recalling the known properties of the Γ-function, we evaluate the last integral as follows:

$$\frac{2}{y^{s+1}}\int_0^{\infty}\frac{(u^2-1)}{(u^2+1)^{2+\frac{s}{2}}}\,du=\frac{2}{y^{s+1}}\int_0^{\infty}\frac{(u-1)}{(u+1)^{2+\frac{s}{2}}}\,\frac{du}{2u^{\frac{1}{2}}}$$

$$=\frac{1}{y^{s+1}}\int_1^{\infty}\frac{(u-2)}{u^{2+\frac{s}{2}}(u-1)^{\frac{1}{2}}}\,du=\frac{1}{y^{s+1}}\int_0^{1}\frac{(1-2u)u^{\frac{s}{2}-\frac{1}{2}}}{(1-u)^{\frac{1}{2}}}\,du$$

$$=\frac{1}{y^{s+1}}\left\{\int_0^1(1-u)^{-\frac{1}{2}}u^{\frac{s}{2}-\frac{1}{2}}du-2\int_0^1(1-u)^{-\frac{1}{2}}u^{\frac{s}{2}+\frac{1}{2}}du\right\}$$

$$=\frac{1}{y^{s+1}}\left\{\frac{\Gamma\left(\frac{1}{2}\right)\Gamma\left(\frac{s+1}{2}\right)}{\Gamma\left(\frac{s}{2}+1\right)}-2\frac{\Gamma\left(\frac{1}{2}\right)\Gamma\left(\frac{s}{2}+\frac{3}{2}\right)}{\Gamma\left(\frac{s}{2}+2\right)}\right\}$$

$$= \frac{\sqrt{\pi}}{y^{s+1}} \left\{ \frac{\Gamma\left(\frac{s+1}{2}\right)}{\Gamma\left(\frac{s}{2}+1\right)} - 2\,\frac{\left(\frac{s+1}{2}\right)\Gamma\left(\frac{s+1}{2}\right)}{\left(\frac{s}{2}+1\right)\Gamma\left(\frac{s}{2}+1\right)} \right\}$$

$$= -\frac{\sqrt{\pi}}{y^{1+s}} \frac{\Gamma\left(\frac{s+1}{2}\right)}{\Gamma\left(\frac{s}{2}+1\right)} \frac{s}{s+2}.$$

Thus, if $\operatorname{Re} s > 0$ and $y = \operatorname{Im} \tau > 0$, then

$$c_{0,m}(\tau, s) = -m^{-1-s} \frac{\sqrt{\pi}}{y^{1+s}} \frac{\Gamma\left(\frac{s+1}{2}\right)}{\Gamma\left(\frac{s}{2}+1\right)} \frac{s}{s+2}. \tag{21}$$

The right side is a meromorphic function in the s-plane which is holomorphic at $s=0$. We now summarize the results (18) and (21) and we obtain

$$\Phi(\tau, s) = 2\zeta(2+s) + 2 \sum_{m \geqq 1} \sum_{k \neq 0} c_{k,m}(\tau, s) - 2\zeta(s+1) \frac{\sqrt{\pi}}{y^{1+s}} \frac{\Gamma\left(\frac{s+1}{2}\right)}{\Gamma\left(\frac{s}{2}+1\right)} \frac{s}{s+2}, \quad \text{for } \operatorname{Im} \tau > 0. \tag{22}$$

By what has been said above, the middle term is an entire function of s and so this representation of $\Phi(\tau, s)$ continues it to a neighborhood of $s=0$. (In fact the right side represents an entire function of s. This follows without difficulty from the properties of the functions ζ and Γ.) *We define $G_2(\tau)$ as the value of the holomorphic function $\Phi(\tau, s)$ at $s=0$:*

$$G_2(\tau) := \Phi(\tau, 0). \tag{23}$$

Since for $S = \begin{pmatrix} a & b \\ c & d \end{pmatrix} \in \Gamma$,

$$\Phi(S\tau, s) = \Phi(\tau, s)(c\tau + d)^2 |c\tau + d|^s,$$

we see that

$$G_2(S\tau) = (c\tau + d)^2 G_2(\tau). \tag{24}$$

We compute the coefficients $c_{k,m}(\tau, 0)$ from the representation (19) of $c_{k,m}(\tau, s)$. By bending the path of integration either up or down and calculating the residue, we obtain

$$c_{k,m}(\tau, 0) = m^{-1} \int_{-\infty}^{\infty} (\tau+u)^{-2} e^{-2\pi i k m u} du = \begin{cases} 0, & k<0, \\ -4\pi^2 k e^{2\pi i k m \tau}, & k>0. \end{cases}$$

Finally we arrive at the expansion

$$G_2(\tau) = \frac{\pi^2}{3} - 8\pi^2 \sum_{n \geqq 1} \sigma_1(n) e^{2\pi i \tau n} - \frac{\pi}{y} \quad \text{for } \operatorname{Im}\tau = y > 0. \tag{25}$$

From the behavior of G_2 under modular transformations we immediately obtain the behavior in $\mathscr{H}$ of the holomorphic function G_2^*,

$$G_2^*(\tau) := G_2(\tau) + \frac{\pi}{y}$$

under modular transformations:

$$G_2^*(S\tau) = G_2(S\tau) + 2\pi i \frac{|c\tau+d|^2}{\tau-\bar{\tau}} = (c\tau+d)^2 G_2(\tau) + 2\pi i \frac{|c\tau+d|^2}{\tau-\bar{\tau}}$$

$$= (c\tau+d)^2 G_2^*(\tau) - 2\pi i \frac{(c\tau+d)^2}{\tau-\bar{\tau}} + 2\pi i \frac{|c\tau+d|^2}{\tau-\bar{\tau}}.$$

Now

$$|c\tau+d|^2 - (c\tau+d)^2 = (c\tau+d)c(\bar{\tau}-\tau).$$

Thus we proved

Theorem 7. *The behavior of*

$$G_2^*(\tau) = \frac{\pi^2}{3} + 2 \sum_{m \geqq 1} \sum_{n \in \mathbb{Z}} (m\tau+n)^{-2}, \qquad \tau \in \mathscr{H}$$

under modular transformations is given by the equation

$$G_2^*(S\tau) = (c\tau+d)^2 G_2^*(\tau) - 2\pi i c(c\tau+d) \quad \text{for } S = \begin{pmatrix} a & b \\ c & d \end{pmatrix} \in \Gamma.$$

§ 3. The Discriminant Δ

We now have a new approach to the discriminant Δ. Integrating the series

$$G_2^*(\tau) = \frac{\pi^2}{3} - 8\pi^2 \sum_{m,k \geqq 1} m e^{2\pi i m k \tau},$$

and then multiplying by $-\frac{1}{4\pi i}$, we arrive at

$$\psi(\tau) := \frac{\pi i \tau}{12} - \sum_{m,k \geqq 1} \frac{1}{k} e^{2\pi i m k \tau}. \tag{26}$$

We want to determine its behavior under modular transformations $S = \begin{pmatrix} a & b \\ c & d \end{pmatrix}$. First we remark that

$$\begin{aligned} \frac{d}{d\tau}(\psi(S\tau) - \psi(\tau)) &= \frac{d\psi(S\tau)}{d\tau} \cdot \frac{dS\tau}{d\tau} - \frac{d\psi(\tau)}{d\tau} \\ &= -\frac{1}{4\pi i}\{G_2^*(S\tau)(c\tau+d)^2 - G_2^*(\tau)\} \\ &= -\frac{1}{4\pi i} \cdot \frac{-2\pi i c}{c\tau+d} = \frac{c}{2} \cdot \frac{1}{c\tau+d}. \end{aligned}$$

Next we integrate to obtain the transformation formula

$$\psi(S\tau) - \psi(\tau) = \tfrac{1}{2}\log(c\tau+d) + A(S),$$

for which we choose the principal branch of the logarithm and where $A(S)$ is independent of τ. We may write this transformation formula in the form

$$\psi(S\tau) - \psi(\tau) = (\operatorname{sgn} c)^2 \frac{1}{2} \log(c\tau+d) - \operatorname{sgn} c \cdot \frac{\pi i}{4} + C(S), \tag{27}$$

which yields $C(S) = C(-S)$ for all $S \in \Gamma$ (in particular for $S = U$). By setting $S = T$ in the transformation formula we obtain the special result

$$C(T) = 0.$$

Likewise, it follows immediately from the expansion (26) for $\psi(\tau)$ that

$$C(U) = \frac{\pi i}{12}.$$

The so-called *Dedekind function*

$$\eta(\tau) := e^{\psi(\tau)}$$

has the transformation formula

$$\eta(S\tau) = \eta(\tau)(\sqrt{c\tau+d})^{s^2} e^{-s\frac{\pi i}{4} + C(S)}, \qquad s = \operatorname{sgn} c, \tag{28}$$

where the square root is taken to have positive real or positive imaginary part. η does not vanish in $\mathscr{H}$ since ψ is holomorphic in $\mathscr{H}$. It follows from (26) that

$$\eta(\tau)=e^{\frac{\pi i\tau}{12}}\prod_{m\geqq 1}(1-e^{2\pi i m\tau}).$$

The relations $C(T)=0$ and $C(U)=\dfrac{\pi i}{12}$ imply that

$$\eta^{24}|_{12}U=\eta^{24}|_{12}T=\eta^{24},$$

thus

$$\eta^{24}|_{12}S=\eta^{24}\quad \text{for all } S\in\Gamma, \tag{29}$$

cf. **II § 4, 1.** Furthermore, $\eta^{24}(\tau)$ is holomorphic in $\mathscr{H}$ and has the expansion

$$\eta^{24}(\tau)=\sum_{n=1}^{\infty}a(n)e^{2\pi i\tau n}.$$

Thus, we have

Theorem 8. *η^{24} is an entire modular form of dimension -12 and has a zero of order 1 at $i\infty$, but has no zeros in $\mathscr{H}$.*

Because $a(1)=1$, it follows from Theorem **7** that

$$\eta^{24}(\tau)=e^{2\pi i\tau}\prod_{n\geqq 1}(1-e^{2\pi i\tau n})^{24}=(2\pi)^{-12}\Delta(\tau). \tag{30}$$

Within the framework of this section this equation may be taken as the definition of Δ. Naturally, $a(n)=\tau(n)$. With this function Δ we have obtained a new way to determine the number of zeros of an entire modular form f_k of dimension $-k$. Indeed,

$$\frac{f_k^{12}}{\Delta^k}$$

is a modular function the numerator of which has as many zeros in $\mathscr{F}$ as the denominator, hence k zeros. This means that f_k has $\dfrac{k}{12}$ zeros in $\mathscr{F}$. This again implies Theorem **II, 13.**

Chapter IV. Subgroups of the Modular Group

In this chapter the questions and results of the first chapter for the homogeneous and inhomogeneous modular group will be carried over and extended to its subgroups. We will restrict ourselves mainly to subgroups of finite index, and turn our particular interest to a special class of subgroups, the so-called congruence groups.

§ 1. Subgroups of the Modular Group

1. Homogeneous and inhomogeneous groups. Let Γ_1 be a subgroup of the homogeneous group Γ. The map

$$\varphi: A \mapsto \bar{A}, \qquad A = \begin{pmatrix} a & b \\ c & d \end{pmatrix} \in \Gamma, \qquad \bar{A}: \tau \mapsto \frac{a\tau+b}{c\tau+d},$$

induces a homomorphism of Γ_1 onto a subgroup

$$\bar{\Gamma}_1 := \varphi(\Gamma_1)$$

of the inhomogeneous modular group $\bar{\Gamma}$. This homomorphism has the kernel $\{\pm I\}$ if $-I \in \Gamma_1$, and is an isomorphism if $-I \notin \Gamma_1$.

Conversely, if a group of transformations from $\bar{\Gamma}$ is given in the form

$$\tau \mapsto \frac{a\tau+b}{c\tau+d}, \qquad a, b, c, d \in \mathbb{Z}, \qquad ad-bc=1,$$

then the set of matrices $\pm\begin{pmatrix} a & b \\ c & d \end{pmatrix}$ forms a group $\Gamma_1 \subset \Gamma$ and the given group is

$$\bar{\Gamma}_1 = \varphi(\Gamma_1).$$

This group Γ_1 contains the matrix $-I$. There exist, however, subgroups of Γ (cf. **§ 2,2**), which do not contain $-I$. For such groups, $\Gamma_1 \cup (-I)\Gamma_1$ is a proper group extension with

$$\bar{\Gamma}_1 = \varphi(\Gamma_1) = \varphi(\Gamma_1 \cup (-I)\Gamma_1).$$

The group $\bar{\Gamma}$ is not the isomorphic image under φ of a subgroup Γ_0 of Γ when Γ_0 does not contain the matrix $-I$. Indeed, it follows from $\varphi(\Gamma_0)=\bar{\Gamma}$, that one of the two matrices $\pm\begin{pmatrix}0 & -1\\ 1 & 0\end{pmatrix}$, and consequently also its square, which equals $-I$, must lie in Γ_0.

Suppose now that Γ_1 is a subgroup of Γ and $-I$ is in Γ_1. The coset decomposition

$$\Gamma=\bigcup_{\nu=1}^{\mu}\Gamma_1 S_\nu, \qquad S_\nu\in\Gamma,$$

implies the coset decomposition

$$\bar{\Gamma}=\bigcup_{\nu=1}^{\mu}\bar{\Gamma}_1 \bar{S}_\nu, \qquad \bar{S}_\nu\in\bar{\Gamma}.$$

For the indices

$$\mu=(\Gamma:\Gamma_1), \qquad \bar{\mu}=(\bar{\Gamma}:\bar{\Gamma}_1)$$

it follows that

$$\bar{\mu}=\mu \quad \text{if} \quad -I\in\Gamma_1.$$

On the other hand,

$$\bar{\mu}=\tfrac{1}{2}\mu \quad \text{if} \quad -I\notin\Gamma_1.$$

If Γ_1 is normal in Γ, then $\bar{\Gamma}_1=\varphi(\Gamma_1)$ is normal in $\bar{\Gamma}$, and conversely. When

$$A\Gamma_1 A^{-1}=\Gamma_1 \quad \text{for all } A\in\Gamma,$$

it follows that

$$\bar{A}\bar{\Gamma}_1\bar{A}^{-1}=\bar{\Gamma}_1 \quad \text{for all } \bar{A}\in\bar{\Gamma},$$

and conversely—this, moreover, is independent of whether or not $-I\in\Gamma_1$.

2. Equivalence classes and fixed points. We carry over to subgroups some of the definitions of Chapter **I**.

Two points $\tau,\tau'\in\mathscr{H}^*$ are called *equivalent* under the subgroup $\bar{\Gamma}_1=\varphi(\Gamma_1)$ of $\bar{\Gamma}$ (for either of the choices of Γ_1 in Γ) if there is an $S\in\Gamma_1$ such that

$$\tau'=S(\tau).$$

An equivalence class under the subgroup $\bar{\Gamma}_1$ is the set of all those points of $\mathscr{H}^*$ which are equivalent to one another under $\bar{\Gamma}_1$.

A *fundamental set* $\mathscr{F}$ for $\bar{\Gamma}_1$ is a set of points of $\mathscr{H}^*$ such that every point of $\mathscr{H}^*$ is equivalent to exactly one point of $\mathscr{F}$.

The *fundamental region* of $\bar{\Gamma}_1$ in the case of finite index $(\bar{\Gamma}:\bar{\Gamma}_1)$ will be discussed in detail in **§ 4**.

A point $\tau_0 \in \mathscr{H}^*$ is called a *fixed point* of $\bar{\Gamma}_1$ if there is an $S \in \Gamma_1$, distinct from $\pm I$, such that

$$S(\tau_0) = \tau_0 .$$

We prove two theorems about points which are equivalent under a subgroup $\bar{\Gamma}_1 = \varphi(\Gamma_1)$ of $\bar{\Gamma}$.

1. If $\tau_0 \in \mathscr{H}^*$ is a fixed point of the substitution $\bar{L} \in \bar{\Gamma}_1$, then $L_1(\tau_0) = \tau_1$ with $L_1 \in \Gamma_1$ is a fixed point of the substitution $\bar{L}_1 \bar{L} \bar{L}_1^{-1} \in \bar{\Gamma}_1$. It follows from

$$L(\tau_0) = \tau_0 ,$$

that

$$(L_1 L L_1^{-1})(L_1(\tau_0)) = L_1(\tau_0) .$$

Of course,

$$(L_1 L L_1^{-1})^f = \pm I \quad \text{if } L^f = \pm I .$$

2. Let r be either a rational number or $i\infty$, and let $r = A(i\infty)$ with $A \in \Gamma$. Further let κ be the smallest natural number k for which

$$A U^k A^{-1} \in \Gamma_1 .$$

If

$$r' = L_1(r) , \qquad L_1 \in \Gamma_1 ,$$

that is, if

$$r' = A'(i\infty) \quad \text{with } A' = L_1 A ,$$

then κ is also the smallest natural number k for which

$$A' U^k A'^{-1} \in \Gamma_1 .$$

It follows from

$$A U^k A^{-1} = L_1^{-1} A' U^k A'^{-1} L_1$$

that

$$A U^k A^{-1} \quad \text{and } A' U^k A'^{-1}$$

simultaneously belong to Γ_1. The transformation $\bar{A} \bar{U}^\kappa \bar{A}^{-1}$ is the generator of the cyclic group of all transformations from $\bar{\Gamma}_1$ with the fixed point $A(i\infty) = r$, for

$$A U^k A^{-1} \in \Gamma_1 \quad \text{is equivalent with } \bar{A} \bar{U}^k \bar{A}^{-1} \in \bar{\Gamma}_1 .$$

§ 2. The Principal Congruence Groups

For a natural number N let z_N denote the residue class of $z \in \mathbb{Z}$ modulo N and let $\mathbb{Z}_N$ denote the ring of all residue classes modulo N. As before Γ denotes the group of homogeneous modular transformations which is isomorphic to the special linear group $SL(2,\mathbb{Z})$. Correspondingly, we use the symbol Γ_N for the group $SL(2,\mathbb{Z}_N)$.

1. The homogeneous principal congruence group of level N. The ring homomorphism $z\mapsto z_N$ of $\mathbb{Z}$ onto $\mathbb{Z}_N$ induces the group homomorphism σ of Γ into Γ_N with

$$\sigma\colon\ \Gamma\to\Gamma_N,\qquad \begin{pmatrix} a & b\\ c & d\end{pmatrix}\mapsto\begin{pmatrix} a_N & b_N\\ c_N & d_N\end{pmatrix}. \tag{1}$$

The kernel

$$\Gamma(N):=\left\{\begin{pmatrix} a & b\\ c & d\end{pmatrix}\in\Gamma\,\middle|\,\begin{pmatrix} a & b\\ c & d\end{pmatrix}\equiv\begin{pmatrix} 1 & 0\\ 0 & 1\end{pmatrix}\bmod N\right\} \tag{2}$$

of σ is a normal subgroup of Γ and is called the *homogeneous principal congruence group of level N*.

We show that the image $\sigma(\Gamma)$ is isomorphic to the full group Γ_N. This is implied by the following

Lemma. *For each solution a,b,c,d in integers of*

$$ad-bc\equiv 1\bmod N$$

there are integers $a'\equiv a$, $b'\equiv b$, $c'\equiv c$, and *$d'\equiv d\bmod N$ with*

$$a'd'-b'c'=1.$$

Proof. The proof proceeds in two steps; first we show that we can assume that the greatest common divisor $(c,d)=1$, i.e. that there are integers c' and d' with $c'\equiv c\bmod N$, $d'\equiv d\bmod N$, and that $(c',d')=1$. By hypothesis $(c,d,N)=1$. If $c=0$, then one may choose $c':=N$ and $d':=d$. Otherwise, form the product $P:=\prod_{p\mid c,\,p\nmid N} p$, where p denotes a prime, and then set $c':=c$, $d':=d+lN$ where l is chosen subject to the condition $d+lN\equiv 1\bmod P$. This choice is possible since $(N,P)=1$. Now $d'\equiv 1\bmod P$, $c'\equiv 0\bmod P$ and so $(c',d')=1$ as desired. In what follows suppose $(c,d)=1$. Then the equation $xd-yc=1$ has integral solutions a_1,b_1, i.e. the matrix $A_1:=\begin{pmatrix} a_1 & b_1\\ c & d\end{pmatrix}$ is an element of Γ. If $A:=\begin{pmatrix} a & b\\ c & d\end{pmatrix}$, then

$$AA_1^{-1}=\begin{pmatrix} a & b\\ c & d\end{pmatrix}\begin{pmatrix} d & -b_1\\ -c & a_1\end{pmatrix}=\begin{pmatrix} ad-bc & a_1b-ab_1\\ 0 & a_1d-cb_1\end{pmatrix}\equiv\begin{pmatrix} 1 & k\\ 0 & 1\end{pmatrix}\bmod N,$$

or $A\equiv U^kA_1\bmod N$, however, $U^kA_1\in\Gamma$. Hence the entries of U^kA_1, yield the required solution: a',b',c', and d'. □

As a corollary we have obtained the desired result which we formulate as

Theorem 1. *The image of Γ under the homomorphism σ is isomorphic to the group $\Gamma_N = SL(2,\mathbb{Z}_N)$:*

$$\sigma(\Gamma) \cong \Gamma/\Gamma(N) \cong \Gamma_N .$$

2. The order of Γ_N. We now determine the order $\mu(N) := |\Gamma_N|$ of Γ_N. Equivalently this is the number of incongruent solutions of

$$ad - bc \equiv 1 \bmod \mathrm{N} .$$

By the Chinese remainder theorem,

$$\mu(N_1 N_2) = \mu(N_1)\mu(N_2)$$

for $(N_1, N_2) = 1$. Consequently we may restrict ourselves to powers of a prime $N = p^\alpha$. There are $\varphi(p^\alpha)$ residue classes $a \bmod p^\alpha$ with $a \not\equiv 0 \bmod p$, where $\varphi(n)$ denotes Euler's function, the number of residue classes $\bmod n$ which are relatively prime to n. To each of these classes for a, the numbers b and c may be choosen arbitrarily modulo p^α. $d \bmod p^\alpha$ is uniquely determined. In this case there are altogether $\varphi(p^\alpha)p^{2\alpha}$ solutions. There are $p^{\alpha-1}$ residue classes $a \bmod \mathrm{p}^\alpha$ with $a \equiv 0 \bmod p$, and, corresponding to each of these, $d \bmod p^\alpha$ may be choosen arbitrarily. Since in this case $(p, bc) = 1$, there are $\varphi(p^\alpha)$ possibilities for $b \bmod p^\alpha$ and $c \bmod p^\alpha$ is again uniquely determined. Hence there are $\varphi(p^\alpha)p^{2\alpha-1}$ additional solutions. Together we obtain

$$\mu(p^\alpha) = \varphi(p^\alpha)\, p^{2\alpha}\left(1 + \frac{1}{p}\right) = p^{3\alpha}\left(1 - \frac{1}{p^2}\right).$$

Hence for arbitrary N we arrive at the result:

$$|\Gamma_N| = \mu(N) = N^3 \prod_{p|N}\left(1 - \frac{1}{p^2}\right), \tag{3}$$

where the product runs over prime divisors of N.

3. The inhomogeneous principal congruence group of level N. We now consider the inhomogeneous case. Suppose as before that $\bar{\Gamma}$ denotes the group of inhomogeneous modular substitutions. By the *inhomogeneous principal congruence group of level N* we understand the group

$$\bar{\Gamma}(N) := \left\{ \tau \to \frac{a\tau + b}{c\tau + d} \,\middle|\, \begin{pmatrix} a & b \\ c & d \end{pmatrix} \in \Gamma, \quad \begin{pmatrix} a & b \\ c & d \end{pmatrix} \equiv \begin{pmatrix} 1 & 0 \\ 0 & 1 \end{pmatrix} \bmod N \right\}. \tag{4}$$

The homogeneous group

$$\Gamma[N] := \Gamma(N) \cup (-I)\Gamma(N) \tag{5}$$

will likewise be called the principal congruence group. We note that $\bar{\Gamma}(N)$ can be obtained as the image of $\Gamma[N]$ under the homomorphism

$$\varphi: A = \begin{pmatrix} a & b \\ c & d \end{pmatrix} \to \bar{A}, \quad \text{where } \bar{A}: \tau \mapsto \frac{a\tau+b}{c\tau+d},$$

which has the kernel $\{\pm I\}$. For $N>2$, $\Gamma[N]$ properly contains $\Gamma(N)$, however, $\bar{\Gamma}[N]=\bar{\Gamma}(N)$. The group $\bar{\Gamma}(N)$ is a normal subgroup of $\bar{\Gamma}$; the factor group

$$\bar{\Gamma}_N := \bar{\Gamma}/\bar{\Gamma}(N)$$

is called the *modulary group of level* N. In view of § **1** and the preceding results, the order of $\bar{\Gamma}_N$ is determined from

$$|\bar{\Gamma}_N| = (\bar{\Gamma}:\bar{\Gamma}(N)) = (\Gamma:\Gamma[N]) = \tfrac{1}{2}(\Gamma:\Gamma(N)) \quad \text{if } N>2.$$

Together with the result for $N=2$ we thus obtain

$$|\bar{\Gamma}_N| = \begin{cases} \mu(2)=6 & \text{for } N=2, \\ \tfrac{1}{2}\mu(N) = \dfrac{N^3}{2}\displaystyle\prod_{p|N}\left(1-\frac{1}{p^2}\right) & \text{for } N>2. \end{cases} \tag{6}$$

The modulary groups $\bar{\Gamma}_N$ of levels $N=2,3,4$ and 5 prove to be isomorphic to the rotation groups of certain regular solids and consequently are isomorphic to certain symmetric groups S_ν or alternating groups A_ν. In particular

$$\begin{aligned} &\bar{\Gamma}_2 \cong \text{triangle group} \cong S_3 \quad \text{and } |\bar{\Gamma}_2| = 6, \\ &\bar{\Gamma}_3 \cong \text{tetrahedral group} \cong A_4 \quad \text{and } |\bar{\Gamma}_3| = 12, \\ &\bar{\Gamma}_4 \cong \text{octahedral group} \cong S_4 \quad \text{and } |\bar{\Gamma}_4| = 24, \\ &\bar{\Gamma}_5 \cong \text{icosahedral group} \cong A_5 \quad \text{and } |\bar{\Gamma}_5| = 60. \end{aligned}$$

The theory of rotation groups is treated in F. Klein [1], the connection with modulary groups is discussed in F. Klein-R. Fricke [1], vol. 1, p. 353ff.

The modulary group $\bar{\Gamma}_p$ of prime level p may obviously be interpreted as the group of projective mappings of the one-dimensional projective space $\mathbb{P}_1$ over the field of p elements F_p. So one has

$$\bar{\Gamma}_p \cong PSL(2,p),$$

where $PSL(2,p)$ denotes the projective group over $\mathbb{P}_1$. Hence $\bar{\Gamma}_p$ may be represented as the permutation group of $p+1$ elements. For prime levels all subgroups of the modulary groups are known, cf. F. Klein-R. Fricke [1], vol. 1, pp. 419–491. For $p \geqq 5$ the modulary groups are simple. The groups $\bar{\Gamma}_5$ and $\bar{\Gamma}_7$ are the non-cyclic simple groups of lowest order.

§ 3. Congruence Groups

In this section we introduce a more general class of subgroups of the modular group.

1. Homogeneous congruence groups. Conductor. A subgroup Γ_1 of Γ is called a *homogeneous congruence group of level* N_1 provided it contains the homogeneous principal congruence group $\Gamma(N_1)$ of level N_1. This number N_1 is not uniquely determined by Γ_1, however, the least level F (for Führer) is well defined and is called the *conductor* of Γ_1. It is more precise to write $\Gamma_1(F)$ for Γ_1. The conductor is characterized by the following property:

Theorem 2. *The conductor F of a congruence group Γ_1 is the smallest natural number N such that for arbitrary $A \in \Gamma$*

$$A\, U^N A^{-1} \in \Gamma_1 .$$

Proof. If N is the smallest natural number such that for all $A \in \Gamma$ the matrix $A\, U^N A^{-1}$ lies in Γ_1, then N divides every level of Γ_1 and consequently divides F. Hence $N \leqq F$. We next show that every $M = \begin{pmatrix} a & b \\ c & d \end{pmatrix} \in \Gamma(N)$ is an element of Γ_1 and so $N \geqq F$. To this end we write $M \in \Gamma(N)$ as a product of matrices in such a way that the fact $M \in \Gamma_1$ is evident. Obviously, $\begin{pmatrix} 1 & gN \\ 0 & 1 \end{pmatrix} \in \Gamma_1$ for all $g \in \mathbb{Z}$. Since $(d, N) = (d, c) = 1$, it is possible to determine an integer g in the equation

$$\begin{pmatrix} a & b \\ c & d \end{pmatrix} \begin{pmatrix} 1 & gN \\ 0 & 1 \end{pmatrix} = \begin{pmatrix} a & b + agN \\ c & d + cgN \end{pmatrix}$$

in such a way that $(d + cgN, F) = 1$. Therefore we assume $(d, F) = 1$ for the element d of M. Furthermore,

$$\begin{pmatrix} 1 & gN \\ 0 & 1 \end{pmatrix} \begin{pmatrix} a & b \\ c & d \end{pmatrix} = \begin{pmatrix} a + cgN & b + dgN \\ c & d \end{pmatrix},$$

and, since $(d, F) = 1$, we can solve

$$b + dgN = \left(\frac{b}{N} + dg\right) N \equiv 0 \bmod F$$

for $g \in \mathbb{Z}$. This means that without loss of generality we may further suppose that $b \equiv 0 \bmod F$. The matrix

$$\begin{pmatrix} 1 & 0 \\ gN & 1 \end{pmatrix} = \begin{pmatrix} 0 & -1 \\ 1 & 0 \end{pmatrix} \begin{pmatrix} 1 & -gN \\ 0 & 1 \end{pmatrix} \begin{pmatrix} 0 & 1 \\ -1 & 0 \end{pmatrix}$$

belongs to Γ_1 as is obvious from this representation. Since $(d, F)=1$, the equation

$$\begin{pmatrix} a & b \\ c & d \end{pmatrix}\begin{pmatrix} 1 & 0 \\ gN & 1 \end{pmatrix}=\begin{pmatrix} a+bgN & b \\ c+dgN & d \end{pmatrix}$$

implies that we may choose g so that $c+dgN\equiv 0 \bmod F$. Thus we may furthermore assume that $c\equiv 0 \bmod F$ and that

$$M=\begin{pmatrix} a & b \\ c & d \end{pmatrix}\equiv\begin{pmatrix} a & 0 \\ 0 & d \end{pmatrix} \bmod F .$$

Since $ad\equiv 1 \bmod F$, we have

$$M=\begin{pmatrix} a & b \\ c & d \end{pmatrix}\equiv\begin{pmatrix} a & ad-1 \\ 1-ad & d(2-ad) \end{pmatrix} \bmod F . \tag{$*$}$$

Consequently there is an $L\in\Gamma(F)\subset\Gamma_1$ with

$$M=M_0 L ,$$

where M_0 denotes the matrix on the right in (*). M_0 may be written as the product

$$M_0=\begin{pmatrix} 1 & 0 \\ 1-d & 1 \end{pmatrix}\begin{pmatrix} a & a-1 \\ 1-a & 2-a \end{pmatrix}\begin{pmatrix} 1 & d-1 \\ 0 & 1 \end{pmatrix},$$

where in view of the remarks above, the first and last factors are in Γ_1, since $d\equiv 1 \bmod N$, while the middle factor is seen to be in Γ_1, since $a\equiv 1 \bmod N$ and

$$\begin{pmatrix} a & a-1 \\ 1-a & 2-a \end{pmatrix}=\begin{pmatrix} 1 & 0 \\ -1 & 1 \end{pmatrix}\begin{pmatrix} 1 & a-1 \\ 0 & 1 \end{pmatrix}\begin{pmatrix} 1 & 0 \\ 1 & 1 \end{pmatrix}.$$

This proves Theorem **2**: $F=N$. □

If we assume that Γ_1 is not a congruence group, but that it is of finite index in Γ, then there is a minimum natural number F, such that $AU^F A^{-1}\in\Gamma_1$ for all $A\in\Gamma$. In this case too, we call F the conductor of Γ_1. Cf. K. Wohlfahrt [1].

2. Special congruence groups. It is customary to give specific designations to three special congruence groups:

$$\Gamma_0(N):=\left\{\begin{pmatrix} a & b \\ c & d \end{pmatrix}\in\Gamma \,\middle|\, c\equiv 0 \bmod N\right\},$$

$$\Gamma^0(N):=\left\{\begin{pmatrix} a & b \\ c & d \end{pmatrix}\in\Gamma \,\middle|\, b\equiv 0 \bmod N\right\}, \tag{7}$$

$$\Gamma_0^0(N):=\left\{\begin{pmatrix} a & b \\ c & d \end{pmatrix}\in\Gamma \,\middle|\, b\equiv c\equiv 0 \bmod N\right\}.$$

It is immediate that the above are groups, and one sees, as in the case of $\Gamma(N)$, that their conductor is N. Obviously the index of $\Gamma_0(N)$ in Γ is equal to the index of $\Gamma^0(N)$ in Γ. To compute the index $(\Gamma:\Gamma_0(N))$, we observe that

$$(\Gamma_0(N):\Gamma(N)) = N\,\varphi(N),$$

because, if $c \equiv 0 \bmod N$, the congruence

$$a\,d - b\,c \equiv 1 \bmod N$$

has exactly $N\varphi(N)$ incongruent solutions. With $\mu(N)$ as in equation (3) we obtain the index

$$\mu_0(N) := (\Gamma:\Gamma_0(N)) = \frac{\mu(N)}{N\,\varphi(N)} = N \prod_{p|N} \left(1 + \frac{1}{p}\right). \tag{8}$$

If $b \equiv c \equiv 0 \bmod N$, then our congruence has exactly $\varphi(N)$ incongruent solutions. Thus

$$(\Gamma:\Gamma_0^0(N)) = \frac{\mu(N)}{\varphi(N)} = N^2 \prod_{p|N} \left(1 + \frac{1}{p}\right).$$

3. Inhomogeneous congruence groups. A subgroup $\bar{\Gamma}_1$ of $\bar{\Gamma}$ is called an *inhomogeneous congruence group of level* N_1 if $\bar{\Gamma}_1$ is the homomorphic image of a homogeneous congruence group Γ_1 of level N_1 under the map

$$\varphi: A \mapsto \bar{A}, \qquad A \in \Gamma_1 .$$

It contains the inhomogeneous principal congruence group $\bar{\Gamma}(N_1)$.

If $\bar{\Gamma}_1$ is an arbitrary subgroup of finite index in $\bar{\Gamma}$, then as *conductor* of $\bar{\Gamma}_1$ we choose the conductor of Γ_1 if $\varphi(\Gamma_1) = \bar{\Gamma}_1$. Obviously, the two (not necessarily different) groups Γ_1 and $\Gamma_1 \cup (-I)\Gamma_1$ have the same conductor. The conductor of the inhomogeneous group $\bar{\Gamma}_1$ has a geometric significance which we shall explain later on (cf. Theorem **5**).

§ 4. Fundamental Region

Let $\bar{\Gamma}_1$ be a subgroup of finite index μ in $\bar{\Gamma}$. In this and the following section of this chapter we assume that $\bar{\Gamma}_1$ is the homomorphic image under φ of the group Γ_1 which contains $-I$. We shall generally omit the bar in the designation of the inhomogeneous groups and of their elements. We trust that this will cause no confusion.

1. Fundamental region. We generalize the definitions of **I, § 5**.

A set $\mathscr{F}_1$ of the extended upper half-plane $\mathscr{H}^*$ is called a *fundamental set* for the group Γ_1 if it contains exactly one representative of each class of points equivalent under Γ_1.

A set $\mathscr{F}_1$ is called a *fundamental region* if $\mathscr{F}_1$ contains a fundamental set, and if

$$\tau \in \mathscr{F}_1, \quad S(\tau) \in \mathscr{F}_1, \quad S \neq I, \quad \text{and } S \in \Gamma_1$$

imply that τ is a boundary point of $\mathscr{F}_1$.

It is easy to show the existence of a fundamental region $\mathscr{F}_1$ for Γ_1. Indeed, if

$$\Gamma = \bigcup_{\nu=1}^{\mu} \Gamma_1 S_\nu, \quad \mu = (\Gamma : \Gamma_1)$$

is a coset decomposition of the modular group (cf. § **1, 1**) and, if $\mathscr{F}$ is a fundamental region of Γ, then

$$\mathscr{H}^* = \bigcup_{\nu=1}^{\mu} \Gamma_1 S_\nu(\mathscr{F}).$$

Hence

$$\mathscr{F}_1 := \bigcup_{\nu=1}^{\mu} S_\nu(\mathscr{F}) \tag{9}$$

contains a fundamental set for the group Γ_1. If—as is always possible—the S_ν are chosen so that

$$S_\nu(\mathscr{F}) \cap S_{\nu'}(\mathscr{F}) = \phi \quad \text{for } \nu \neq \nu',$$

then the Γ_1-equivalent points of $\mathscr{F}_1$ lie on the boundary of $\mathscr{F}_1$.

Moreover, we prove

Theorem 3. *There exists a fundamental region $\mathscr{F}_1$ for Γ_1 which is constructed from modular triangles in such a way that every pair of them can be connected by a chain of modular triangles with a common side. This fundamental region $\mathscr{F}_1$ is simply connected.*

Proof. Proceeding from a fixed modular triangle we always add to our region $\mathscr{B}$ a modular triangle that has a side in common with $\mathscr{B}$ and that is not equivalent to any other modular triangle in $\mathscr{B}$. If in this way one obtains a representative of every equivalence class of points under Γ_1, then $\mathscr{B}$ is a connected fundamental region $\mathscr{F}_1$ for Γ_1. If one does not obtain representatives of all equivalence classes, then for every fundamental region $\mathscr{F}_1 \supset \mathscr{B}$ there is a modular triangle Δ not in $\mathscr{B}$. However, this is impossible. For suppose we join an interior point of Δ with an interior point of a modular triangle of $\mathscr{B}$ by a polygonal path not passing through a point Γ-equivalent to $i\infty$, i or ρ. Then the first modular triangle along this path from $\mathscr{B}$ which is not equivalent to any triangle of $\mathscr{B}$, is necessarily equivalent to a triangle which has a side in common with a triangle in $\mathscr{B}$. This contradicts the construction of $\mathscr{B}$.

A fundamental region $\mathscr{F}_1$ constructed in this way from modular triangles is simply connected. Otherwise there is a simple closed curve $\mathscr{C}$ in $\mathscr{F}_1$ that cannot be continuously shrunk in $\mathscr{F}_1$ to a point, i.e. $\mathscr{C}$ encloses an obviously finite set $\mathscr{L}$ of modular triangles whose union does not intersect $\mathscr{F}_1$. In this case there exists a substitution $V \in \Gamma_1$ with

$$\tilde{\mathscr{F}}_1 := V(\mathscr{F}_1) \subseteqq \mathscr{L}.$$

Also $\tilde{\mathscr{F}}_1$ is not simply connected. The repetition of this process leads to the conclusion that $\mathscr{L}$ consists of infinitely many modular triangles. This contradicts the above observation and proves Theorem **3**. □

Unless otherwise stated we shall assume in what follows that $\mathscr{F}_1$ is composed of whole modular triangles that are connected along their sides. We then have:

Two distinct fundamental regions that are related by a substitution $L \in \Gamma_1$ have at most one connected set of sides in common.

As in the proof of the simple connectedness of $\mathscr{F}_1$, this may be shown by assuming that $\mathscr{F}_1 \cup L(\mathscr{F}_1)$ is the union of two sets of modular triangles with more than one common connected set of sides.

2. The boundary of the fundamental region. Now we consider the boundary of a fixed fundamental region $\mathscr{F}_1$ and observe: to each modular triangle Δ not in $\mathscr{F}_1$, but with a side in common with $\mathscr{F}_1$, there is a transformation $V \in \Gamma_1$ such that $V(\Delta)$ is a boundary triangle of $\mathscr{F}_1$. Such maps together with their inverses are called *boundary substitutions*. $V(\mathscr{F}_1)$ and $\mathscr{F}_1$ have only boundary points in common. Since the boundary substitutions permit us to map $\mathscr{F}_1$ on every neighboring fundamental region, we have

Theorem 4. *The boundary substitutions generate the inhomogeneous group Γ_1.*

We want to determine the nature of the boundary of $\mathscr{F}_1$ at the points that are equivalent to ρ, i and $i\infty$ under Γ. If τ, equivalent to i or $\rho = e^{2\pi i/3}$ under Γ, is a fixed point of Γ_1, then it is a proper boundary point of $\mathscr{F}_1$, i.e. τ is not an interior point of $\mathscr{F}_1$, for each choice of $\mathscr{F}_1$. In this case, at a point τ_i equivalent to i under Γ there is only one modular triangle in $\mathscr{F}_1$ touching τ_i. On the other hand at τ_ρ, equivalent to ρ under Γ, either two modular triangles abut at τ_ρ or $\mathscr{F}_1^-$ contains exactly one other point equivalent to τ_ρ under Γ_1. In either of these last two cases the sum of the interior angles of the triangles of $\mathscr{F}_1$ at the *cycle* of points equivalent to τ_ρ under Γ_1 is $\dfrac{2\pi}{3}$. A *cycle* of the fixed point $\tau_0 \in \mathscr{H}^*$ is the set of points $\{L(\tau_0) \mid L \in \Gamma_1, L(\tau_0) \in \mathscr{F}_1^-\}$.

If τ is not a fixed point of Γ_1, then all of the modular triangles abutting at τ are inequivalent under Γ_1 and so there is a complete neighborhood $\mathcal{U}$ of τ in $\mathcal{H}$ composed of points inequivalent under Γ_1. If $\tau \in \partial \mathcal{F}_1$, then the set of points in $\mathcal{F}_1$ equivalent to points of $\mathcal{U}$ is not connected. Corresponding to the points $L(\tau)$ belonging to the cycle of τ, we have partial sets $L(\mathcal{U}) \cap \mathcal{F}_1$, the sum of whose angles is 2π.

The rational cusps of the fundamental region are always fixed points and thus are proper boundary points. The modular triangles touching at a cycle of points equivalent under Γ_1 to a fixed τ_∞ can be transformed by the substitutions of Γ_1 to take the form of a fan at τ_∞. This is clear, since the transformations of Γ with the fixed point τ_∞ form a cyclic group, as do those of Γ_1. If, say, $A U^\kappa A^{-1}$, $\kappa > 0$, is the generating element of this cyclic group, then the triangles $A U^\nu(\mathcal{F})$ for $\nu = 0, 1, \ldots, \kappa - 1$, form a fan consisting of all Γ_1-inequivalent triangles touching τ_∞. We say: the fundamental region has *fan width* κ at the cusp τ_∞. Theorem **2** and the 2nd Theorem from **§1, 2** imply

Theorem 5. *The conductor of a group Γ_1 is equal to the least common multiple of the fan widths at the rational cusps.*

3. Conjugate subgroups. Normal subgroups. A fundamental region of a subgroup conjugate to Γ_1 can be obtained from the fundamental region $\mathcal{F}_1$ of Γ_1 by applying

Theorem 6. *If $\mathcal{F}_1$ is a fundamental region for Γ_1 and if $L \in \Gamma$, then*

$$\mathcal{F}' := L^{-1}(\mathcal{F}_1)$$

is a fundamental region for the subgroup

$$\Gamma' := L^{-1} \Gamma_1 L$$

conjugate to Γ_1.

Proof. Let $\mathcal{F}_1$ be a fundamental region of Γ_1. Then $\mathcal{H}^* = \Gamma_1(\mathcal{F}_1)$, or equivalently

$$\mathcal{H}^* = (L^{-1} \Gamma_1 L) L^{-1}(\mathcal{F}_1),$$

and so $L^{-1}(\mathcal{F})$ contains a fundamental set of Γ'. Points equivalent under Γ_1' correspond to similar equivalent points under Γ_1. □

Before we consider the case when Γ_1 is normal in Γ, we introduce the oftentimes useful notion of a *half-triangle.* This is a region $A(\mathcal{F}^*)$ with $A \in \Gamma^*$, where Γ^* denotes the modular group extended by reflections and $\mathcal{F}^*$ denotes the corresponding fundamental region described in **II, §2**.

If Γ_1 is normal in Γ it follows from Theorem **6** that, if $\mathcal{F}_1$ is a fundamental region for Γ_1, then so is $L(\mathcal{F}_1)$ for all $L \in \Gamma$.

Theorem 7. *If Γ_1 is normal in Γ, then equally many non-equivalent half-triangles abut at the points equivalent to ρ, i or $i\infty$, respectively.*

Proof. Suppose that τ_0 is such a point with $L(\tau_0)=\tau_0$ for some $L\in\Gamma_1$ and that $M\in\Gamma$ maps τ_0 to τ_1, then

$$LM^{-1}(\tau_1)=M^{-1}(\tau_1), \quad \text{so } MLM^{-1}(\tau_1)=\tau_1 .$$

Since Γ_1 was assumed to be normal, $MLM^{-1}\in\Gamma_1$. However, this proves our claim. (See also § **1,2** 1.) □

We denote half the numbers of Γ_1-inequivalent half-triangles which abut at the points equivalent under Γ to ρ, i or $i\infty$ by n_ρ, n_i or n_∞ respectively, and call the triple (n_i, n_ρ, n_∞) the *branch schema* of the normal subgroup Γ_1. Obviously $n_i=1$ or 2, $n_\rho=1$ or 3, and n_∞ is the least positive integer for which $U^{n_\infty}\in\Gamma_1$. Let σ_i, σ_ρ, and σ_∞ denote the maximal number of Γ_1-inequivalent points which are Γ-equivalent to i, ρ, and $i\infty$, respectively. If Γ_1 has the branch schema (n_i, n_ρ, n_∞) and index $\mu=(\Gamma:\Gamma_1)$, then

$$\sigma_i=\frac{\mu}{n_i}, \qquad \sigma_\rho=\frac{\mu}{n_\rho}, \qquad \text{and } \sigma_\infty=\frac{\mu}{n_\infty}. \tag{10}$$

In conclusion we indicate the branch schema for the principal congruence subgroup $\Gamma[N]$. Obviously $n_\infty=N$, and for $N\geqq 2$ neither $\pm T=\begin{pmatrix}0 & -1\\ 1 & 0\end{pmatrix}$ with $T(i)=i$ nor $\pm R=\pm\begin{pmatrix}0 & -1\\ 1 & 1\end{pmatrix}$ with $R(\rho)=\rho$ is congruent to the identity modulo N. Thus $n_i=2$ and $n_\rho=3$. The branch schema for $\Gamma[N]$, $N\geqq 2$ has therefore the form

$$(n_i, n_\rho, n_\infty)=(2, 3, N).$$

The branch schema for $\Gamma(1)$ is $(1, 1, 1)$.

§ 5. Fundamental Regions for Special Subgroups

1. Fundamental region for the principal congruence group $\Gamma[2]$. In the last section we deduced that $n_\rho=3$. Thus all of the half-triangles abutting at a point equivalent to ρ under Γ are inequivalent under $\Gamma[2]$. Since $\mu=6$, we may choose as fundamental region the six modular triangles abutting at $-\bar{\rho}$. The boundary correspondences are given by U^2, $TU^{-2}T$, and $(UT)U^{-2}(UT)^{-1}$ (cf. Fig. 12):

$$e_6=U^2(e_1), \qquad e_3=TU^{-2}T(e_2), \qquad e_5=UTU^{-2}TU^{-1}(e_4).$$

We obtain simpler generators for $\bar{\Gamma}(2)$ from the boundary subsituations if we modify the fundamental region to that of Fig. 13. We

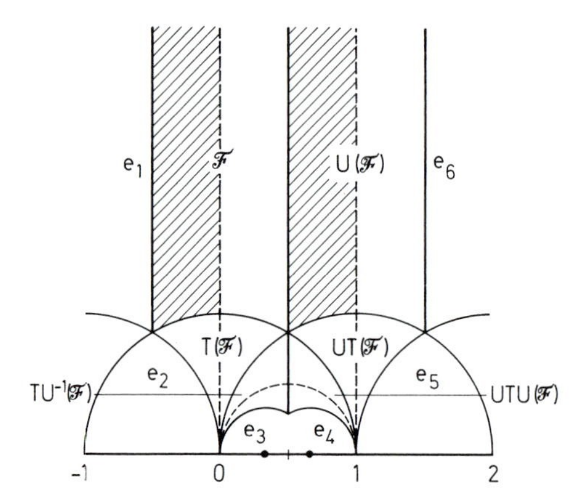

Fig. 12

see that corresponding to this new fundamental region there are only two distinct boundary substitutions, namely,

$$\bar{U}^2: \tau \mapsto \tau+2 \quad \text{and} \quad \bar{S}: \tau \mapsto \frac{\tau}{2\tau+1}.$$

Likewise these generate $\bar{\Gamma}[2]$. We state without proof that $\bar{\Gamma}[2]$ is the free group on $\bar{U}^2$ and $\bar{S}$.

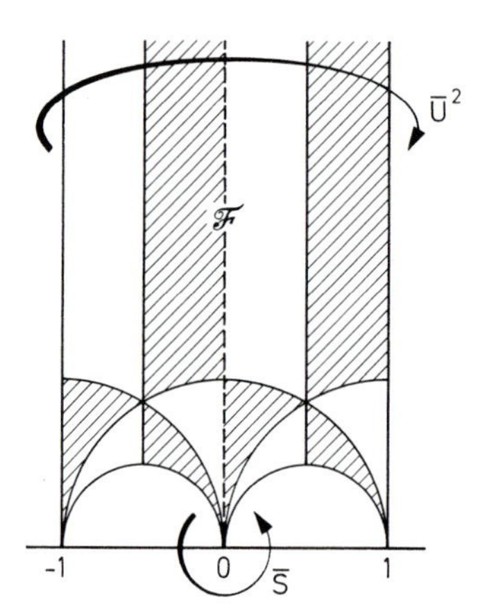

Fig. 13

2. Fundamental region of the theta group $\Gamma_1 := \Gamma[2] \cup \Gamma[2]\, T$. Since $T^2 = -I$, $\Gamma[2]$ is a subgroup of index $(\Gamma_1 : \Gamma[2]) = 2$ in Γ_1, and therefore the index $(\Gamma : \Gamma_1)$ is 3. Since U^2 and T are in Γ_1, its fundamental region is given as in Fig. 14. This figure verifies that Γ_1 is not normal in Γ due to its behavior at the fixed points of Γ. As boundary correspond-

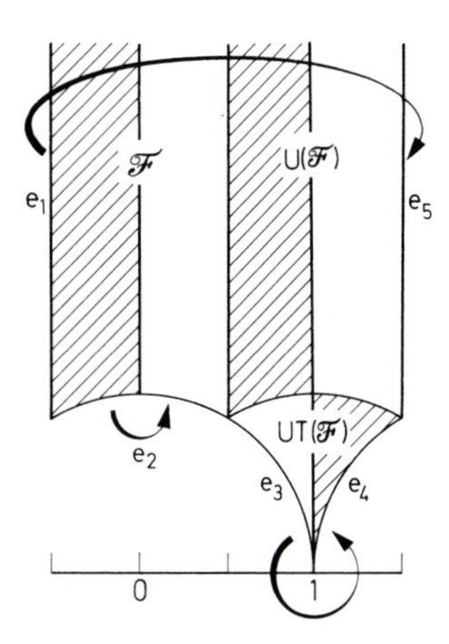

Fig. 14

ences we have

$$e_5 = U^2(e_1), \quad e_2 = T(e_2), \quad e_4 = (U\,T)\,U^{-1}(U\,T)^{-1}\,(e_3).$$

The second fundamental region, Fig. 15, leads to the simpler system of generators for $\overline{\Gamma}_1$, namely, $\overline{U}^2$ and $\overline{T}$. Because of its connection with the transformation formulas of the *theta function:*

$$\vartheta(\tau) := \sum_{m=-\infty}^{\infty} e^{\pi i m^2 \tau},$$

Γ_1 is also called the *theta group.*

The groups conjugate to the theta group $\Gamma_1 = \Gamma_\vartheta$ are

$$\Gamma^0(2) = U^{-1}\,\Gamma_\vartheta\,U,$$

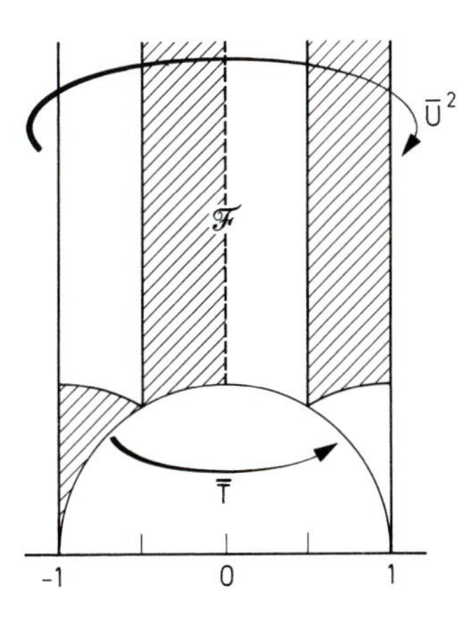

Fig. 15

with fundamental region

$$U^{-1}(\mathcal{F}) \cup \mathcal{F} \cup T(\mathcal{F}),$$

and boundary correspondences

$$U^2,\ U^{-1}TU,\ UTU,$$

and the group

$$\Gamma_0(2) = (UT)^{-1}\,\Gamma_\vartheta\, UT.$$

3. Fundamental region of the group $\bar{\Gamma}_1 := \Gamma[2] \cup \Gamma[2]\,UT \cup \Gamma[2](UT)^2$. Besides the theta group and its two conjugates there is only one other group which contains $\Gamma[2]$, namely $\bar{\Gamma}_1$ defined above. It has index 2 in Γ and thus is normal in Γ. Since $U \notin \bar{\Gamma}_1$ and $T \notin \bar{\Gamma}_1$, one may choose either of the two accompanying figures 16 or 17 as a fundamental region. The boundary correspondences can be verified as above.

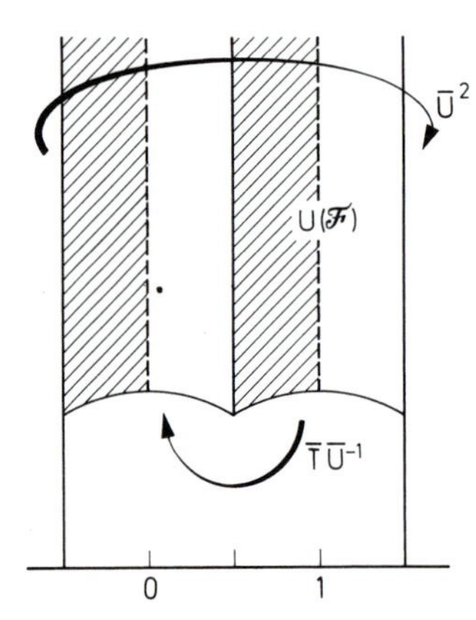

Fig. 16

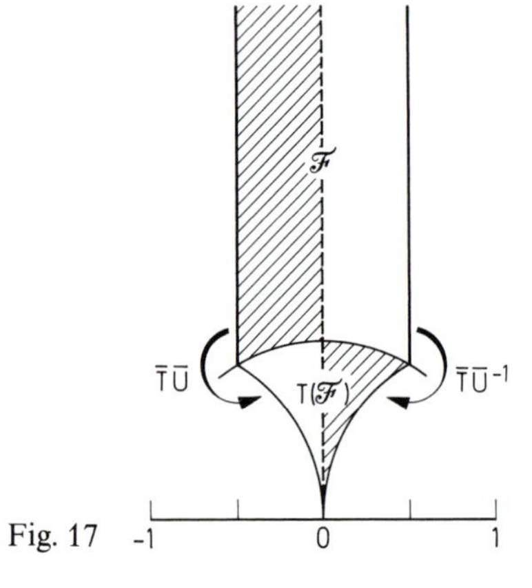

Fig. 17

4. The system of inequivalent cusps for the principal congruence subgroup $\Gamma[N]$. The number of inequivalent cusps is known: by (10) it is $\dfrac{\mu}{n_\infty}$, where $n_\infty = N$ and where $\mu = |\Gamma_N|$ is given in (6). Now let $\dfrac{r}{s}$ and $\dfrac{r_1}{s_1}$ be two rational numbers in reduced form. Corresponding to these there are two matrices

$$L := \begin{pmatrix} r & b \\ s & d \end{pmatrix}, \quad L_1 := \begin{pmatrix} r_1 & b_1 \\ s_1 & d_1 \end{pmatrix} \tag{11}$$

in Γ for which

$$L(i\infty) = \frac{r}{s} \quad \text{and} \quad L_1(i\infty) = \frac{r_1}{s_1}.$$

The transformations

$$V := L_1 U^k L^{-1}, \qquad k \in \mathbb{Z},$$

are the modular transformations for which

$$V\left(\frac{r}{s}\right)=\frac{r_1}{s_1}.$$

Among these transformations V there is one in $\Gamma[N]$ if and only if the matrix congruence

$$L_1^{-1}L\equiv \pm U^k \bmod N, \quad \text{for some } k\in\mathbb{Z},$$

can be satisfied. This is equivalent to the following three congruences:

$$\begin{aligned} d_1 r-b_1 s&\equiv \pm 1 \mod N, && (1') \\ -s_1 r+r_1 s&\equiv 0 \mod N, && (2') \quad (A) \\ -s_1 b+r_1 d&\equiv \pm 1 \mod N. && (3') \end{aligned}$$

It follows from (1′) and (2′) that

$$r\equiv \pm r_1, \quad \text{and } s\equiv \pm s_1 \bmod N \tag{12}$$

and that consequently (3′) is also satisfied, where on the right sides the same sign is to be taken.

Conversely, if the congruences (12) hold, then the entries of the matrices of (11) satisfy (A).

Since, as was shown in § **2, 1**, corresponding to any two integers $r,s\in\mathbb{Z}$ with $(r,s,N)=1$, $N>2$, there are two integers $r_1\equiv r$ and $s_1\equiv s$ modulo N with $(r_1,s_1)=1$, and since in (12) only one sign can be taken, we have

Theorem 8. *If* $N>2$, *the number* $\sigma_\infty(N)$ *of inequivalent rational cusps for* $\Gamma[N]$ *is equal to one half the number of pairs* (a,b) *incongruent* $\bmod N$ *with* $(a,b,N)=1$. *If* $N=2$, $\sigma_\infty(N)$ *is equal to the number of these pairs.*

A fundamental region for $\Gamma[N]$ may be constructed by taking a fan of width N at each point of a maximal system of $\Gamma[N]$-inequivalent rational numbers.

We give an example of a complete system of inequivalent rational cusps in the case $N=5$:

$$\tfrac{1}{0},\ \tfrac{2}{5},\ \tfrac{0}{1},\ \tfrac{1}{1},\ \tfrac{2}{1},\ \tfrac{3}{1},\ \tfrac{4}{1},\ \tfrac{5}{2},\ \tfrac{1}{2},\ \tfrac{7}{2},\ \tfrac{3}{2},\ \tfrac{9}{2}.$$

5. The fundamental region of the group $\Gamma^0(q)$, q a prime number. Here $-I\in\Gamma^0(q)$. For $q>2$ it is easy to prove that Γ has the coset decomposition

$$\Gamma=\bigcup_{-\frac{q+1}{2}<\nu\leq\frac{q+1}{2}}\Gamma^0(q)\,S_\nu$$

with

$$S_\nu = U^\nu \quad \text{for } |\nu| \leqq \frac{q-1}{2}, \quad \text{and } S_{\frac{q+1}{2}} = T.$$

Thus

$$\mathscr{F}_1 := \bigcup_{-\frac{q+1}{2} < \nu \leqq \frac{q+1}{2}} S_\nu(\mathscr{F})$$

is a fundamental region for $\Gamma^0(q)$. To study the boundary correspondences, for integral ν, $0 < |\nu| \leqq \frac{q-1}{2}$, we denote by $\mathscr{E}_\nu$ the circular arc of radius 1 centered at ν. There are two maps that transform $\mathscr{E}_\nu$ to $\mathscr{E}_{\nu'}$, namely, $U^{\nu'-\nu}$ and $U^{\nu'} T U^{-\nu}$. Since, however, for $\nu \neq \nu'$, $U^{\nu'-\nu} \notin \Gamma^0(q)$, there remains at most the map

$$U^{\nu'} T U^{-\nu} = \begin{pmatrix} \nu' & -1-\nu\nu' \\ 1 & -\nu \end{pmatrix}.$$

This is an element of $\Gamma^0(q)$ if

$$\nu\nu' + 1 \equiv 0 \bmod q.$$

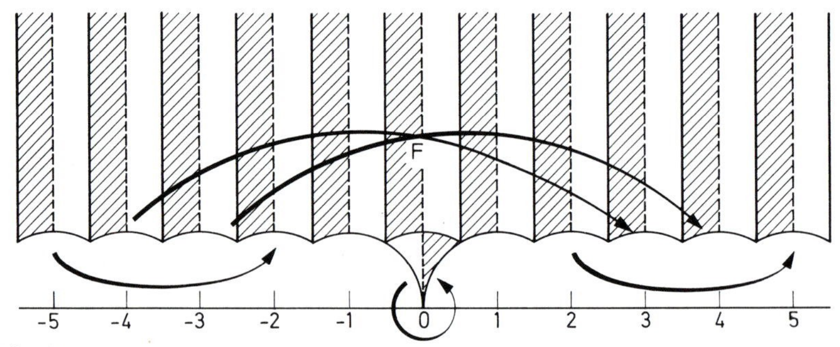

Fig. 18

We thus obtain:

Two circular arcs $\mathscr{E}_\nu$ and $\mathscr{E}_{\nu'}$ are equivalent under $\Gamma^0(q)$ with $q > 2$ if and only if

$$\nu\nu' + 1 \equiv 0 \bmod q.$$

We consider the example $q = 11$, which in view of the above theorem has the following pairs of equivalent sides:

$$\mathscr{E}_{-5} \leftrightarrow \mathscr{E}_{-2}, \quad \mathscr{E}_5 \leftrightarrow \mathscr{E}_2, \quad \mathscr{E}_{-4} \leftrightarrow \mathscr{E}_3, \quad \mathscr{E}_4 \leftrightarrow \mathscr{E}_{-3}, \quad \mathscr{E}_{-1} \leftrightarrow \mathscr{E}_1.$$

Correspondingly, for $q=5$, one has

$$\mathscr{E}_{-2} \leftrightarrow \mathscr{E}_{-2}, \quad \mathscr{E}_2 \leftrightarrow \mathscr{E}_2, \quad \mathscr{E}_{-1} \leftrightarrow \mathscr{E}_1 .$$

The group $\Gamma^0(2)$ was handled in subsection **2.**

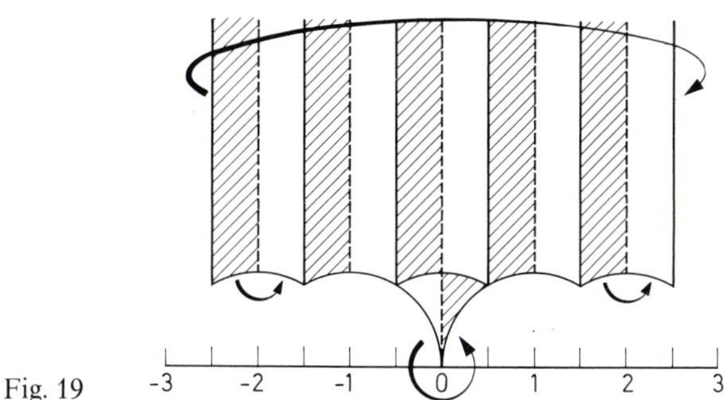

Fig. 19

§ 6. The Quotient Space $\mathscr{H}^*/\Gamma_1$

In this section let Γ_1 again be a subgroup of finite index in Γ and $-I \in \Gamma_1$.

1. The topological spaces $\mathscr{H}^*$ and $\mathscr{H}^*/\Gamma_1$. We form the quotient space

$$\mathfrak{R} = \mathscr{H}^*/\Gamma_1 ,$$

consisting of the elements

$$\langle\tau\rangle := \{M\tau \mid M \in \Gamma_1\} .$$

We denote by σ the canonical map

$$\sigma \colon \mathscr{H}^* \to \mathfrak{R}, \qquad \tau \mapsto \langle\tau\rangle . \tag{13}$$

We obtain a representation of $\mathfrak{R}$ if we identify Γ_1-equivalent points of the closure $\mathscr{F}_1^-$ of the fundamental region for Γ_1.

Before we consider a topological structure for $\mathfrak{R}$, we introduce a topology on $\mathscr{H}^*$. To do so we choose for each $\tau \in \mathscr{H}^*$ a family of sets B_τ in the following way:

1) For a point $\tau \in \mathscr{H}$ let B_τ be the family of sets

$$\mathscr{U}_{\tau,r} := \left\{ v \in \mathscr{H} : \left| \frac{v-\tau}{v-\bar{\tau}} \right| < r \right\} \tag{14}$$

with $0<r<1$.

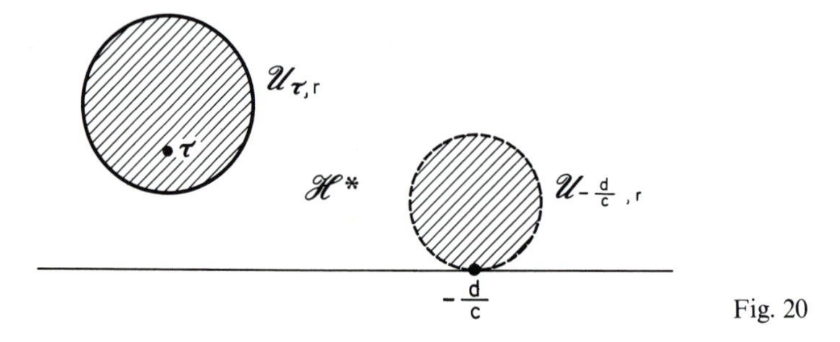

Fig. 20

2) If $\tau = -\dfrac{d}{c}$, $(c, d)=1$, is a point equivalent to $i\infty$ under Γ, we define B_τ to be the family of sets

$$\mathscr{U}_{\tau,r} := \left\{v\in\mathscr{H} : \operatorname{Im}\frac{av+b}{cv+d} > \frac{1}{r}\right\} \cup \{\tau\} \tag{15}$$

with an arbitrary choice of $\begin{pmatrix} a & b \\ c & d \end{pmatrix} \in \Gamma$ and with $0<r<1$. A set $\mathscr{M}$ of $\mathscr{H}^*$ is said to be open if and only if $\mathscr{M} = \bigcup \mathscr{U}_{\tau,r}$. Thus $\mathscr{H}^*$ is a topological space, and the sets $\mathscr{U}_{\tau,r} \in \mathsf{B}_\tau$ form a neighborhood basis. From now on we shall view $\mathscr{H}^*$ as a topological space rather than a point set. We prove

Theorem 9. *$\mathscr{H}^*$ is a connected Hausdorff space. $\mathscr{H}^*$ is not locally compact.*

Proof. The first assertion is immediately clear. For the proof of the second we show that the rational points do not have compact neighborhoods. If $\mathscr{U}$ were a compact neighborhood of $\tau\in\mathbb{Q}$, then $\mathscr{U}$ would contain a closed and thus compact disk $\mathscr{U}_{\tau,r}^-$. Now suppose that $\{\tau_\nu\}$ is a sequence on the boundary of $\mathscr{U}_{\tau,r}^-$ that converges to τ in the Euclidean topology. Then each neighborhood (15) would contain almost all of the τ_ν. This, however, is impossible for $\mathscr{U}_{\tau,r'}$ with $r' \leqq r$. $\square$

Now we endow $\mathfrak{R}$ with the quotient topology. A set $\mathfrak{U}$ in $\mathfrak{R}$ is open if and only if the pre-image $\sigma^{-1}(\mathfrak{U})$ is open in $\mathscr{H}^*$. Then we have

Theorem 10. *The canonical map σ is both continuous and open. $\mathfrak{R}$ is a connected, compact Hausdorff space.*

Proof. By definition of the quotient topology σ is continuous. If $\mathscr{A}$ is an open set in $\mathscr{H}^*$, then

$$\sigma^{-1}\sigma(\mathscr{A}) = \bigcup_{M\in\Gamma_1} M(\mathscr{A})$$

is open, and so σ is an open map.

$\mathscr{H}^*$ contains a connected compact set that intersects all equivalence classes under Γ_1, namely the closure of $\mathscr{F}_1$. It follows that $\mathfrak{R}=\mathscr{H}^*/\Gamma_1$ is likewise connected and compact. Since the separation axiom is obviously satisfied in $\mathfrak{R}$, this completes the proof. □

2. The surface $\mathscr{H}^*/\Gamma_1$. We recall the following definition: a connected Hausdorff space together with a covering by open sets which are homeomorphic to open sets in the Euclidean plane is called a *surface.* An element of this covering together with its corresponding homeomorphism is called a *chart*; the collection of charts is called an *atlas.*

We now want to give $\mathfrak{R}$ the structure of a surface. We obtain certain neighborhoods of the point $\langle\tau\rangle$ of $\mathfrak{R}$ by taking

$$\mathfrak{U}_{\langle\tau\rangle}:=\sigma(\mathscr{U}_{\tau,r}). \tag{16}$$

For these we choose r so small that either $\mathscr{U}_{\tau,r}$ is contained entirely in a modular triangle or is contained in the union of half-triangles abutting at τ. We use the same r for all $\tau\in\langle\tau\rangle$. Then $\mathfrak{U}_{\langle\tau\rangle}$ is a well defined neighborhood of the class. Since σ is open and surjective, the family of all $\mathfrak{U}_{\langle\tau\rangle}$ is an open cover for $\mathfrak{R}$. Since we fixed r, for short we write $\mathscr{U}_\tau$ in place of $\mathscr{U}_{\tau,r}$.

We now construct a homeomorphism $\phi_{\langle\tau\rangle}$ that maps the neighborhood $\mathfrak{U}_{\langle\tau\rangle}$ on an open set $\mathscr{E}_{\langle\tau\rangle}$ of the Euclidean plane. If σ_τ denotes the restriction of σ to $\mathscr{U}_\tau$, then the pre-image of $\mathfrak{U}_{\langle\tau\rangle}$ is $\mathscr{U}_\tau=\sigma_\tau^{-1}(\mathfrak{U}_{\langle\tau\rangle})$. We need a continuous open map

$$\phi_\tau\colon \mathscr{U}_\tau\to\mathscr{E}_\tau \tag{17}$$

into the Euclidean plane such that

$$\phi_{\langle\tau\rangle}:=\phi_\tau\circ\sigma_\tau^{-1}\colon \mathfrak{U}_{\langle\tau\rangle}\to\mathscr{E}_{\langle\tau\rangle}$$

is bijective. Here the correspondence σ_τ^{-1} is not necessarily a mapping! We continue to write $\mathscr{E}_{\langle\tau\rangle}$ and $\phi_{\langle\tau\rangle}$, since we shall prove that for suitable choice of ϕ_τ the set $\mathscr{E}_\tau$ does not depend on the representative τ, and that likewise the homeomorphism $\phi_{\langle\tau\rangle}$ is a well defined class function.

We distinguish two cases:

1) τ is not a parabolic fixed point of Γ_1:

We fix a representative $\tau_0\in\langle\tau\rangle$. Then each $\tau\in\langle\tau\rangle$ is an image $\tau=S(\tau_0)$ for some $S\in\Gamma_1$. If $S=\begin{pmatrix} a & b\\ c & d\end{pmatrix}$, we define

$$\phi_\tau(v):=\left(\frac{v-\tau}{v-\bar\tau}\cdot\frac{c\tau_0+d}{c\bar\tau_0+d}\right)^k, \qquad v\in\mathscr{U}_\tau. \tag{18}$$

In the above we assume $k=1$ if τ is not a fixed point, and $k=2$ or 3 in case τ is an elliptic fixed point. Then the factor $\left(\dfrac{c\tau_0+d}{c\bar\tau_0+d}\right)^k$ is uniquely determined by τ_0 and τ. In case $k=2$ or $k=3$ it has the value 1. (See **I**, (9) and I § **3**,**1**.) σ_τ^{-1} assigns to every element of $\mathfrak{U}_{\langle\tau\rangle}$ exactly k values of $\mathcal{U}_\tau$. These are mapped onto a single image in $\mathcal{E}_{\langle\tau\rangle} := \phi_\tau(\mathcal{U}_\tau)$. The product $\phi_\tau \circ \sigma_\tau^{-1}$ is thus bijective. As a holomorphic function ϕ_τ is continuous and open; by the preceding theorem σ_τ is likewise continuous and open. Consequently $\phi_{\langle\tau\rangle} = \phi_\tau \circ \sigma_\tau^{-1}$ is continuous, open, and bijective, and thus is a homeomorphism. Since for $S = \begin{pmatrix} a & b \\ c & d \end{pmatrix}$,

$$\frac{v-\tau}{v-\bar{\tau}} = \frac{v_0-\tau_0}{v_0-\bar{\tau}_0} \cdot \frac{c\tau_0+d}{c\bar{\tau}_0+d}, \qquad v = S(v_0), \tag{18'}$$

$\mathcal{E}_{\langle\tau\rangle}$ and $\phi_{\langle\tau\rangle}$ are well defined.

2) τ is a parabolic fixed point of Γ_1:

In this case $\tau = -\dfrac{d}{c}$ with $(d, c) = 1$ and we choose

$$\phi_\tau(v) := e^{\frac{2\pi i}{\kappa} A(v)}, \qquad v \in \mathcal{U}_\tau, \tag{19}$$

where κ is the fan width and $A = \begin{pmatrix} a & b \\ c & d \end{pmatrix} \in \Gamma$. As was explained above we find that $\phi_\tau \circ \sigma_\tau^{-1}$ is a homeomorphism. The cusp τ determines ϕ_τ up to a multiple of a power of $e^{2\pi i/\kappa}$. ϕ_τ is fixed by the choice of A. If we subject τ and v to the substitution $S \in \Gamma_1$, then, with $A_1 := AS^{-1}$, we have

$$e^{\frac{2\pi i}{\kappa} A_1(S(v))} = e^{\frac{2\pi i}{\kappa} A(v)}. \tag{19'}$$

The choice of A_1 is admissible since $A_1(S\tau) = i\infty$. Thus $\mathcal{E}_{\langle\tau\rangle}$ as well as $\phi_{\langle\tau\rangle}$ are independent of the choice of representative. In conclusion we have

Theorem 11. *$\mathfrak{R}$ is a surface. An atlas for $\mathfrak{R}$ is the system*

$$\mathsf{A} := \{(\mathfrak{U}_{\langle\tau\rangle}, \phi_{\langle\tau\rangle}) : \langle\tau\rangle \in \mathfrak{R}\}.$$

3. The Riemann surface $\mathcal{H}^*/\Gamma_1$. As usual we define: A surface is called a *Riemann surface* if its charts are holomorphically consistent, that is, if the images in the Euclidean plane of the intersection of two elements of the covering are related by a holomorphic function.

We show that the surface $\mathfrak{R}$ with atlas A is a Riemann surface. Let $\mathfrak{D}$ denote the intersection of $\mathfrak{U}_{\langle\tau_1\rangle}$ and $\mathfrak{U}_{\langle\tau_2\rangle}$, $\langle\tau_1\rangle \neq \langle\tau_2\rangle$, and let $\mathcal{F}_1 := \phi_{\langle\tau_1\rangle}(\mathfrak{D})$, $\mathcal{F}_2 := \phi_{\langle\tau_2\rangle}(\mathfrak{D})$ be the images of $\mathfrak{D}$ in the Euclidean plane. Then

$$\phi_{\langle\tau_2\rangle} \circ \phi_{\langle\tau_1\rangle}^{-1} = \phi_{\tau_2} \circ \sigma_{\tau_2}^{-1} \circ \sigma_{\tau_1} \circ \phi_{\tau_1}^{-1}$$

restricted to $\mathscr{F}_1$ is a map from $\mathscr{F}_1$ to $\mathscr{F}_2$. This map is holomorphic because $\mathscr{F}_1$ contains no images of points equivalent under Γ_1 to i, ρ or $i\infty$, and because the restriction of $\sigma_{\tau_2}^{-1} \circ \sigma_{\tau_1}$ to $\phi_{\tau_1}^{-1}(\mathscr{F}_1)$ is the restriction of an element of Γ_1. Thus we have

Theorem 12. $\mathfrak{R}$ *is a Riemann surface with atlas* A.

§ 7. Genus of the Fundamental Region

1. Euler's theorem on polyhedra. The Riemann surface $\mathfrak{R}$ constructed in the last section is compact and is obviously also orientable—as is every Riemann surface. For a compact orientable surface $\mathfrak{F}$ we define:

A homeomorphic image of the closed unit interval of the reals in $\mathfrak{F}$ is called a *simple arc* on $\mathfrak{F}$. A *curve* on $\mathfrak{F}$ is a finite sucession of simple arcs where the endpoint of one arc coincides with the initial point of the next. If the end point of a curve coincides with the initial point, we speak of a *closed curve.* A homeomorphic image in $\mathfrak{F}$ of a closed disk of the Euclidean plane is called a *polygon* on $\mathfrak{F}$. Suppose that a system $\mathscr{L}$ of finitely many curves with finitely many points of intersection is given on $\mathfrak{F}$. Assume further that on the one hand the closures of connected components of the complement of $\mathscr{L}$ are polygons and on the other hand that their intersection consists either of a single point, a single edge, or the empty set. Such a system of curves is called a *polygonal decomposition* of $\mathfrak{F}$.

The meaning of *vertex*, *edge*, and *face* of a polygonal decomposition is then clear. We denote their numbers by v, e, and f respectively. With these notions we now formulate

Euler's Theorem on polyhedra. *For each polygonal decomposition of* $\mathfrak{F}$ *the number* $v-e+f$ *has the same value.*

For a proof of the polyhedral theorem we refer to H. Behnke and F. Sommer [1] p. 507 or C. L. Siegel [2], vol. 1, p. 124.

Likewise we cite:

The number $g := 1 - \dfrac{v-e+f}{2}$ *is the maximal number of non-intersecting closed curves that do not decompose* $\mathfrak{F}$.

This important topological invariant is called the *genus* of $\mathfrak{F}$.

2. Applications of the theorem on polyhedra to subgroups of the modular group. We now return to the Riemann surface $\mathfrak{R}$ of a subgroup Γ_1 of finite index in the modular group Γ. The edges of the images of the half-triangles under the canonical map σ, (13), obviously form a polygonal decomposition of $\mathfrak{R}$. For this decomposition $f=2\mu$. If we denote the vertices by $V_1, V_2, \ldots, V_v$ and the number of half-triangles abutting at V_ν by $2\kappa_\nu$, then

$$e=\sum_{\nu=1}^{v} \kappa_\nu, \quad v=\sum_{\nu=1}^{v} 1 .$$

Thus, we obtain the first form of the genus formula

$$g=1-\mu+\frac{1}{2}\sum_{\nu=1}^{v}(\kappa_\nu-1). \tag{20}$$

We also call g the *genus of the fundamental region* for the group Γ_1.

If Γ_1 is normal in Γ, then we know by Theorem **7** that there are equally many half-triangles abutting at each of the vertices V_ν of $\mathfrak{R}$ which are images under σ of points Γ-equivalent to ρ, i, or $i\infty$ respectively. We have denoted these numbers by $2n_i$, $2n_\rho$, and $2n_\infty$. In this notation we obtained $\frac{\mu}{n_i}$, $\frac{\mu}{n_\rho}$, and $\frac{\mu}{n_\infty}$, respectively, for the number of vertices of $\mathfrak{R}$ that correspond to i, ρ and $i\infty$. Hence:

If Γ_1 is normal in Γ, then

$$g=1-\mu+\frac{\mu}{n_i}\,\frac{n_i-1}{2}+\frac{\mu}{n_\rho}\cdot\frac{n_\rho-1}{2}+\frac{\mu}{n_\infty}\,\frac{n_\infty-1}{2}. \tag{21}$$

The principal congruence subgroup $\Gamma[N]$, $N\geqq 2$, has the branch schema $(2, 3, N)$. Thus its fundamental region has the genus

$$g=1+\frac{\mu}{N}\,\frac{N-6}{12} \quad \text{with } \mu=\frac{\mu(N)}{2} \quad \text{for } N>2, \text{ and } \mu=\mu(2) \quad \text{for } N=2. \tag{22}$$

A formula for $\mu(N)$ is given in (3). In particular for

$N=2$	3	4	5	6	7	8	9	10	11	12,
$g=0$	0	0	0	1	3	5	10	13	26	25.

3. Branch schema and the corresponding normal subgroup. The branch schema of a normal subgroup can only be of the form $(1, 1, n)$, $(2, 1, n)$, $(1, 3, n)$ or $(2, 3, n)$. We discuss these cases:

By (21), since the genus is a non-negative integer,

$$(n_i, n_\rho, n_\infty)=(1, 1, n)$$

leads to

$$g=1-\mu+\frac{\mu}{n}\cdot\frac{n-1}{2}=1-\frac{\mu}{n}\cdot\frac{n+1}{2}=0 .$$

Because $\frac{\mu}{n} \in \mathbb{N}$, so also is $\frac{2}{n+1}$, and thus $n=1$ and $\mu=1$. There is one and only one subgroup with the branch schema (1, 1, 1), namely the full modular group.

As above

$$(n_i, n_\rho, n_\infty) = (2, 1, n)$$

yields

$$g = 1 - \mu + \frac{\mu}{2} \cdot \frac{1}{2} + \frac{\mu}{n} \cdot \frac{n-1}{2} = 1 - \frac{\mu}{n} \cdot \frac{n+2}{4} = 0$$

and because $\frac{\mu}{n} \in \mathbb{N}$, it follows that $n=\mu=2$. Thus at most the schema (2, 1, 2) can occur. However, in **§ 5,3** we met a group corresponding to this schema, namely,

$$\Gamma_1 = \bigcup_{\nu=0}^{2} \Gamma[2](UT)^\nu .$$

Γ_1 is the only group with this branch schema, since the fundamental region given there together with boundary correspondences are already determined by the branch schema. Since the index of Γ_1 in Γ is 2, Γ_1 is a normal subgroup.

The case

$$(n_i, n_\rho, n_\infty) = (1, 3, n)$$

implies

$$g = 1 - \mu + \frac{\mu}{3} \cdot \frac{2}{2} + \frac{\mu}{n} \cdot \frac{n-1}{2} = 1 - \frac{\mu}{n} \cdot \frac{n+3}{6} = 0$$

and thus $n=\mu=3$. If there is a group with this branch behavior, the fundamental region must appear as indicated in figure 21. The resulting boundary substitutions are

$$S_1 := U^3, \quad S_2 := T, \quad S_3 := UTU^{-1}, \quad S_4 := U^{-1}TU .$$

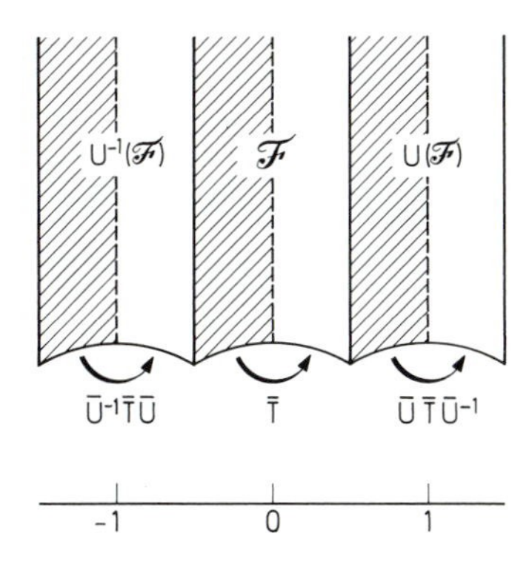

Fig. 21

Now, as is easily proved,

$$\Gamma_1 := \bigcup_{\nu=1}^{4} \Gamma[3] S_\nu$$

is a group, and is in fact normal in Γ. Since $(\Gamma:\Gamma_1)=3$, and since I, U, and U^{-1} are pairwise incongruent modulo Γ_1, the figure represents a fundamental region for Γ_1. Therefore Γ_1 has branch schema (1, 3, 3) and by construction is uniquely determined by it.

There remains the case

$$(n_i, n_\rho, n_\infty) = (2, 3, n).$$

This branch schema is realized by the principal congruence subgroup $\Gamma[n]$ for every $n \geqq 2$.

For this schema we consider the normal subgroups of genus $g=0$ and $g=1$. By (21)

$$g = 1 + \frac{\mu}{n} \, \frac{n-6}{12}.$$

From $g=0$ it follows that $-\dfrac{12}{n-6} \in \mathbb{N}$. Hence we obtain the following possibilities for n and corresponding μ:

n	2	3	4	5
μ	6	12	24	60

In F. Klein and R. Fricke [1] it is proved that besides the principal congruence groups $\Gamma[N]$, $N \in \{2,3,4,5\}$, there are no other groups in these four cases that have the same branch schema.

If $g=1$, then $n=6$ and μ is undetermined. Beside the principal congruence group $\Gamma[6]$ with $\mu=72$ there are yet other groups with this branch behavior. We give an example. Let us consider Γ as a matrix group and Γ' as its commutator subgroup. From **I**, Theorem **11** it follows that $-I \notin \Gamma'$ and $(\Gamma:\Gamma')=12$. Let

$$\Gamma^+ := \Gamma' \cup (-I)\Gamma',$$

then Γ^+ is normal in Γ and $\mu=(\Gamma:\Gamma^+)=6$. In terms of the generators R and T of the modular group Γ we see that

$$\overline{I}, \overline{T}, \overline{R}, \overline{R\,T}, \overline{R^2}, \overline{R}^2\,\overline{T}$$

are representatives of $\overline{\Gamma}^+$ in $\overline{\Gamma}$. Making use of $\overline{R}=\overline{T}\,\overline{U}$ and $\overline{R}^3=\overline{I}$, we find that

$$\overline{I}, \overline{U}, \overline{U}^2, \overline{U}^3, \overline{U}^4, \overline{U}^5$$

is an equivalent system of representatives. Thus it follows that $n_\infty = 6$. Since $\bar{T} \notin \bar{\Gamma}^+$, it follows that $n_i = 2$, and since $\bar{R} \notin \bar{\Gamma}^+$, it follows that $n_\rho = 3$. Thus one obtains the fundamental region for $\bar{\Gamma}^+$ and corresponding boundary substitutions as illustrated. We see immediately from the branch schema (2, 3, 6) that the genus g is 1. One easily confirms that $\bar{\Gamma}^+$ is a congruence group and that naturally the conductor $F = 6$.

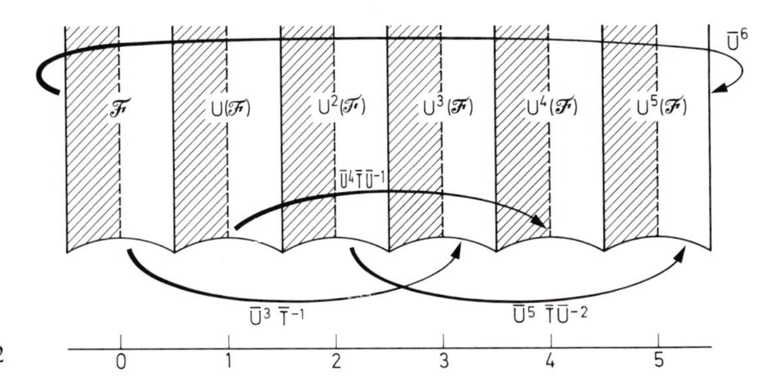

Fig. 22

4. Investigation of the genus formula. We transform the general genus formula

$$g = 1 - \mu + \frac{1}{2} \sum_{\nu=1}^{v} (\kappa_\nu - 1) \tag{20}$$

to a form which is of interest in its own right and that allows the calculation of the genus in concrete cases.

Suppose the subgroup Γ_1 of Γ contains the matrix $-I$, and that

$$\Gamma = \bigcup_{\nu=1}^{\mu} \Gamma_1 S_\nu, \tag{23}$$

where the matrices S_ν run through a system of pairwise incongruent matrices modulo Γ_1. As in **§ 7,2** let

$$V_1, V_2, \ldots, V_v$$

be the vertices of the polygonal decomposition of $\mathfrak{R} = \mathcal{H}^*/\Gamma_1$. If V_ν is the image under σ of a point of the class $\langle \rho \rangle \bmod \Gamma$, then

$\kappa_\nu = 1$, if V_ν is the image of a fixed point of Γ_1, and

$\kappa_\nu = 3$, whenever this is not the case.

If V_ν is an image of a point of the class $\langle i \rangle \bmod \Gamma$, then

$\kappa_\nu = 1$, if V_ν is the image of a fixed point of Γ_1, and

$\kappa_\nu = 2$, whenever this is not the case.

We denote by ε_ρ the number of the V_ν corresponding to $\langle\rho\rangle \bmod \Gamma$ that have $\kappa_\nu=1$, and by α_ρ the number of those that have $\kappa_\nu=3$. Let the corresponding numbers of the V_ν belonging to the class $\langle i\rangle$ be denoted by ε_i and α_i. The contribution of the σ_∞ vertices V_ν with pre-image in the class $\langle i\infty\rangle$ under Γ to the sum

$$\sum_{\nu=1}^{v}(\kappa_\nu-1)$$

is $\mu-\sigma_\infty$. Hence we derive

$$g=1-\mu+\alpha_\rho+\tfrac{1}{2}\alpha_i+\tfrac{1}{2}\mu-\tfrac{1}{2}\sigma_\infty .$$

Since

$$2\varepsilon_\rho+6\alpha_\rho=2\varepsilon_i+4\alpha_i=2\mu ,$$

we obtain the desired formula for the genus:

$$g=1+\frac{\mu}{12}-\frac{\varepsilon_\rho}{3}-\frac{\varepsilon_i}{4}-\frac{\sigma_\infty}{2}. \tag{24}$$

In the above formula ε_ρ is the number of Γ_1-inequivalent fixed points of $\langle\rho\rangle$ in the fundamental region $\mathscr{F}_1$ for Γ_1 in the form

$$\mathscr{F}_1=\bigcup_{\nu=1}^{\mu} S_\nu(\mathscr{F}),$$

and ε_i and σ_∞ are the corresponding numbers for $\langle\rho\rangle$ and $\langle i\infty\rangle$.

We investigate the numbers ε_ρ a little further. A point $S(\rho)$, $S\in\Gamma$, is a fixed point for Γ_1 if and only if there is an $L\in\Gamma_1$ $L\neq\pm I$, with

$$LS(\rho)=S(\rho), \quad \text{i.e,} \quad S^{-1}LS(\rho)=\rho ,$$

thus

$$S^{-1}LS=\pm R^\alpha \quad \text{or} \quad SR^\alpha S^{-1}\in\Gamma_1, \qquad \alpha=1 \text{ or } 2 .$$

The last conclusion holds since $-I\in\Gamma_1$. Since, however, $SRS^{-1}\in\Gamma_1$ is equivalent with $SR^2S^{-1}\in\Gamma_1$, it follows that $S(\rho)$ is a fixed point for Γ_1 if and only if

$$SRS^{-1}\in\Gamma_1 .$$

If $S(\rho)$ is a fixed point for Γ_1 and $S_1=g_1S$ with $g_1\in\Gamma_1$, then $S_1(\rho)$ is also a fixed point for Γ_1 and we have

$$SRS^{-1}\in\Gamma_1 \quad \text{and} \quad S_1RS^{-1}\in\Gamma_1 .$$

It follows that

$$S_1RS^{-1}=(S_1S^{-1})(SRS^{-1})\in\Gamma_1 ,$$

and so

$$S_1S^{-1}\in\Gamma_1 , \quad \text{or } S_1 \text{ congruent } S \text{ modulo } \Gamma_1 .$$

If, in addition, we take for granted the analogous behavior at $\langle i \rangle$ and use the known behavior at $\langle i\infty \rangle$, we obtain the following

Theorem 14. *The number ε_ρ of inequivalent fixed points under Γ_1 in the class $\langle \rho \rangle$ is equal to the number of incongruent solutions S modulo Γ_1 of*

$$S R S^{-1} \in \Gamma_1 . \tag{25}$$

The number ε_i is equal to the number of incongruent solutions S modulo Γ_1 of

$$S T S^{-1} \in \Gamma_1 . \tag{26}$$

The number σ_∞ is the number of incongruent solutions S modulo Γ_1 of

$$S U^F S^{-1} \in \Gamma_1 ,$$

where F is the conductor of Γ_1. Moreover

$$\sigma_\infty = \sum_{\nu=1}^{\mu} \frac{1}{\kappa_\nu} , \tag{27}$$

where κ_ν is the fan width at the cusp $S_\nu(i\infty)$, the smallest natural number k with

$$S_\nu U^k S_\nu^{-1} \in \Gamma_1 .$$

§ 8. The Genus of the Fundamental Region of $\Gamma_0(N)$

We calculate this genus from formula (24). In **§ 3,2** we introduced the group

$$\Gamma_0(N) = \left\{ \begin{pmatrix} a & b \\ c & d \end{pmatrix} \in \Gamma : c \equiv 0 \bmod N \right\} ,$$

whose index in Γ is

$$\mu_0(N) = N \cdot \prod_{p|N} \left(1 + \frac{1}{p} \right) .$$

We remark that $-I \in \Gamma_0(N)$.

1. Congruence mod $\Gamma_0(N)$. Two matrices $A = \begin{pmatrix} a & b \\ c & d \end{pmatrix}$ and $A_1 = \begin{pmatrix} a' & b' \\ c' & d' \end{pmatrix}$ from Γ are congruent modulo $\Gamma_0(N)$ if $A A_1^{-1}$ is an element of $\Gamma_0(N)$, i.e. if

$$c d' - c' d \equiv 0 \bmod N . \tag{28}$$

On the other hand, any two number pairs (c, d), (c', d') with $(c, d, N) = (c', d', N) = 1$ satisfying (28) can after a possible change mod N always be completed to matrices of Γ congruent modulo $\Gamma_0(N)$. The condition (28) is equivalent to the existence of a $\delta \in \mathbb{Z}$ for which

$$c \equiv \delta c' \bmod N \quad \text{and} \quad d = \delta d' \bmod N . \tag{29}$$

Indeed, because of $(c,d,N)=(c',d',N)=1$, there are two matrices $A=\begin{pmatrix} a & b \\ c_1 & d_1 \end{pmatrix}$, $A_1=\begin{pmatrix} a' & b' \\ c'_1 & d'_1 \end{pmatrix}$ in Γ with $c_1\equiv c, d_1\equiv d, c'_1\equiv c', d'_1\equiv d' \bmod N$. If $c\,d'-c'\,d\equiv 0 \bmod N$, then

$$A\,A_1^{-1}\equiv\begin{pmatrix} \alpha & \beta \\ 0 & \delta \end{pmatrix} \bmod N$$

with certain $\alpha, \beta, \delta\in\mathbb{Z}$. From

$$\begin{pmatrix} a & b \\ c & d \end{pmatrix}\equiv\begin{pmatrix} \alpha & \beta \\ 0 & \delta \end{pmatrix}\begin{pmatrix} a' & b' \\ c' & d' \end{pmatrix} \bmod N$$

we therefore obtain the congruence (29). Conversely, we see immediately that (29) implies (28). We shall call two number pairs (c,d) and (c',d') *proportional* $\bmod N$ if (29) is satisfied.

By what we just proved above, we obtain one representative of each class of incongruent matrices of Γ modulo $\Gamma_0(N)$ if we complete the number pairs (c,d) with

1) $(c,d,N)=1$,

2) (c,d) pairwise non-proportional $\bmod N$,

to matrices of Γ.

After these preparatory remarks we calculate the values ε_ρ, ε_i, and σ_∞ for $\Gamma_0(N)$. Here we use the theorem about factoring rational primes in a quadratic field. See H. Hasse [1], E. Hecke [1] or I. Niven and H.S. Zuckerman [1], p. 215ff.

2. Determination of ε_ρ and ε_i. In §7 we recognized that ε_ρ is the maximal number of incongruent $A\in\Gamma$ modulo $\Gamma_0(N)$ with $A\,R\,A^{-1}\in\Gamma_0(N)$. For $A=\begin{pmatrix} a & b \\ c & d \end{pmatrix}$ this is equivalent to 1), 2) (above), and

3) $c^2-c\,d+d^2\equiv 0 \bmod N$.

These three conditions have a simple meaning if we associate to the number pair (c,d) the integer $\omega=c-d\rho$ in the quadratic number field $\mathbb{Q}(\rho)$. Then 1), 2) and 3) become

1′) (ω, N) has no divisors in $\mathbb{Z}\setminus\{\pm 1\}$,

2′) $\omega'\neq\omega$ implies $\omega'\not\equiv g\,\omega \bmod N$ for any $g\in\mathbb{Z}\setminus\{0\}$

and

3′) $\mathcal{N}(\omega)\equiv 0 \bmod N$, where $\mathcal{N}(\omega)$ denotes the norm of ω.

Suppose N has the prime factorization

$$N=3^a\prod_{\nu=1}^{r} p_\nu^{m_\nu}$$

with non-negative a and positive m_ν.

If among the p_ν there is a prime p for which Legendre's symbol $\left(\frac{-3}{p}\right) = -1$, then $\mathcal{N}(\omega) = \omega\bar{\omega} \equiv 0 \bmod p$ implies that p divides (ω, N). This contradicts 1'). If $a \geqq 2$ then $\mathcal{N}(\omega) = \omega\bar{\omega} \equiv 0 \bmod 9$ implies that $\omega \equiv 0 \bmod 3$, from which it follows that $3|(\omega, N)$.

Thus in both of these cases $\varepsilon_\rho = 0$.

In the remaining cases $a=0$ or $a=1$ and $\left(\frac{-3}{p_\nu}\right) = 1$ for all ν. Thus each rational prime $p_\nu \neq 3$ dividing N has a factorization $p_\nu = \pi_\nu \cdot \bar{\pi}_\nu$ in the ring of integers $\mathbb{Z}(\rho)$ of $\mathbb{Q}(\rho)$ with non-equivalent primes π_ν and $\bar{\pi}_\nu$. Suppose the number ω satisfies conditions 1') and 3'). If $\omega \equiv 0 \bmod \pi_\nu \bar{\pi}_\nu$, then $p_\nu = \pi_\nu \cdot \bar{\pi}_\nu$ divides ω as well as N, which contradicts 1'). Thus one can so choose π_ν and $\bar{\pi}_\nu$ so that

$$\omega = \lambda \cdot \gamma \quad \text{with } \lambda \in \mathbb{Z}(\rho) \quad \text{and } (\lambda, \bar{\gamma}) = 1, \tag{30}$$

where

$$\gamma := \sqrt{-3}^{\,a}\, \pi_1^{m_1} \ldots \pi_r^{m_r}.$$

All numbers ω of the form (30) satisfy the conditions 1') and 3'). On the other hand 2') shows us that the factor λ has no influence on the number of simultaneous solutions of 1'), 2'), and 3'). For if $\lambda' \in \mathbb{Z}(\rho)$ with $(\lambda', \bar{\gamma}) = 1$ is given, then $\lambda' \equiv g\lambda \bmod \bar{\gamma}$ is solvable for $g \in \mathbb{Z}(\rho)$. The residue classes $\bmod \bar{\gamma}$ can be represented by rationals, since for each $z \in \mathbb{Z}$ with $z \equiv 0 \bmod \bar{\gamma}$, we also have $z \equiv 0 \bmod N$. Thus we can choose g in $\mathbb{Z}$. By multiplying by γ one obtains from $\lambda' \equiv g\lambda \bmod \bar{\gamma}$ that

$$\omega' \equiv g\,\omega \bmod N.$$

Since, however, $\omega = \lambda\gamma$ and $\omega' \equiv g\,\omega \bmod N$ for $g \in \mathbb{Z}$ imply $\omega' = \lambda'\gamma$ with the same γ, we see that the different simultaneous solutions of 1'), 2') and 3') are determined by the different possibilities for γ. Their number, ε_ρ, is the number of r-tuples $(\pi_1', \ldots, \pi_r')$ with $\pi_i' \in \{\pi_i, \bar{\pi}_i\}$. Consequently $\varepsilon_\rho = 2^r$. Together with our earlier results we have

$$\varepsilon_\rho = \begin{cases} 0 & \text{if } 9|N, \text{ and} \\ \prod\limits_{p|N} \left(1 + \left(\frac{-3}{p}\right)\right) & \text{otherwise}. \end{cases} \tag{31}$$

Since the calculation of ε_i follows analogously when $\mathbb{Q}(\rho)$ is replaced by the Gaussian number field $\mathbb{Q}(i)$, we only state the final result

$$\varepsilon_i = \begin{cases} 0 & \text{if } 4|N, \text{ and} \\ \prod\limits_{p|N} \left(1 + \left(\frac{-1}{p}\right)\right) & \text{otherwise}. \end{cases} \tag{32}$$

3. Evaluation of σ_∞. From § 7 the number of $\Gamma_0(N)$-inequivalent rational cusps is

$$\sigma_\infty = \sum_{\nu=1}^{\mu_0(N)} \frac{1}{\kappa_\nu}.$$

In this formula κ_ν is the smallest natural number k for which the matrix $A_\nu U^k A_\nu^{-1} \in \Gamma_0(N)$, where $A_\nu = \begin{pmatrix} a_\nu & b_\nu \\ c_\nu & d_\nu \end{pmatrix} \in \Gamma$ and $\frac{a_\nu}{c_\nu}$ is a rational cusp of $\mathscr{H}^*$ which is a pre-image of the νth vertex V_ν under σ. From $A_\nu U^{\kappa_\nu} A_\nu^{-1} \equiv \begin{pmatrix} \alpha & \beta \\ 0 & \delta \end{pmatrix} \bmod N$ we obtain $c_\nu^2 \kappa_\nu \equiv 0 \bmod N$. Thus

$$\kappa_\nu = \frac{N}{(c_\nu^2, N)}.$$

If we introduce $t := (c, N)$ and $t_\nu := (c_\nu, N)$, then $(c_\nu^2, N) = t_\nu \cdot \left(t_\nu, \frac{N}{t_\nu}\right)$ and

$$\sigma_\infty = \frac{1}{N} \sum_{\nu=1}^{\mu_0(N)} t_\nu \left(t_\nu, \frac{N}{t_\nu}\right).$$

Now we count how often a fixed $c \bmod N$ occurs among the c_ν. To each $c \bmod N$ there correspond exactly $\varphi(t)$ values of $d \bmod t$. Thus there are $\frac{N}{t} \varphi(t)$ incongruent values for d modulo N. Since we already included the pairs (c, d) which are proportional $\bmod N$, we must divide this sum by $\varphi(N)$ to obtain σ_∞. Thus we have found

$$\sigma_\infty = \frac{1}{\varphi(N)} \sum_{c \bmod N} \left(t, \frac{N}{t}\right) \varphi(t).$$

Each divisor t of N appears for $\varphi\left(\frac{N}{t}\right)$ different $c \bmod N$, thus

$$\sigma_\infty = \frac{1}{\varphi(N)} \sum_{t|N} \left(t, \frac{N}{t}\right) \varphi(t) \varphi\left(\frac{N}{t}\right).$$

Since

$$\varphi(n_1)\varphi(n_2) = \varphi(n_1 n_2) \frac{\varphi((n_1, n_2))}{(n_1, n_2)},$$

we have

$$\sigma_\infty = \sum_{t|N} \varphi\left(\left(t, \frac{N}{t}\right)\right). \tag{33}$$

We collect these results in

Theorem 15. *The genus of the fundamental region of the congruence group* $\Gamma_0(N)$ *is given by*

$$g = 1 + \frac{\mu_0(N)}{12} - \frac{\varepsilon_\rho}{3} - \frac{\varepsilon_i}{4} - \frac{\sigma_\infty}{2},$$

where the values of $\mu_0(N)$, ε_ρ, ε_i, *and* σ_∞ *are claculated from* (8), (31), (32), *and* (33).

For $N \leqq 25$ we obtain from these formulas the genus of the groups $\Gamma_0(N)$:

$$\begin{aligned} g&=0 \quad \text{for } N=1,\ldots,10, 12, 13, 16, 25;\\ g&=1 \quad \text{for } N=11, 14, 15, 17,\ldots, 21;\\ g&=2 \quad \text{for } N=22, 23;\\ g&=3 \quad \text{for } N=24. \end{aligned}$$

For § **8** cf. W. Maak [1].

Chapter V. Function Theory for the Subgroups of Finite Index in the Modular Group

In this chapter we introduce functions and forms of integral dimension for subgroups Γ_1 of finite index in Γ, and as in chapter **II**, at first without regard to their existence, we investigate their basic properties. Then again, independent of questions of existence, we focus on modular forms of dimension -2 and their connection with certain integrals. The existence of such functions and forms will be derived from the Riemann-Roch Theorem of the theory of algebraic functions. In order to understand this theorem we compile certain facts. In conclusion we apply the Riemann-Roch Theorem to the calculation of the $\mathbb{C}$-dimension of the vector space of entire modular forms of fixed dimension.

§ 1. Functions for Subgroups

The subgroups Γ_1 and $\Gamma_1 \cup (-I)\Gamma_1$ support the same functions. We therefore assume in this section that $-I \in \Gamma_1$. We denote the finite index of Γ_1 in Γ by μ.

1. Definition. A function f on $\mathscr{H}^*$ is called a *modular function for* Γ_1 if the following hold:

a) f is meromorphic on $\mathscr{H}$;

b) for all $S \in \Gamma_1$ and $\tau \in \mathscr{H}^*$, $f(S(\tau)) = f(\tau)$;

c) at a rational cusp $-\dfrac{d}{c}$, $(c, d) = 1$, f has an expansion of the form

$$f(\tau) = \sum_{\nu \geqq \nu_0} c_\nu e^{\frac{2\pi i}{\kappa_1} A(\tau)\nu}, \qquad A = \begin{pmatrix} a & b \\ c & d \end{pmatrix} \in \Gamma, \tag{1}$$

where κ_1 is a natural number. This expansion is valid for $\tau \in \mathscr{H}$ with imaginary part of $A(\tau)$ sufficiently large.

$f\left(-\dfrac{d}{c}\right)$ is defined in the usual way by (1).

The cusp $-\frac{d}{c}$ determines the matrix A only up to a factor of $\pm U^n$ with arbitrary $n \in \mathbb{Z}$ and thus determines $e^{\frac{2\pi i}{\kappa_1} A(\tau)}$ to within a κ_1^{-th} root of unity. One can choose $\kappa_1 = \kappa$ the fan width of the fundamental region for Γ_1 at $-\frac{d}{c}$. This follows from the expansion (1), since

$$A^{-1} U^{\kappa} A \in \Gamma_1$$

and by b) the left side in (1) is invariant under the transformation

$$\tau \mapsto A^{-1} U^{\kappa} A(\tau) \quad \text{or} \quad A(\tau) \mapsto U^{\kappa} A(\tau).$$

2. First properties. We enumerate some properties of modular functions for Γ_1 which follow directly from Definition **1**:

1. The modular functions for Γ_1 form a field.

2. If f is a modular function for Γ_1 and if Γ_2 is a subgroup of finite index in Γ_1, then f is a modular function for Γ_2.

3. If f is a modular function for Γ_1 and is invariant under the transformations of the group extension $\Gamma_2 \subset \Gamma$, then f is a modular function for Γ_2.

4. If f is a modular function for Γ_1 and $S \in \Gamma$, then

$$f|S : \mathscr{H}^* \to \hat{\mathbb{C}}, \qquad \tau \mapsto f(S(\tau)),$$

is a modular function for $S^{-1} \Gamma_1 S$.

Concerning 4. we remark that the first two conditions are obviously fulfilled. The validity of c) follows from the existence of an expansion

$$f(\tau) = \sum_{\nu \geqq \nu_0} c_\nu e^{\frac{2\pi i}{\kappa_1} A S^{-1}(\tau) \nu}$$

at the point $S A^{-1}(\infty)$ when τ is replaced by $S(\tau)$.

$f|S$ coincides formally even for real $S = \begin{pmatrix} a & b \\ c & d \end{pmatrix}$, $ad - bc > 0$, with $f|_k S$ for $k = 0$ as defined in **II, §4,1** when $\mathscr{H}^*$ replaces $\mathscr{H}$. We prove

Theorem 1. *Each modular function f for Γ_1 satisfies an algebraic equation of degree μ,*

$$G(f) = \sum_{\nu=0}^{\mu} R_\nu(J) f^\nu = 0, \tag{2}$$

where the $R_\nu(J)$ are rational functions of J over $\mathbb{C}$.

Proof. Let $\Gamma = \bigcup_{\nu=1}^{\mu} \Gamma_1 S_\nu$ be a coset decomposition of Γ with certain $S_\nu \in \Gamma$. The coefficients of the polynomial in X,

$$G(X) := \prod_{\nu=1}^{\mu} (x - f|S_\nu), \tag{3}$$

are symmetric functions of the $f|S_\nu$. For each $A \in \Gamma$ and for each ν there is a ν' and a substitution $G_\nu \in \Gamma_1$ such that $S_\nu A = G_\nu S_{\nu'}$. For fixed A, ν and ν' both run through all μ indices. Thus the coefficients of $G(X)$ are invariant under Γ. The coefficients are obviously meromorphic functions of τ in $\mathscr{H}$ and, since by 4. each $f|S_\nu$ has an expansion of the form (1), so do the coefficients of $G(x)$. Thus they are rational functions of J. Since one of the S_ν's lies in Γ_1, it follows that $G(f) = 0$. □

As a purely algebraic consequence of Theorem **1** we deduce

Theorem 2. *The field of modular functions for Γ_1 is an algebraic extension of $\mathbb{C}(J)$ of degree $\mu_1 \leqq \mu$.*

We denote this field by K_{Γ_1}. From the existence theorems of function theory it will follow that the equality sign holds—see **§ 4**.

3. Meromorphic functions on $\mathfrak{R}$. The modular functions f for Γ_1 have an invariant power series expansion in terms of the special uniformizing variables ϕ_τ as introduced in **IV, § 6**, (18) and (19). For fixed $\tau \in \mathscr{H}^*$,

$$f(v) = \sum_{\nu \geqq \nu_0} c_\nu \phi_\tau^\nu(v)$$

for all v in a neighborhood $\mathscr{U}_\tau$ of τ. Now let

$$S \in \Gamma_1, \quad \tau' = S(\tau), \quad \text{and } v' = S(v),$$

then by **IV**, (18′) and (19′) it follows that

$$\phi_\tau(v) = \phi_{\tau'}(v'),$$

and that

$$f(v) = \sum_{\nu \geqq \nu_0} c_\nu \phi_{\tau'}^\nu(v'), \quad v \in \mathscr{U}_\tau.$$

Since

$$f(v) = f(v'),$$

we conclude that

$$f(v') = \sum_{\nu \geqq \nu_0} c_\nu \phi_{\tau'}^\nu(v), \quad v' \in S(\mathscr{U}_\tau).$$

In particular, we obtain from this last equation

Theorem 3. *As measured in local variables, the orders of zeros, poles, and, more generally, the orders of c-points of a modular function for Γ_1 are the same at equivalent points under Γ_1.*

Consequently the function f permits the definition of a function F on the Riemann surface $\mathfrak{R}$ associated with Γ_1 by means of the equation

$$F(\langle v\rangle) := f(v). \tag{4}$$

On account of the local expansion

$$F(\langle v\rangle) = \sum_{\nu \geqq \nu_0} c_\nu \phi_\tau^\nu(v),$$

F is meromorphic on $\mathfrak{R}$. Conversely, by means of

$$f(v) := F(\langle v\rangle),$$

and by virtue of the expansions for F on $\mathfrak{R}$, every meromorphic function F on $\mathfrak{R}$ gives rise to a modular function f for Γ_1. The indicated correspondence maps the meromorphic functions on $\mathfrak{R}$ one-to-one onto the modular functions for Γ_1. Their orders at the corresponding points $\langle v\rangle$ and v are the same.

Theorem 4. *Each function on $\mathfrak{R}$ corresponding to a non-constant modular function for Γ_1 takes every value equally often.*

This is a theorem about algebraic functions. The theorem also has a proof similar to the special case $\Gamma = \Gamma_1$ already treated. We do not wish to carry it through.

4. $\mathfrak{R}$ as a covering surface. Let $w = J(\tau)$ and $J(\langle\tau\rangle) := J(\tau)$. The Riemann surface $\mathfrak{R} = \mathscr{H}^*/\Gamma_1$ is mapped by J onto a μ-sheeted covering surface of the w-sphere in a $1-1$ manner. (See also **II, § 3,3.**) We investigate the analytic character of this map.

1) If $\tau_0 \in \mathscr{H}$ is not Γ-equivalent to either ρ or i, then a sufficiently small neighborhood $\mathscr{U}_{\tau_0, r}$ of τ_0 is mapped conformally by J onto a schlicht neighborhood of a point over $w_0 = J(\tau_0)$. A modular function f for Γ_1 can be expanded in a series

$$f(\tau) = \sum_{\nu \geqq \nu_0} c_\nu (\tau - \tau_0)^\nu$$

for τ in a sufficiently small neighborhood of τ_0. On the other hand

$$w = J(\tau) = \sum_{\nu \geqq 0} a_\nu (\tau - \tau_0)^\nu, \qquad a_1 \neq 0,$$

and, since

$$\tau - \tau_0 = \sum_{\nu \geqq 0} d_\nu (w - w_0)^\nu, \qquad d_1 \neq 0, \tag{6}$$

it follows that $w - w_0$ is a local variable for $\langle \tau_0 \rangle$ on $\mathfrak{R}$.

2) If τ_ρ is equivalent to ρ under Γ and is a fixed point for Γ_1, then for a suitably chosen fundamental region $\tilde{\mathscr{F}}_1$ for Γ_1 the intersection $\mathscr{U}_{\tau_\rho,r} \cap \tilde{\mathscr{F}}_1$ is mapped by J onto a schlicht neighborhood over $J(\tau_\rho)=0$. The function f has the expansion

$$f(\tau)=\sum_{\nu \geqq \nu_0} c_\nu \left(\frac{\tau-\tau_\rho}{\tau-\bar{\tau}_\rho}\right)^{3\nu}.$$

From

$$J(\tau)=\sum_{\nu \geqq 1} a_\nu \left(\frac{\tau-\tau_\rho}{\tau-\bar{\tau}_\rho}\right)^{3\nu}, \quad a_1 \neq 0,$$

we obtain

$$\left(\frac{\tau-\tau_\rho}{\tau-\bar{\tau}_\rho}\right)^{3}=\sum_{\nu \geqq 1} d_\nu w^\nu, \quad d_1 \neq 0. \tag{7}$$

Thus w is a local variable at $\langle\tau_\rho\rangle$ on $\mathfrak{R}$.

If τ_ρ is not a fixed point of Γ_1, then a suitably chosen fundamental region $\tilde{\mathscr{F}}_1$ contains a set $\mathscr{U}_{\tau_\rho,r}$. This set is mapped by J onto a neighborhood over $w=0$, and in this neighborhood each $w \neq 0$ is covered exactly three times, i.e. over $w=0$ is a branch point of order 2. A local variable at τ_ρ is

$$\frac{\tau-\tau_\rho}{\tau-\bar{\tau}_\rho}=\sqrt[3]{\sum_{\nu=1}^{\infty} d_\nu w^\nu}=\sqrt[3]{w}\sum_{\nu=0}^{\infty} b_\nu w^\nu, \quad b_0 \neq 0. \tag{8}$$

We then have $\sqrt[3]{w}$ as the local variable at $\langle\tau_\rho\rangle$ on the Riemann surface $\mathfrak{R}$.

3) Correspondingly, a point τ_i, Γ-equivalent to i, is mapped on a branch point over $w=1$ of order 0 or of order 1 according as τ_i is or is not a fixed point of Γ_1. As local variable at $\langle\tau_i\rangle$ one uses analogously either $w-1$ or $\sqrt{w-1}$, respectively.

4) If $\tau_\infty=-\dfrac{d}{c}$ is a rational point, then τ_∞ is mapped onto a branch point over $J(i\infty)=\infty$ of order $\kappa-1$. If we take

$$t_{\tau_\infty}=e^{\frac{2\pi i}{\kappa}A(\tau)}, \quad A=\begin{pmatrix} a & b \\ c & d \end{pmatrix} \in \Gamma,$$

as the local variable, then J has the expansion

$$J(\tau)=\sum_{\nu \geqq -1} a_\nu t_{\tau_\infty}^{\kappa\cdot\nu}, \quad a_{-1} \neq 0,$$

or

$$J(\tau)^{-1}=\sum_{\nu \geqq 1} b_\nu t_{\tau_\infty}^{\kappa\cdot\nu}, \quad b_1 \neq 0.$$

Hence $t_{\tau\infty}^{\kappa}$ can be developed in a power series

$$t_{\tau\infty}^{\kappa} = \sum_{\nu \geqq 1} d_\nu w^{-\nu}, \quad d_1 \neq 0, \tag{9}$$

and $\sqrt[\kappa]{\frac{1}{w}}$ is a local variable for $\langle\tau_\infty\rangle$ on $\mathfrak{R}$.

We say that the point $\langle\tau\rangle$ of $\mathfrak{R}$ lies over $J(\tau)$. Accordingly, there are ε_ρ, ε_i, and σ_∞ points lying over 0, 1 and ∞, respectively. If τ is not Γ-equivalent to ρ, i or $i\infty$, then there are exactly μ points over $J(\tau)$.

§ 2. Modular Forms for Subgroups

In this section we do not assume that $-I \in \Gamma_1$.

1. Definitions and properties. A function h of the two complex parameters ω_1 and ω_2 is called a *homogeneous modular form for* Γ_1 if it is defined for $\operatorname{Im}(\tau)>0$ with $\tau = \frac{\omega_1}{\omega_2}$ and if it satisfies the following four conditions:

a) $h(\lambda\omega_1, \lambda\omega_2) = \lambda^{-k} h(\omega_1, \omega_2)$ for all $\lambda \neq 0$ and a $k \in \mathbb{Z}$,

b) $h(a\omega_1 + b\omega_2,\ c\omega_1 + d\omega_2) = h(\omega_1, \omega_2)$ for $\begin{pmatrix} a & b \\ c & d \end{pmatrix} \in \Gamma_1$,

c) $\omega_2^k h(\omega_1, \omega_2) = h(\tau, 1)$ is meromorphic in $\mathscr{H}$, and

d) for each rational point $-\frac{d}{c}$, $(c, d)=1$, there is an expansion of the form

$$(c\omega_1 + d\omega_2)^k h(\omega_1, \omega_2) = \sum_{\gamma \geqq \nu_0} c_\nu e^{\frac{2\pi i}{\kappa_1} A(\tau)\nu}$$

with $A = \begin{pmatrix} a & b \\ c & d \end{pmatrix} \in \Gamma$, $\kappa_1 \in \mathbb{N}$, that is valid for τ with $\operatorname{Im} A(\tau)$ sufficiently large.

The number $-k$ is called the *dimension* of the modular form h.

A function $f(\tau)$ is called an *inhomogeneous modular form for* Γ_1 *of dimension* $-k$ if $\omega_2^{-k} f\left(\frac{\omega_1}{\omega_2}\right)$ is a homogeneous modular form of dimension $-k$ for Γ_1. For inhomogeneous modular forms the conditions b) and d) become

b') $$f(A(\tau)) = (c\tau + d)^k f(\tau) \quad \text{for } A = \begin{pmatrix} a & b \\ c & d \end{pmatrix} \in \Gamma_1,$$

and

d') $$(c\tau + d)^k f(\tau) = \sum_{\nu \geqq \nu_0} c_\nu e^{\frac{2\pi i}{\kappa_1} A(\tau)\nu}, \quad A = \begin{pmatrix} a & b \\ c & d \end{pmatrix} \in \Gamma.$$

Conversely, the conditions b′) and d′) for inhomogeneous modular forms f imply the conditions b) and d) for the homogeneous form $\omega_2^{-k} f\left(\frac{\omega_1}{\omega_2}\right)$. If $h(\omega_1, \omega_2)$ is a homogeneous modular form, then $h(\tau, 1)$ is an inhomogeneous modular form.

Here also one can choose $\kappa_1 = \kappa$ the fan width of the fundamental region for Γ_1 at $-\frac{d}{c}$. Indeed, as is easily verified, the left side in d′) remains invariant under the transformation

$$\tau \mapsto A^{-1} U^{\kappa} A(\tau) \quad \text{or} \quad A(\tau) \mapsto U^{\kappa} A(\tau).$$

An inhomogeneous modular form f for Γ_1 that is holomorphic in $\mathscr{H}$ and whose expansion for $(c\tau + d)^k f(\tau)$ at all cusps $-\frac{d}{c}$ begins with an index $\nu_0 \geqq 0$, is called an *entire modular form* for Γ_1. If the expansion at $-\frac{d}{c}$ begins with $\nu_0 \geqq 1$, we say that f vanishes at the cusp $-\frac{d}{c}$. If an entire form f vanishes at all cusps we call it a *cusp form.*

For an arbitrary function f on $\mathscr{H}$ and for real $A = \begin{pmatrix} a & b \\ c & d \end{pmatrix}$ with $|A| = ad - bc > 0$ we recall the definition

$$f|_k A: \mathscr{H} \to \hat{\mathbb{C}}, \qquad \tau \mapsto |A|^{\frac{k}{2}} \frac{f(A(\tau))}{(c\tau + d)^k} \tag{10}$$

given in **II, §4, 1.**

As a direct consequence of the definitions of homogeneous and inhomogeneous forms we have

Theorem 5. *The set of modular forms, the entire modular forms, and the cusp forms, each of the same dimension for Γ_1, form vector spaces over the complex field.*

The product and quotient of two modular forms for Γ_1 of dimensions $-k_1$ and $-k_2$ are modular forms for Γ_1 of dimensions $-(k_1 + k_2)$ and $-(k_1 - k_2)$, respectively.

Modular forms of dimension 0 are modular functions if their values at the cusps are defined so as to be consistent with the expansion d′).

The following two properties are immediate from the definition of modular forms.

1. If f is a modular form for Γ_1 and if Γ_2 is a subgroup of finite index in Γ_1, then f is a modular form for Γ_2.
2. If f is a homogeneous modular form for Γ_1 and if f is invariant under the transformations of the group extension $\Gamma_2 \subset \Gamma$, then f is a homogeneous modular form for Γ_2.

We prove a further property of modular forms in

Theorem 6. *If f is a modular form of dimension $-k$ for Γ_1 and if* $S=\begin{pmatrix}\alpha & \beta\\ \gamma & \delta\end{pmatrix}\in\Gamma$, *then $f|_k S$ is a modular form of dimension $-k$ for $S^{-1}\Gamma_1 S$.*

Proof. We prove only the existence of an expansion d′) for $f|_k S$. The rest is clear. If $AS^{-1}=\begin{pmatrix}a_1 & b_1\\ c_1 & d_1\end{pmatrix}$, then f has the expansion

$$(c_1\tau+d_1)^k f(\tau)=\sum_{\nu\geqq\nu_0} c_\nu e^{\frac{2\pi i}{\kappa}AS^{-1}(\tau)\nu}$$

at the point $SA^{-1}(i\infty)$. If we replace τ by $S(\tau)$, then the claim for the expansion follows from

$$c_1 S(\tau)+d_1=\frac{c\tau+d}{\gamma\tau+\delta}.$$

Thus

$$(c\tau+d)^k\frac{f(S(\tau))}{(\gamma\tau+\delta)^k}=\sum_{\nu\geqq\nu_0} c_\nu e^{\frac{2\pi i}{\kappa}A(\tau)\nu}.$$

Hence $f|_k S$ is a modular form for the group $S^{-1}\Gamma_1 S$. ☐

Obviously, if f is an entire form or a cusp form, then $f|_k S$ is likewise an entire form or a cusp form.

2. Expansions of a modular form at the vertices of a fundamental region. First, if $-I\in\Gamma_1$, it follows as in the case of the full modular group, that there are no modular forms $f\neq 0$ of odd dimension for Γ_1.

If τ_ρ is Γ-equivalent to $\rho=e^{2\pi i/3}$ and is a fixed point for Γ_1, then by **I, § 3,4,** if $\tau_\rho=\frac{\rho-d}{c}$, say, we have $A(\tau_\rho)=\tau_\rho$, where

$$A=\begin{pmatrix}-(1+d), & -\dfrac{d^2+d+1}{c}\\ c & d\end{pmatrix}.$$

If $-A$ is in Γ_1 then, because $(-A)^3=-I$ is in Γ_1, there are no modular forms of odd dimension. The function $(\tau-\bar\tau_\rho)^k f(\tau)$ has an expansion

$$(\tau-\bar\tau_\rho)^k f(\tau)=\sum_{\nu\geqq\nu_0} c_\nu\left(\frac{\tau-\tau_\rho}{\tau-\bar\tau_\rho}\right)^\nu$$

at τ_ρ. If we apply the substitution A to the left side, we obtain

$$(A(\tau)-\bar{\tau}_\rho)^k f(A(\tau)) = \rho^k(\tau-\bar{\tau}_\rho)^k f(\tau),$$

since

$$A(\tau)-\bar{\tau}_\rho = A(\tau)-A(\bar{\tau}_\rho) = \frac{\tau-\bar{\tau}_\rho}{c\tau+d}\cdot\rho.$$

On the right side we obtain

$$\sum_{\nu\geqq\nu_0} c_\nu\left(\frac{A(\tau)-\tau_\rho}{A(\tau)-\bar{\tau}_\rho}\right)^\nu = \sum_{\nu\geqq\nu_0} c_\nu\bar{\rho}^{2\nu}\left(\frac{\tau-\tau_\rho}{\tau-\bar{\tau}_\rho}\right)^\nu,$$

and therefore together we have

$$\sum_{\nu\geqq\nu_0} c_\nu\left(\frac{\tau-\tau_\rho}{\tau-\bar{\tau}_\rho}\right)^\nu = (\tau-\bar{\tau}_\rho)^k f(\tau) = \sum_{\nu\geqq\nu_0} c_\nu\bar{\rho}^{2\nu+k}\left(\frac{\tau-\tau_\rho}{\tau-\bar{\tau}_\rho}\right)^\nu.$$

It follows that only the indices with $\nu\equiv k \bmod 3$ appear in this expansion. We have proved:

In a neighborhood of the fixed point τ_ρ, f has the expansion

$$(\tau-\bar{\tau}_\rho)^k f(\tau) = \sum_{\substack{\nu\geqq\nu_0\\ \nu\equiv k \bmod 3}} c_\nu\left(\frac{\tau-\tau_\rho}{\tau-\bar{\tau}_\rho}\right)^\nu. \tag{11}$$

If τ_i is Γ-equivalent to i and is a fixed point of Γ_1, say $\tau_i = \dfrac{i-d}{c}$, then

$$B = \begin{pmatrix} -d & -\dfrac{d^2+1}{c} \\ c & d \end{pmatrix}$$

fixes τ_i, i.e. $B(\tau_i)=\tau_i$. Since $(\pm B)^2=-I$, there are no modular forms of odd dimension for Γ_1. As above we obtain:

In a neighborhood of the fixed point τ_i, f has the expansion

$$(\tau-\bar{\tau}_i)^k f(\tau) = \sum_{\substack{\nu\geqq\nu_0\\ 2\nu\equiv k \bmod 4}} c_\nu\left(\frac{\tau-\tau_i}{\tau-\bar{\tau}_i}\right)^\nu. \tag{12}$$

The order of zeros and poles of a modular form for Γ_1 will be measured as in the case of Γ (cf. **II § 4, 3**), i.e. they will be measured at the cusps and elliptic fixed points of a fundamental region for Γ_1 using **1**, d), (11) and (12).

The two expansions (11) and (12) reveal that, as measured in the local variable t_i or t_ρ, the orders of the zeros or poles of f at τ_ρ or τ_i can be fractional. However, for forms of the same dimension their orders at these points are congruent modulo 1.

3. Expansions of forms at points equivalent under Γ_1. Let f be a modular form of dimension $-k$ for the group Γ_1. Corresponding to each $\tau \in \mathscr{H}^*$, $\tau \neq i\infty$, there is an expansion

$$(v-\bar{\tau})^k f(v) = \sum_{\nu \geqq \nu_0} c_\nu \psi_\tau^\nu(v), \tag{13}$$

valid for v in a neighborhood $\mathscr{U}_\tau$ of τ, where

$$\psi_\tau(v) = \frac{v-\tau}{v-\bar{\tau}} \, \frac{c\tau_0+d}{c\bar{\tau}_0+d} \quad \text{for } \tau \in \mathscr{H}$$

with τ_0 a fixed representative of the class of τ under Γ_1 (cf. **IV, 6, 2**), and where

$$\psi_\tau(v) = e^{\frac{2\pi i}{\kappa} A(v)} \quad \text{for } \tau = -\frac{d}{c} \in \mathbb{Q}$$

with $A = \begin{pmatrix} a & b \\ c & d \end{pmatrix} \in \Gamma$. Since

$$\psi_\tau(v) = \psi_{\tau'}(v'), \qquad v' = S(v), \qquad \tau' = S(\tau), \qquad S = \begin{pmatrix} \alpha & \beta \\ \gamma & \delta \end{pmatrix} \in \Gamma_1,$$

(cf. **IV, § 6, 2**) we also have

$$(v-\bar{\tau})^k f(v) = \sum_{\nu \geqq \nu_0} c_\nu \psi_{\tau'}^\nu(v'), \qquad v \in \mathscr{U}_\tau.$$

First of all, we suppose that τ and τ' are different from $i\infty$. Then from

$$v' - \bar{\tau}' = \frac{v-\bar{\tau}}{(\gamma v+\delta)(\gamma\bar{\tau}+\delta)}$$

and from the transformation formula we have

$$(v'-\bar{\tau}')^k f(v') = (\gamma\bar{\tau}+\delta)^{-k}(v-\bar{\tau})^k f(v).$$

Thus f has the expansion

$$(v'-\bar{\tau}')^k f(v') = (\gamma\bar{\tau}+\delta)^{-k} \sum_{\nu \geqq \nu_0} c_\nu \psi_{\tau'}^\nu(v') \tag{14}$$

If on the other hand $\tau = i\infty$, $\tau' = \frac{\alpha}{\gamma}$, say, then we have

$$f(v) = \sum_{\nu \geqq \nu_0} c_\nu \psi_{i\infty}^\nu(v), \tag{15}$$

and also

$$f(v) = \sum_{\nu \geqq \nu_0} c_\nu \psi_{\tau'}^\nu(v').$$

We obtain from

$$f(v') = (\gamma v+\delta)^k f(v) \text{ and } v' = \frac{\alpha v+\beta}{\gamma v+\delta} = \frac{\alpha}{\gamma} - \frac{1}{\gamma(\gamma v+\delta)}.$$

that

$$(v'-\tau')^k f(v') = (-\gamma)^{-k} \sum_{\nu \geqq \nu_0} c_\nu \psi_{\tau'}^{\nu}(v'). \tag{16}$$

In particular our expansions prove

Theorem 7. *The orders of zeros and poles of an inhomogeneous modular form for Γ_1 are the same at Γ_1-equivalent points.*

4. The number of zeros of an entire modular form of dimension $-k$ for Γ_1. We recall that the discriminant Δ is a cusp form of dimension -12 for Γ and thus for Γ_1 as well. We show:

There are no entire modular forms of positive dimension for Γ_1.

If f_k were an entire modular form of dimension $k>0$ for Γ_1, then $f_k^{12}\Delta^k$ would have dimension 0 and would thus be a modular function for Γ_1 without poles, and hence would be a constant C. Thus $f_k^{12} = C\Delta^{-k}$, $C \neq 0$, which contradicts the fact that f_k was to be entire.

Using Δ, we can determine the number of zeros of an entire modular form f_{-k} of dimension $-k$ for Γ_1 in a fundamental set $\mathscr{F}_1$. As a modular form for Γ, the discriminant has an expansion

$$\left(\tau+\frac{d}{c}\right)^{12} \Delta(\tau) = \sum_{\nu \geqq 1} c_\nu e^{\frac{2\pi i}{\kappa} A(\tau)\cdot \nu \cdot \kappa}, \qquad A = \begin{pmatrix} a & b \\ c & d \end{pmatrix} \in \Gamma,$$

at the cusp $-\frac{d}{c}$, where κ is the fan width of the fundamental region for Γ_1 at $-\frac{d}{c}$. Thus Δ has a zero of order κ at $-\frac{d}{c}$. Hence the total number of zeros of Δ in $\mathscr{F}_1$ is $\bar{\mu}$, the index of $\Gamma_1 \cup (-I)\Gamma_1$ in Γ. Since $f_{-k}^{12}\Delta^{-k}$ is a modular function, f_{-k}^{12} and Δ^k have the same number of zeros. Thus we have proved

Theorem 8. *As measured in the local variable, an entire modular form $f \neq 0$ of dimension $-k$ for Γ_1 has $\frac{\bar{\mu}k}{12}$ zeros in $\mathscr{F}_1$, where $\bar{\mu}$ is the index $(\Gamma : \Gamma_1 \cup (-I)\Gamma_1)$.*

For an arbitrary modular form it is obvious that the number of zeros minus the number of poles in $\mathscr{F}_1$ is $\frac{\bar{\mu}k}{12}$.

As a corollary to this theorem we have

Theorem 9. *The vector space of entire modular forms of dimension $-k$ for Γ_1 has finite $\mathbb{C}$-dimension*

$$h \leqq 1 + \frac{\bar{\mu}k}{12}.$$

Proof. Let $\Phi_s, s=1,2,\ldots,h$ be linearly independent entire modular forms of dimension $-k$ for Γ_1. Further, let

$$\Phi_s(\tau)=\sum_{\nu\geqq 0} c_\nu^{(s)} e^{\frac{2\pi i}{\kappa}\cdot\tau\cdot\nu}, \qquad s=1,2,\ldots,h,$$

be their expansions at $i\infty$. Consider the function Φ:

$$\Phi(\tau)=\sum_{s=1}^{h} x_s\Phi_s(\tau)=\sum_{\nu\geqq 0}\sum_{s=1}^{h} x_s c_\nu^{(s)} e^{\frac{2\pi i}{\kappa}\cdot\tau\cdot\nu}.$$

The homogeneous system of linear equations

$$\sum_{s=1}^{h} x_s c_\nu^{(s)}=0, \qquad \nu=0,\ldots,h-2$$

has a non-trivial solution $x_1^*,\ldots,x_h^*$. Consequently

$$\Phi^*(\tau):=\sum_{s=1}^{h} x_s^*\Phi_s(\tau) \tag{17}$$

has a zero of order at least $h-1$ at $i\infty$. If $h-1>\frac{\bar{\mu}k}{12}$, then clearly $\Phi^*=0$. So the maximal number of linearly independent entire modular forms of dimension $-k$ for Γ_1 satisfies

$$h\leqq 1+\frac{\bar{\mu}k}{12}. \quad \square$$

The $\mathbb{C}$-dimension h will be investigated further in **§ 4, 4.**

§ 3. Modular Forms of Dimension -2 and Integrals

We again assume that $-I\in\Gamma_1$. Here this will lead to no loss of generality.

1. Integrals for Γ_1. The derivative with respect to τ of a modular function F for Γ_1,

$$f(\tau):=\frac{dF(\tau)}{d\tau},$$

is a modular form of dimension -2. We prove this in a more general setting.

Definition. A meromorphic function F on $\mathscr{H}$ is called an *integral for Γ_1* if it has expansions

$$F(\tau)=c_r^* A(\tau)+\sum_{\nu\geqq\nu_0} c_\nu t_r^\nu, \quad t_r=e^{\frac{2\pi i}{\kappa}A(\tau)}, \quad A=\begin{pmatrix} a & b\\ c & d\end{pmatrix}\in\Gamma \tag{18}$$

at the cusps $r=-\frac{d}{c}$ and satisfies

$$F(S(\tau))=F(\tau)+\pi_S, \qquad S\in\Gamma_1, \qquad \tau\in H.$$

π_S is a constant independent of τ. The number π_S is called a *period of the integral.*

Naturally $\pi_{-S}=\pi_S$.

If $c_r^* \neq 0$, we say that F has a *logarithmic singularity* at the point $-\frac{d}{c}$, because

$$A(\tau)=\frac{\kappa}{2\pi i}\log t_r.$$

We subdivide the integrals into three classes: If an integral F for Γ_1 is holomorphic in H^*, then F is called an *integral of the* 1st*-kind.* If all the coefficients $c_r^*=0$, i.e. if F has no logarithmic singularities, then F is an *integral of the* 2nd*-kind*; otherwise it is an *integral of the* 3rd*-kind.* We shall not consider a more general concept of the integral; we shall restrict ourselves to single-valued integrals on $\mathscr{H}$. Obviously modular functions for Γ_1 are integrals for Γ_1. We now derive

Theorem 10. *The derivative* $f(\tau):=\frac{dF(\tau)}{d\tau}$ *of an integral* F *for* Γ_1 *is a modular form of dimension* -2 *for* Γ_1.

Proof. For $f(\tau)$ and $S=\begin{pmatrix} a & b \\ c & d \end{pmatrix}\in\Gamma_1$ we have

$$f(S\tau)=\frac{dF(S(\tau))}{dS(\tau)}=\frac{dF(S(\tau))}{d\tau}\left(\frac{dS(\tau)}{d\tau}\right)^{-1}=f(\tau)(c\tau+d)^2. \tag{19}$$

If $r=-\frac{d}{c}$, then from (18) we obtain the expansion

$$f(\tau)=c_r^*(c\tau+d)^{-2}+\sum_{\nu\geqq\nu_0} d_r t_r^\nu (c\tau+d)^{-2}$$

at $\tau=r$. The function f is meromorphic in $\mathscr{H}$, since F is meromorphic there. Thus f is a modular form of dimension -2. □

Next we prove

Theorem 11. *If* f *is a modular form of dimension* -2 *for* Γ_1, *and if the residues of* f *vanish at all* $\tau\in\mathscr{H}$, *then the function* F, *where*

$$F(\tau):=\int_{\tau_0}^{\tau} f(v)\,dv$$

is an integral for Γ_1.

Proof. F is obviously meromorphic in $\mathcal{H}$. At $r=-\frac{d}{c}$, f has the expansion

$$f(\tau)=\sum_{\nu\geqq\nu_0} c_\nu e^{\frac{2\pi i}{\kappa}\cdot A(\tau)\cdot\nu}(c\tau+d)^{-2}, \qquad A=\begin{pmatrix} a & b \\ c & d \end{pmatrix}\in\Gamma,$$

from which it follows that

$$F(\tau)=\int_{\tau_0}^{\tau} f(v)\,dv=\sum_{\nu\geqq\nu_0,\,\nu\neq 0} c_\nu \frac{\kappa}{2\pi i\nu} e^{\frac{2\pi i}{\kappa}A(\tau)\cdot\nu}+\int_{\tau_0}^{\tau} c_0(cv+d)^{-2}\,dv \tag{20}$$

$$=c_0 A(\tau)+\sum_{\nu\geqq\nu_0,\,\nu\neq 0} c_\nu \frac{\kappa}{2\pi i\nu} e^{\frac{2\pi i}{\kappa}A(\tau)\cdot\nu}+C$$

with a constant C. Moreover, if $S\in\Gamma_1$,

$$F(S(\tau))=\int_{\tau_0}^{S(\tau)} f(v)\,dv=\int_{\tau_0}^{\tau} f(v)\,dv+\int_{\tau}^{S(\tau)} f(v)\,dv.$$

Since

$$\frac{d}{d\tau}\left(\int_{\tau}^{S(\tau)} f(v)\,dv\right)=f(S(\tau))\frac{d}{d\tau}S(\tau)-f(\tau)\equiv 0,$$

the integral $\int_{\tau}^{S(\tau)} f(v)\,dv$ is a constant π_S. □

The rational points r, for which the coefficient c_r^* in the expansion (18) is 0, are to be adjoined to the domain of the integral F. In particular, Theorem **11** is true for entire modular forms of dimension -2 for Γ_1. For entire functions f expansion (20) shows that F is an integral of the 1st-kind if f is a cusp form, and F is an integral of the 3rd-kind if f is not a cusp form.

The most general integrals are obtained if one does not require that the residues of f vanish in $\mathcal{H}$.

The modular forms of dimension -2 are designated as *integrands of the* 1st, 2nd *or* 3rd*-kind* according as the corresponding integrals are of the 1st, 2nd or 3rd-kind, respectively.

2. Periods of Integrals. Applying two substitutions $A,B\in\Gamma_1$ to the functional equation of an integral F for Γ_1 yields

$$F(AB(\tau))=F(B(\tau))+\pi_A=F(\tau)+\pi_A+\pi_B=F(\tau)+\pi_{AB}.$$

Thus the periods satisfy

$$\pi_{AB}=\pi_A+\pi_B \tag{21}$$

and also satisfy

Theorem 12. *The periods of an integral for Γ_1 give a representation of degree 1 for the factor group $\bar{\Gamma}_1/\bar{\Gamma}_1'$, where $\bar{\Gamma}'$ is the commutator subgroup of $\bar{\Gamma}_1$.*

Here we have taken the inhomogeneous groups since $-I$ is not in the commutator subgroup Γ' of Γ and hence is not in the commutator subgroup Γ_1' of Γ_1.

As a simple consequence of (21) we have:

If $S \in \Gamma_1$ has finite order, then

$$\pi_S = 0 . \tag{22}$$

An integral for Γ_1 has an expansion of the form

$$F(\tau) = \sum_{\nu \geqq \nu_0} c_\nu \left(\frac{\tau - \tau_0}{\tau - \bar{\tau}_0} \right)^{j_{\tau_0} \cdot \nu} \tag{23}$$

at $\tau_0 \in \mathscr{H}$, where $j_{\tau_0} = 1, 2$, or 3 depending on whether τ_0 is not a fixed point of Γ_1, is an elliptic fixed point Γ-equivalent to i, or to ρ.

We prove:

If S is a parabolic substitution of Γ_1 with fixed point r, then

$$\pi_S = c_r^* \cdot \kappa \cdot n \quad \text{with an } n \in \mathbb{Z} \text{ depending on } S . \tag{24}$$

By definition F has the expansion

$$F(\tau) = c_r^* A(\tau) + \sum_{\nu \geqq \nu_0} c_\nu e^{\frac{2\pi i}{\kappa} \cdot A(\tau) \cdot \nu}, \qquad A = \begin{pmatrix} a & b \\ c & d \end{pmatrix} \in \Gamma$$

at $r = -\dfrac{d}{c}$. The matrix S can be written in the form $S = \pm A^{-1} U^{\kappa n} A$, from which it follows that

$$\begin{aligned} F(S\tau) &= c_r^* U^{\kappa \cdot n} A(\tau) + \sum_{\nu \geqq \nu_0} c_\nu e^{\frac{2\pi i}{\kappa} U^{\kappa \cdot n} A(\tau) \cdot \nu} \\ &= c_r^* (A(\tau) + \kappa \cdot n) + \sum_{\nu \geqq \nu_0} c_\nu e^{\frac{2\pi i}{\kappa} A(\tau) \nu} \\ &= c_r^* \cdot \kappa \cdot n + F(\tau) . \end{aligned}$$

This proves the result.

3. Zeros. Let f be an entire modular form of dimension -2 for Γ_1. It follows from the expansions (11) and (12) that f has a zero of order at least $\frac{2}{3}$ at the fixed points τ_ρ of Γ_1, and f has a zero of order at least $\frac{1}{2}$ at the fixed point τ_i of Γ_1. Let

$$F(\tau) = \int_{\tau_0}^{\tau} f(v) \, dv .$$

We shall now compare the orders of the zeros of $\frac{dF}{d\tau}=f(\tau)$ with those of $\frac{dF}{dt}$, where t is the local variable. Their order coincides at points of $\mathscr{H}$ not fixed by elements of Γ_1. On the other hand, in the neighborhood of a fixed point τ_ρ of Γ_1, we obtain

$$\frac{dF}{d\tau}=\frac{dF}{dt_\rho}\cdot\frac{dt_\rho}{d\tau}=\frac{dF}{dt_\rho}\cdot 3\left(\frac{\tau-\tau_\rho}{\tau-\overline{\tau}_\rho}\right)^2\cdot\frac{\tau_\rho-\overline{\tau}_\rho}{(\tau-\overline{\tau}_\rho)^2},$$

and so

$$(\tau-\overline{\tau}_\rho)^2\frac{dF}{d\tau}=C\left(\frac{\tau-\tau_\rho}{\tau-\overline{\tau}_\rho}\right)^2\frac{dF}{dt_\rho}, \quad C\neq 0. \tag{25}$$

Hence at τ_ρ the order of the zero of $\frac{dF}{d\tau}$ is larger by $\frac{2}{3}$ than the (non-negative integral) order of $\frac{dF}{dt_\rho}$. Correspondingly in the neighborhood of a fixed τ_i of Γ_1 we obtain

$$(\tau-\overline{\tau}_i)^2\frac{dF}{d\tau}=C\left(\frac{\tau-\tau_i}{\tau-\overline{\tau}_i}\right)\frac{dF}{dt_i}, \quad C\neq 0, \tag{26}$$

i. e. at τ_i the order of the zero of $\frac{dF}{d\tau}$ is larger by $\frac{1}{2}$ than the order of the zero of $\frac{dF}{dt_i}$. At the cusp $r=-\frac{d}{c}$,

$$\frac{dF}{d\tau}=\frac{dF}{dt_r}\cdot\frac{dt_r}{d\tau}=\frac{dF}{dt_r}\cdot\frac{2\pi i}{\kappa}\cdot(c\tau+d)^{-2}\cdot t_r.$$

Thus

$$(c\tau+d)^2\frac{dF}{d\tau}=\frac{2\pi i}{k}\cdot t_r\cdot\frac{dF}{dt_r}. \tag{27}$$

Hence a cusp form $\frac{dF}{d\tau}$ has a zero at $\tau=r$ whose order exceeds that of $\frac{dF}{dt_r}$ by 1, i. e. $\frac{dF}{dt_r}$ has a pole of order 1 at r in case $\frac{dF}{d\tau}$ does not vanish at r, otherwise the order of its zero is smaller by 1.

We consider the case of a cusp form. The aforementioned zeros of orders $\frac{2}{3}$, $\frac{1}{2}$ and 1 are called *inherent zeros* and the remaining ones are called the *essential zeros*. For the number of essential zeros, which coincides with the number of zeros of $\frac{dF}{dt}$, we obtain

$$\frac{2\overline{\mu}}{12}-\frac{2}{3}\varepsilon_\rho-\frac{1}{2}\varepsilon_i-\sigma_\infty. \tag{28}$$

By **IV, § 7,** (24) this expression is equal to $2g-2$, where g is the genus of the fundamental region for Γ_1. We have proved

Theorem 13. *A cusp form of dimension* -2 *for* Γ_1 *has* $2g-2$ *essential zeros on a fundamental set.*

This theorem implies that there are no cusp forms of dimension -2 when $g=0$.

§ 4. The Riemann-Roch Theorem and Applications

In this section we formulate the Riemann-Roch Theorem. We collect the necessary facts to understand it and to apply it. For a proof we refer to a book on algebraic functions, or say G. Springer [1]. From the Riemann-Roch Theorem we derive in particular that the degree of the field K_{Γ_1} of modular functions for Γ_1 over the field K_Γ of modular functions for Γ is the same as the index of the group $\Gamma_1 \cup (-I)\Gamma_1$ in Γ (cf. Theorem **2**). Further, from it we derive a formula for the $\mathbb{C}$-dimension h of the vector space of entire even-dimensional modular forms for Γ_1 (cf. inequality (17)).

We let $\mathfrak{R}$ be an arbitrary compact Riemann surface, $\mathfrak{p}$ a point of $\mathfrak{R}$ and $t_\mathfrak{p}$ the local variable at $\mathfrak{p}$.

1. Differentials on a Riemann surface. Suppose that for every $\mathfrak{p} \in \mathfrak{R}$, and for every local variable $t_\mathfrak{p}$ at $\mathfrak{p}$ there is given a map

$$(\mathfrak{p}, t_\mathfrak{p}) \mapsto \Phi(t_\mathfrak{p}) \tag{29}$$

where $\Phi(t_\mathfrak{p})$ is a meromorphic function of $t_\mathfrak{p}$. Suppose in addition, if two such maps

$$(\mathfrak{p}, t_\mathfrak{p}) \mapsto \Phi(t_\mathfrak{p}), \qquad (\mathfrak{p}, \tilde{t}_\mathfrak{p}) \mapsto \tilde{\Phi}(\tilde{t}_\mathfrak{p}) \tag{30}$$

are given then

$$\tilde{\Phi}(\tilde{t}_\mathfrak{p}) = \Phi(t_\mathfrak{p}) \frac{d t_\mathfrak{p}}{d \tilde{t}_\mathfrak{p}} \tag{31}$$

is satisfied. The class of such maps associated with the points $\mathfrak{p}$ of $\mathfrak{R}$ is called a *differential* on $\mathfrak{R}$ and is denoted by ω.

By the *order* of the differential ω at the point $\mathfrak{p} \in \mathfrak{R}$ we understand the order of $\Phi(t_\mathfrak{p})$ at $t_\mathfrak{p}=0$. By (31) this order is well defined; it is 0 except for finitely many points of $\mathfrak{R}$.

In order to explain the line integral $\int_{\mathfrak{L}} \omega$ over the curve $\mathfrak{L}$ on $\mathfrak{R}$ we choose a chart $\mathfrak{U}$ on $\mathfrak{R}$ with the homeomorphism α from $\mathfrak{U}$ onto an open set $\mathscr{E}$ in the complex t-plane. t is then a local variable for $\mathfrak{U}$

(cf. **IV, § 6**). Suppose $\phi(t)$ is a representation of ω in the local variable t. Suppose that $\mathfrak{L}$ is a curve in $\mathfrak{U}$ and $\alpha(\mathfrak{L})=\mathscr{L}$ is rectifiable. Assume that $\phi(t)$ is holomorphic on $\mathscr{L}$. Then the line integral is defined by

$$\int_{\mathfrak{L}} \omega := \int_{\alpha(\mathfrak{L})} \phi(t)\,dt$$

and in the general case as a sum of such integrals. Because of (31) the line integral is independent of the choice of the local variable.

The *residue* of the differential ω at the point $\mathfrak{p}$ is defined by

$$\operatorname*{Res}_{\mathfrak{p}} \omega := \frac{1}{2\pi i} \int_{\mathfrak{p}} \omega\,,$$

where the integration is performed by traversing a sufficiently small circle about $\mathfrak{p}$ in the positive direction.

If F is a function and if ω is a differential on $\mathfrak{R}$, then $F\omega$ is obviously a differential on $\mathfrak{R}$.

If ω_1 and ω_2 are differentials with the representatives

$$(\mathfrak{p}, t_{\mathfrak{p}}) \mapsto \Phi_1(t_{\mathfrak{p}}) \quad \text{and} \quad (\mathfrak{p}, t_{\mathfrak{p}}) \mapsto \Phi_2(t_{\mathfrak{p}})\,,$$

then the class of maps

$$(\mathfrak{p}, t_{\mathfrak{p}}) \mapsto \Phi_1(t_{\mathfrak{p}}) + \Phi_2(t_{\mathfrak{p}})$$

defines a differential ω that is called the sum of the differentials and is denoted by

$$\omega = \omega_1 + \omega_2\,.$$

By an analogous definition the quotient of two differentials is a function on $\mathfrak{R}$.

If F is a function on $\mathfrak{R}$, then the class of maps

$$(\mathfrak{p}, t_{\mathfrak{p}}) \mapsto \frac{dF(t_{\mathfrak{p}})}{dt_{\mathfrak{p}}}$$

is a differential which we denote by dF.

2. Divisors on Riemann surfaces. A formal product

$$\boldsymbol{d} := \prod_{\mathfrak{p}} \mathfrak{p}^{n_{\mathfrak{p}}}\,,$$

independent of the order of the factors and taken over all points $\mathfrak{p}$ of a Riemann surface $\mathfrak{R}$ with only finitely many non-zero exponents $n_{\mathfrak{p}} \in \mathbb{Z}$, is called a *divisor on* $\mathfrak{R}$. Under formal commutative multiplication the set of divisors becomes an Abelian group $\boldsymbol{D}$.

The *degree of a divisor* is defined by the sum

$$\deg \boldsymbol{d} := \sum_{\mathfrak{p}} n_{\mathfrak{p}} .$$

A divisor is *integral* if $n_{\mathfrak{p}} \geqq 0$ for all $\mathfrak{p}$. Moreover, the degree function is additive:

$$\deg \boldsymbol{d}_1 \boldsymbol{d}_2 = \deg \boldsymbol{d}_1 + \deg \boldsymbol{d}_2 .$$

We assign divisors to functions F and differentials ω on $\mathfrak{R}$ by

$$\boldsymbol{d}(F) := \prod_{\mathfrak{p}} \mathfrak{p}^{\operatorname{ord}_{\mathfrak{p}}(F)}, \qquad \boldsymbol{d}(\omega) := \prod_{\mathfrak{p}} \mathfrak{p}^{\operatorname{ord}_{\mathfrak{p}}(\omega)}, \tag{32}$$

where $\operatorname{ord}_{\mathfrak{p}}(F)$ and $\operatorname{ord}_{\mathfrak{p}}(\omega)$ are not 0 for finitely many $\mathfrak{p} \in \mathfrak{R}$. We call $\boldsymbol{d}(F)$ the *divisor of the function* F and call $\boldsymbol{d}(\omega)$ *the divisor of the differential* ω. The divisors of functions are called *principal divisors*. They form a subgroup $\boldsymbol{D}_0$ of $\boldsymbol{D}$, and if $\boldsymbol{d}_0 \in \boldsymbol{D}_0$, then $\deg \boldsymbol{d}_0 = 0$. The cosets of $\boldsymbol{D}$ by $\boldsymbol{D}_0$ are called *divisor classes* and all elements of the same class have the same degree. We speak of the *degree of a class*.

Obviously, the divisors of differentials form a divisor class which we denote by $\boldsymbol{D}_1$.

Definition. A divisor $\boldsymbol{d}_1 = \prod_{\mathfrak{p}} \mathfrak{p}^{n_{\mathfrak{p}}}$ divides a divisor $\boldsymbol{d}_2 = \prod_{\mathfrak{p}} \mathfrak{p}^{m_{\mathfrak{p}}}$, in symbols $\boldsymbol{d}_1 | \boldsymbol{d}_2$, if and only if $n_{\mathfrak{p}} \leqq m_{\mathfrak{p}}$ for all $\mathfrak{p} \in \mathfrak{R}$. The set of all functions F whose divisors $\boldsymbol{d}(F)$ are multiples of a divisor $\boldsymbol{d}$, form a vector space over $\mathbb{C}$ whose dimension is called the *dimension of the divisor* $\boldsymbol{d}$:

$$\dim \boldsymbol{d} := \dim \{F : \boldsymbol{d} | \boldsymbol{d}(F)\} . \tag{33}$$

The divisors of a divisor class have the same dimension, so that the dimension of a class is well defined. Clearly, if $\deg \boldsymbol{d} > 0$, then $\dim \boldsymbol{d} = 0$.

After these preparations we formulate the Riemann-Roch Theorem.

3. The Riemann-Roch Theorem. *The dimension of a divisor* $\boldsymbol{d}^{-1}$ *on a Riemann surface* $\mathfrak{R}$ *of genus* g *is finite and satisfies*

$$\dim \boldsymbol{d}^{-1} = \deg \boldsymbol{d} - (g-1) + \dim(\boldsymbol{d} \boldsymbol{D}_1^{-1}) . \tag{34}$$

As we remarked earlier, we will not prove this theorem.

The dimension of the divisor $\boldsymbol{d} \boldsymbol{D}_1^{-1}$ is at the same time the maximal number of linearly independent differentials whose divisors are multiples of $\boldsymbol{d}$.

As a consequence of the Riemann-Roch Theorem we have:

There are non-constant functions on $\mathfrak{R}$ that have a pole at an arbitrarily given point $\mathfrak{p}$ of order at most $g+1$ and that are otherwise holomorphic.

For a proof one only has to substitute the divisor $\boldsymbol{d}=\mathfrak{p}^{g+1}$ in (34). Then

$$\dim \boldsymbol{d}^{-1} \geqq (g+1)-(g-1)=2\,.$$

If in the Riemann-Roch Theorem one takes a divisor $\boldsymbol{d}_1$ of the class $\boldsymbol{D}_1$ of divisors of differentials, then

$$\deg \boldsymbol{D}_1 - \dim \boldsymbol{D}_1^{-1} = g-2\,.$$

If one uses the result

$$\deg \boldsymbol{D}_1 = 2g-2 \tag{35}$$

from the theory of algebraic functions, one obtains

$$\dim \boldsymbol{D}_1^{-1} = g \tag{36}$$

and the converse. These two assertions, which precede the Riemann-Roch Theorem in the development of the theory, will be used later.

We now shall prove the previously mentioned supplement to Theorem **2**. Suppose that $-I \in \Gamma_1$ and $\Gamma = \bigcup_{\nu=1}^{\mu_1} \Gamma_1 S_\nu$ is the coset decomposition of Γ. If $\mathfrak{R}$ is the Riemann surface over the J-sphere associated with Γ_1 and if f is a modular function for Γ_1 that—as a meromorphic function on $\mathfrak{R}$—has exactly one pole, then the functions $f|S_\nu$, $\nu=1,2,\dots,\mu_1$ are pairwise distinct. So the function f has degree μ_1 over $K_\Gamma = \mathbb{C}(J)$ and thus is a generator for K_{Γ_1} over K_Γ (cf. **VI,** Theorem **3**).

In conclusion we have

Theorem 14. *The field K_{Γ_1} of modular functions for Γ_1 is algebraic over the field K_Γ of modular functions for Γ. The degree of the extension K_{Γ_1} over K_Γ is the index of Γ_1 in Γ.*

4. The $\mathbb{C}$-dimension of the space of entire modular forms of dimension $-k$. In some cases the Riemann-Roch Theorem enables us to determine explicitly the $\mathbb{C}$-dimension of the vector space of entire modular forms of dimension $-k$, $k>0$, for the group Γ_1. It is not assumed that $-I \in \Gamma_1$. First we introduce *divisors of modular forms.* As we have already seen the orders of zeros and poles of modular forms for Γ_1 are the same at Γ_1-equivalent points and differ from 0 at only finitely many classes of equivalent points. Hence the divisor of a modular form is well-defined by:

$$\boldsymbol{d}(f) := \prod_{\mathfrak{p}} \mathfrak{p}^{[\mathrm{ord}_{\mathfrak{p}}(f)]}, \tag{37}$$

where $\mathfrak{p}$ runs through a complete set of inequivalent points of $\mathscr{H}^*$ under Γ_1. This definition includes the case of modular functions.

If there exists a modular form f_{-k} of dimension $-k$, not necessarily entire, then because of the association of functions for Γ_1 with functions on $\mathfrak{R}$, the $\mathbb{C}$-dimension of the vector space of entire modular forms is $\dim \boldsymbol{d}_k^{-1}$, where $\boldsymbol{d}_k = \boldsymbol{d}(f_{-k})$. We prove

Theorem 15. *Let Γ_1 be a group whose fundamental region has genus g. If Γ_1 has a modular form of dimension $-k \leqq -1$, then the $\mathbb{C}$-dimension of the vector space of entire modular forms of dimension $-k$ for Γ_1 is*

$$(k-1)(g-1) + \left[\frac{k}{4}\right]\varepsilon_i + \left[\frac{k}{3}\right]\varepsilon_\rho + \frac{k}{2}\sigma_\infty + A_k, \tag{38}$$

where $A_k = 0$ for $k \geqq 2$ and $A_1 = \dim(\boldsymbol{d}_1 \boldsymbol{D}_1^{-1}) \geqq 0$. The numbers ε_i, ε_ρ, and σ_∞ are defined in **IV, § 7, 4.**

Proof. By the Riemann-Roch Theorem

$$\dim(\boldsymbol{d}_k^{-1}) = \deg \boldsymbol{d}_k - (g-1) + \dim(\boldsymbol{d}_k \boldsymbol{D}_1^{-1}).$$

From the definition, (37), the supplement to Theorem **8**, (11) and (12) one obtains

$$\deg \boldsymbol{d}_k = \frac{\bar{\mu} k}{12} - \left(\frac{k}{4} - \left[\frac{k}{4}\right]\right)\varepsilon_i - \left(\frac{k}{3} - \left[\frac{k}{3}\right]\right)\varepsilon_\rho .$$

In view of the genus formula **IV,** (24) and (35) in the form

$$\deg \boldsymbol{D}_1^{-1} = -2(g-1),$$

we conclude that

$$\deg \boldsymbol{d}_k = k(g-1) + \left[\frac{k}{4}\right]\varepsilon_i + \left[\frac{k}{3}\right]\varepsilon_\rho .$$

Thus

$$D := \deg(\boldsymbol{d}_k \boldsymbol{D}_1^{-1}) = (k-2)(g-1) + \left[\frac{k}{4}\right]\varepsilon_i + \left[\frac{k}{3}\right]\varepsilon_\rho + \frac{k}{2}\sigma_\infty .$$

If $g>1$ and $k \geqq 2$, or $g=1$ and $k \geqq 1$ or $g=0$ and $k \leqq 2$, then we have the inequality

$$D \geqq \frac{k}{2}\sigma_\infty > 0 .$$

If $g=0$ and $k>2$ we have

$$D \geqq 2-k + \frac{k-3}{4}\varepsilon_i + \frac{k-3}{2}\varepsilon_\rho + \frac{k-3}{2}\sigma_\infty + \frac{3}{2}\sigma_\infty$$
$$= 2-k+(k-3)\left(1+\frac{\bar{\mu}}{12}\right) + \frac{3}{2}\sigma_\infty \geqq \frac{3}{2}\sigma_\infty - 1 > 0 .$$

In the remaining case $g>1$ and $k=1$,

$$D = \tfrac{1}{2}\sigma_\infty - (g-1)$$

and is possibly negative. Thus

$$\begin{aligned}\dim \boldsymbol{d}_k^{-1} &= \deg \boldsymbol{d}_k - (g-1) + A_k \\ &= (k-1)(g-1) + \left[\frac{k}{4}\right]\varepsilon_i + \left[\frac{k}{3}\right]\varepsilon_\rho + \frac{k}{2}\sigma_\infty + A_k,\end{aligned} \tag{39}$$

where $A_k = 0$ for $k \geqq 2$ and $A_1 = \dim(\boldsymbol{d}_1 \boldsymbol{D}_1^{-1})$. □

If there are modular forms of odd dimension for Γ_1 then the number of cusps must be even since $\dim \boldsymbol{d}_k^{-1}$ is an integer.

Further we state

Theorem 15'. *There always exist modular forms for Γ_1 of arbitrary even dimension.*

The derivatives of modular functions are modular forms f_{-2} of dimension -2. Modular forms of arbitrary even dimension are obtained as powers of f_{-2}. If $k=2$ it follows from (39) that the dimension of the vector space of entire forms of dimension -2 for Γ_1 is

$$g + \sigma_\infty - 1. \tag{39'}$$

Cf. J. Lehner [1].

5. The connection between differentials and modular forms of dimension -2. We have seen in **§ 3** that the integral

$$F(\tau) = \int_{\tau_0}^{\tau} f(v)\,dv$$

of a modular form f of dimension -2 for Γ_1 for which the residues are 0 at all points of $\mathscr{H}$ defines a single-valued integral for Γ_1. This function is meromorphic in $\mathscr{H}$ and has at most logarithmic singularities at the cusps.

The power series expansions that we derived in **§ 1,3** for modular functions obviously also hold for integrals for Γ_1, however, now there also appears a logarithmic term in the expansion at the cusps. The expansions at Γ_1-equivalent points differ by an additive constant. We map the path of integration $\mathfrak{L}$ on $\mathscr{H}^*$ by the map σ of **IV, § 6,** (13) onto the path $\sigma(\mathscr{L})$ on $\mathfrak{R}$. Then we obtain a function that is no longer single-valued on $\mathfrak{R}$, but is single-valued on a covering surface lying over $\mathfrak{R}$. We see that this is $\mathscr{H}^*$. The different expansions at a point of $\mathfrak{R}$ differ by an additive constant. Thus the classes

$$(\mathfrak{p}, t_\mathfrak{p}) \mapsto \frac{dF}{dt_\mathfrak{p}}, \qquad \mathfrak{p} \in \mathfrak{R},$$

represent a differential which as in the case of functions we again denote by dF. The residue of dF may be different from 0 only at the images of the cusps. Thus, with the modular form f of dimension -2 is associated a differential on $\mathfrak{R}$.

Conversely, for an arbitrary differential ω on $\mathfrak{R}$, the integral

$$F = \int_{\mathfrak{p}_0}^{\mathfrak{p}} \omega \quad \text{for fixed } \mathfrak{p}_0$$

has an expansion

$$F = c_{\mathfrak{p}_1} \log t_{\mathfrak{p}_1} + \sum_{\nu \geqq \nu_0} c_\nu t_{\mathfrak{p}_1}^{\nu} \tag{40}$$

in the local variable $t_{\mathfrak{p}_1}$ at each point $\mathfrak{p}_1$ which depends on the path from $\mathfrak{p}_0$ to $\mathfrak{p}_1$. Moreover, $c_{\mathfrak{p}_1}$ is the residue of ω at $\mathfrak{p}_1$. Two distinct expansions at the same point differ by an additive constant. If we consider the expansion (40) on $\mathscr{H}^*$, and if we assume that $\operatorname{Res}_{\mathfrak{p}} \omega = 0$ for all $\mathfrak{p} \in \mathfrak{R}$ that are not σ-images of rational cusps, then due to the simple connectedness of $\mathscr{H}$, we obtain a single-valued meromorphic function F on $\mathscr{H}$ whose expansions at the cusps may contain a logarithmic term. Also, since two paths $\mathfrak{L}_1', \mathfrak{L}_2'$ on $\mathfrak{R}$ from $\mathfrak{p}_0$ to $\mathfrak{p}$ are the continuous images under σ of two paths $\mathscr{L}_1, \mathscr{L}_2$ on $\mathscr{H}^*$, which both start at the same point and end at Γ_1-equivalent points, F satisfies the defining properties of an integral and $\dfrac{dF}{d\tau}$ is a modular form for Γ_1. Thus associated with the differential ω on $\mathfrak{R}$ is a modular form f of dimension -2 for Γ_1. The correspondence between forms of dimension -2 and differentials is obviously invertible:

$$\omega \mapsto f \mapsto \omega .$$

One calls the differential ω on $\mathfrak{R}$ a *differential of the 1st kind* if it is everywhere holomorphic, a *differential of the 2nd kind* if all the residues vanish and a *differential of the 3rd kind* otherwise. Their integrals depending on the nature of the differentials are called *integrals of the 1st, 2nd, or 3rd kind.* Hence, this classification for forms of dimension -2 and their corresponding integrals for Γ_1 agrees with the classification given in **§3, 1.**

The $\mathbb{C}$-dimension of the space of integrals of the 1st kind or the dimension of cusp forms of dimension -2 for Γ_1 is thus equal to the genus of the fundamental region of Γ_1 or to the genus of the Riemann surface $\mathfrak{R}$. It follows from (39′) that there are $\sigma_\infty - 1$ linearly independent entire modular forms of dimension -2 which span a space containing no cusp forms.

Chapter VI. Fields of Modular Functions

We begin this chapter by proving some general theorems about fields of modular functions. We then turn to the investigation of special fields. Here we will not make use of existence theorems of function theory. We start with the absolute invariant J for which we have a representation as the quotient of modular forms. This representation may also be considered as its definition. By various processes we shall derive new functions from functions of level 1. We shall then determine their behavior under modular substitutions. The same will hold for forms of level 1. The so-called division fields, generated from the Weierstrass $\wp$-function of the theory of elliptic functions, should in a certain sense be discussed here, but they will be more conveniently treated in the next chapter.

§ 1. Algebraic Field Extensions of $\mathbb{C}(J)$

Let Γ_1 be a subgroup of finite index μ in Γ, $-I \in \Gamma_1$ and let

$$\Gamma = \bigcup_{\nu=1}^{\mu} \Gamma_1 S_\nu \tag{1}$$

be a coset decomposition of Γ. As in **V, § 1** we denote the field of modular functions for Γ_1 by K_{Γ_1}.

Suppose f is a function defined on $\mathscr{H}^*$, is meromorphic in $\mathscr{H}$, and has expansions (cf. **V, § 1**) of the form

$$f(\tau) = \sum_{\nu \geqq \nu_0} c_\nu e^{\frac{2\pi i}{\kappa_1} A(\tau)\nu}, \quad A = \begin{pmatrix} a & b \\ c & d \end{pmatrix} \in \Gamma, \quad \kappa_1 \in \mathbb{N},$$

at all rational points $-\dfrac{d}{c}$. The group of all modular substitutions under which f is invariant is called the *invariance group* of f and is denoted by I_f.

Theorem 1. *Let* $f \in K_{\Gamma_1}$. *If the polynomial*

$$G(x) = \prod_{\nu=1}^{\mu} (x - f|S_\nu) \tag{2}$$

in $K_\Gamma[x]$ *is irreducible, then* f *generates* K_{Γ_1},

$$K_{\Gamma_1} = K_\Gamma(f),$$

and conversely.

Proof. Recall Theorem **1** of Chapter **V**. The first part of this theorem follows already from **V**, Theorem **2**; however, the converse requires **V**, Theorem **14**. □

Moreover, from **V**, Theorem **14** we also obtain: If two groups Γ_1 and Γ_2 both contain $-I$ and if the fields K_{Γ_1} and K_{Γ_2} are the same, then so are the groups Γ_1 and Γ_2.

Lemma. *If* $f \in K_{\Gamma_1}$ *and* if

$$g(x) = \sum_{n=0}^{m} R_n(J)\, x^n$$

is a polynomial in x *with coefficients* $R_n(J)$ *in* K_Γ *with* f *as root, i.e.*

$$g(f) = 0,$$

then

$$g(f|S) = 0 \quad \textit{for all } S \in \Gamma.$$

Proof. Indeed, if

$$g(f(\tau)) = \sum_{n=0}^{m} R_n(J(\tau))\, f^n(\tau) = 0,$$

then

$$g(f|S)(\tau) = g(f(S\tau)) = 0,$$

i. e.

$$g(f|S) = 0 \quad \text{for all } S \in \Gamma. \quad □$$

As an immediate consequence of the lemma we have

Theorem 2. *The algebraic conjugates of* $f \in K_{\Gamma_1}$ *over* K_Γ *are the functions* $f|S$. *If*

$$K_{\Gamma_1} = K_\Gamma(f),$$

then

$$K_{S^{-1}\Gamma_1 S} = K_\Gamma(f|S).$$

Further, our lemma and **V**, Theorem **2** imply

Theorem 3. *If* $f \in K_{\Gamma_1}$ *and if the functions* $f|S_\nu$, $\nu = 1, 2, \ldots, \mu$ *are pairwise distinct, then the polynomial* (2) *over* K_Γ *is irreducible and* $K_{\Gamma_1} = K_\Gamma(f)$.

This theorem was used to prove **V**, Theorem **14**.

Theorem 4. *If Γ_1 is a normal subgroup of finite index in Γ, then K_{Γ_1} is a Galois extension of K_Γ and conversely.*

Proof. The first part of the theorem follows immediately from Theorem **2.**

To prove the converse let K_{Γ_1} be a Galois extension of K_Γ. Then

$$K_{\Gamma_1} = K_\Gamma(f)$$

with a certain $f \in K_\Gamma$ as follows from **V**, Theorem **14.** By Theorem **2** of this chapter the conjugates $f|S$ generate the fields $K_{S^{-1}\Gamma_1 S}$ which by assumption coincide with K_{Γ_1}. By the remark following Theorem **1** the groups also coincide, i. e. Γ_1 is normal in Γ. ▯

All isomorphisms between K_{Γ_1} and its conjugates are of the form

$$I_S: K_{\Gamma_1} \to K_{S^{-1}\Gamma_1 S}, \quad f \mapsto f|S \quad \text{for all } f \in K_{\Gamma_1}. \tag{3}$$

Here $I_S = I_{S'}$ if and only if $S' \in \Gamma_1 S$. Under the usual definition of product, the automorphisms among these I_S's form a group G. In the case of a Galois field K_{Γ_1}, G is isomorphic to the factor group Γ modulo its normal subgroup Γ_1:

$$G \cong \Gamma/\Gamma_1. \tag{4}$$

Theorem 5. *If f is a function with invariance group I_f of finite index in Γ, then*

$$K_{I_f} = K_\Gamma(f). \tag{5}$$

Proof. Obviously

$$K_\Gamma(f) \subset K_{I_f}.$$

The functions $f|S_\nu$, $\nu = 1, \ldots, \mu$ are pairwise distinct when the S_ν come from a coset decomposition (1). Hence the degree

$$[K_\Gamma(f): K_\Gamma] = \mu.$$

Moreover, the degree

$$[K_{I_f}: K_\Gamma] = \mu.$$

This completes the proof. ▯

We do not intend to go into uniformization theory, but we want to prove

Theorem 6. *Let $f \in K_{\Gamma_1}$ be holomorphic in $\mathscr{H}$, $\Gamma = \bigcup_{\nu=1}^{\mu} \Gamma_1 S_\nu$ be a coset decomposition of Γ_1 and*

$$G(x) = \prod_{\nu=1}^{\mu} (x - f|S_\nu) = \sum_{m,n} a_{m,n} x^m J^n$$

with constant coefficients $a_{m,n}$. Then all solutions of the equation

$$Q(x,y) := \sum_{m,n} a_{m,n} x^m y^n = 0 \tag{6}$$

are given in parametric representation by

$$x = f(\tau), \quad y = J(\tau), \quad \tau \in \mathcal{H}. \tag{7}$$

Proof. All pairs of the form (7) are obviously solutions. If $Q(x_0, y_0) = 0$ and if

$$y_0 = J(\tau_0),$$

where τ_0 determined up to equivalence under Γ, then

$$Q(x, y_0) = \prod_{\nu=1}^{\mu} (x - f(S_\nu(\tau_0))) = 0$$

has the solutions

$$x_0^{(\nu)} = f(S_\nu(\tau_0)), \quad \nu = 1, \ldots, \mu.$$

The μ pairs $(x_0^{(\nu)}, y_0^{(\nu)})$ of solutions of $Q(x,y) = 0$ may be represented in the form

$$x_0^{(\nu)} = f(\tau_\nu), \quad y_0^{(\nu)} = y_0 = J(\tau_\nu), \quad \tau_\nu = S_\nu(\tau_0),$$

i. e. all solutions of (6) have the form

$$x = f(\tau), \quad y = J(\tau). \quad \square$$

§ 2. The Fields $\mathbb{C}(\sqrt{J-1})$ and $\mathbb{C}(\sqrt[3]{J})$

1. The field $\mathbb{C}(\sqrt{J-1})$. The function $J-1$ has no singularities in $\mathcal{H}$; its zeros are of order 2 and are located at the points Γ-equivalent to i. Hence there is a function

$$f(\tau) = \sqrt{J(\tau) - 1} \tag{8}$$

holomorphic in $\mathcal{H}$. It follows from

$$f^2 | S = (f|S)^2 = f, \quad S \in \Gamma,$$

that

$$f|S = \varepsilon_S f \quad \text{for all } S \in \Gamma, \tag{9}$$

with a multiplier ε_S depending only on S and satisfying

$$\varepsilon_S^2 = 1, \quad \varepsilon_{SS'} = \varepsilon_S \varepsilon_{S'} \quad \text{for all } S, S' \in \Gamma. \tag{10}$$

It follows from the q-expansion of $J(\tau)$ that

$$f|U = \varepsilon_U f \quad \text{with } \varepsilon_U = -1,$$

and from

$$UT(-\bar{\rho}) = -\bar{\rho}, \qquad f(-\bar{\rho}) \neq 0,$$

that

$$f|UT = \varepsilon_{UT} f \quad \text{with } \varepsilon_{UT}=1.$$

Since U and UT generate Γ, (10) and

$$-\varepsilon_U = \varepsilon_{UT} = 1$$

imply that f is invariant under the subgroup of Γ of index 2 generated by U^2 and UT (cf. **V**, **§ 5,3** and **§ 7,3**). As follows from the corresponding expansions for J, f has the required expansions at the cusps, hence the invariance group of f is given by

$$I_f = \{U^2, UT\}.$$

As measured in the local variable $e^{\pi i \tau}$, f has a pole of order 1 at $i\infty$ and is holomorphic in $\mathcal{H}$. Thus the field $\mathbb{C}(f)$ of rational functions in f is identical with K_{I_f}.

2. The field $\mathbb{C}(\sqrt[3]{J})$. Since J is holomorphic in $\mathcal{H}$ and since its zeros are of order 3, the function

$$g = \sqrt[3]{J} \tag{11}$$

is holomorphic in $\mathcal{H}$. As before,

$$g|S = \varepsilon_S \cdot g \quad \text{for all } S \in \Gamma \tag{12}$$

with a multiplier ε_S which satisfies

$$\varepsilon_S^3 = 1, \qquad \varepsilon_{SS'} = \varepsilon_S \varepsilon_{S'} \quad \text{for all } S, S' \in \Gamma. \tag{13}$$

It follows from the q-expansion for J that

$$g|U = \varepsilon_U g \quad \text{with } \varepsilon_U = e^{\frac{2\pi i}{3}},$$

and from

$$T(i) = i, \qquad g(i) \neq 0,$$

that

$$g|T = \varepsilon_T g \quad \text{with } \varepsilon_T = 1.$$

Entirely analogous to the case of the field $\mathbb{C}(\sqrt{J-1})$ we now obtain: the invariance group I_g is generated by U^3 and T, i.e.

$$I_g = \{U^3, T\},$$

and has index 3 in Γ. Moreover,

$$\mathbb{C}(g) = K_{I_g}$$

(for the group I_g see **V**, **§ 7, 3**).

§ 3. Transformation Groups of Order n

In this section we consider a situation which relates group theory, algebra, function theory, and number theory; however, we will not go too deeply into these connections which are the matter of the so-called theory of "Complex Multiplication". See M. Deuring [1].

1. Transformations of order n. Let $M=\begin{pmatrix}\alpha & \beta\\ \gamma & \delta\end{pmatrix}$ be a matrix of order n, i. e. a matrix with rational integral entries and with positive determinant

$$|M|=n\geqq 1.$$

We assume that the entries of M are relatively prime,

$$(\alpha,\beta,\gamma,\delta)=1.$$

The linear transformation corresponding to M will be called a *transformation of order n*. For fixed n we will let $\mathcal{M}_n$ denote both the set of all such matrices and their corresponding transformations.

By a *transformation group Γ_M of order n* we mean the group

$$\Gamma_M:=\Gamma\cap M^{-1}\Gamma M, \tag{14}$$

which may be considered either as a group of matrices or as a group of linear transformations. Alternatively, we may express the group Γ_M by

$$\Gamma_M=\{A\in\Gamma\colon MAM^{-1}\in\Gamma\}. \tag{14'}$$

If $M=M_0=\begin{pmatrix}n & 0\\ 0 & 1\end{pmatrix}$, then because of

$$\frac{1}{n}\begin{pmatrix}n & 0\\ 0 & 1\end{pmatrix}\begin{pmatrix}a & b\\ c & d\end{pmatrix}\begin{pmatrix}1 & 0\\ 0 & n\end{pmatrix}=\begin{pmatrix}a & nb\\ \dfrac{c}{n} & d\end{pmatrix},$$

we obtain

$$\Gamma_{M_0}=\Gamma_0(n),$$

one of the groups introduced in **IV**, § **3**, **7**. If follows from a calculation that the principal congruence group $\Gamma[n]$ is contained in all Γ_M. The groups Γ_M are thus congruence groups as follows also from Theorem **10** of this chapter.

2. Equivalent transformations. We now want to survey all transformations and transformation groups of order n. It follows from (14) that

$$\Gamma_{SMS_1}=S_1^{-1}\Gamma_M S_1 \quad \text{for all } S,S_1\in\Gamma. \tag{15}$$

Correspondingly we define: two transformations M, M' from $\mathcal{M}_n$ are called *equivalent* or *congruent* modulo Γ, in symbols

$$M_1 \sim M \quad \text{or } M_1 \equiv M \bmod \Gamma,$$

if there is an S in Γ such that

$$M_1 = S\,M\,.$$

This defines an equivalence relation, and

$$M_1 \sim M \quad \text{implies } \Gamma_{M_1} = \Gamma_M\,.$$

In order to determine the class number of $\mathcal{M}_n$ modulo Γ we prove

Theorem 7.

1. *For each* $M = \begin{pmatrix} \alpha & \beta \\ \gamma & \delta \end{pmatrix} \in \mathcal{M}_n$ *there is an* $M_1 = \begin{pmatrix} \alpha_1 & \beta_1 \\ 0 & \delta_1 \end{pmatrix} \in \mathcal{M}_n$ *such that* $M \sim M_1$.

2. *Two transformations from* $\mathcal{M}_n$,

$$M = \begin{pmatrix} \alpha & \beta \\ 0 & \delta \end{pmatrix}, \quad M_1 = \begin{pmatrix} \alpha_1 & \beta_1 \\ 0 & \delta_1 \end{pmatrix},$$

are equivalent if and only if

$$\alpha = t\alpha_1, \quad \delta = t\delta_1, \quad \beta \equiv t\beta_1 \bmod \delta, \quad t \in \{1, -1\}. \tag{16}$$

3. *A complete system of representatives of the equivalence classes of* $\mathcal{M}_n \bmod \Gamma$ *is given by the set of the matrices*

$$\begin{pmatrix} \alpha_\nu & \beta_\nu \\ 0 & \delta_\nu \end{pmatrix} \quad \text{with } \alpha_\nu > 0,\ \alpha_\nu \delta_\nu = n,\ \beta_\nu \bmod \delta_\nu,\ (\alpha_\nu, \beta_\nu, \delta_\nu) = 1. \tag{17}$$

Proof. 1. Let $M = \begin{pmatrix} \alpha & \beta \\ \gamma & \delta \end{pmatrix} \in \mathcal{M}_n$. We must produce $\begin{pmatrix} a & b \\ c & d \end{pmatrix} \in \Gamma$ such that

$$\begin{pmatrix} a & b \\ c & d \end{pmatrix} \begin{pmatrix} \alpha & \beta \\ \gamma & \delta \end{pmatrix} = \begin{pmatrix} * & * \\ 0 & * \end{pmatrix}.$$

For this purpose we determine c, d so that $c\alpha + d\gamma = 0$, $(c,d) = 1$, and then determine a, b such that $ad - bc = 1$.

2. The equivalence of M and M_1 or more precisely the equation

$$\begin{pmatrix} \alpha_1 & \beta_1 \\ 0 & \delta_1 \end{pmatrix} = \begin{pmatrix} a & b \\ c & d \end{pmatrix} \begin{pmatrix} \alpha & \beta \\ 0 & \delta \end{pmatrix} = \begin{pmatrix} a\alpha & a\beta + b\delta \\ c\alpha & c\beta + d\delta \end{pmatrix}$$

for suitable $\begin{pmatrix} a & b \\ c & d \end{pmatrix} \in \Gamma$ implies that

$$c = 0, \quad a = d = \pm 1, \quad \text{thus } \alpha_1 = \alpha, \quad \delta_1 = \delta, \quad \text{or } \alpha_1 = -\alpha, \quad \delta_1 = -\delta\,,$$

and hence that

$$\beta_1 = \pm\beta + b\delta \equiv \pm\beta \bmod \delta .$$

Conversely, if (16) is satisfied, then

$$M_1 M^{-1} = \frac{1}{n}\begin{pmatrix}\alpha_1 & \beta_1\\ 0 & \delta_1\end{pmatrix}\begin{pmatrix}\delta & -\beta\\ 0 & \alpha\end{pmatrix} = \begin{pmatrix}\dfrac{\alpha_1\delta}{n}, & \dfrac{-\alpha_1\beta+\beta_1\alpha}{n}\\ 0, & \dfrac{\alpha\delta_1}{n}\end{pmatrix},$$

hence

$$M_1 = SM, \qquad S \in \Gamma .$$

3. The third assertion is an immediate consequence of the first two. □

3. The number of classes of equivalent transformations. We prove

Theorem 8. *The number $\psi(n)$ of equivalence classes of $\mathcal{M}_n \bmod \Gamma$ is*

$$\psi(n) = n\prod_{p\mid n}\left(1+\frac{1}{p}\right).$$

Proof. We have to determine the number of solutions

$$\alpha > 0, \qquad \alpha\delta = n, \qquad \beta \bmod \delta, \qquad (\alpha,\beta,\delta)=1. \tag{18}$$

For fixed positive δ the number of $\beta \bmod \delta$ that are relatively prime to a divisor t of δ is

$$\frac{\delta}{t}\,\varphi(t),$$

where φ denotes Euler's function. If we set $t=(\alpha,\delta)=\left(\frac{n}{\delta},\delta\right)$ and sum over all positive divisors δ of n, we obtain

$$\psi(n) = \sum_{\delta\mid n}\frac{\delta}{\left(\frac{n}{\delta},\delta\right)}\,\varphi\left(\left(\frac{n}{\delta},\delta\right)\right). \tag{19}$$

Since φ is multiplicative, so is ψ:

$$\psi(n_1 n_2) = \psi(n_1)\psi(n_2) \quad \text{if } (n_1,n_2)=1 .$$

Thus it suffices to calculate

$$\psi(p^k), \qquad p \text{ prime}, \qquad k \in \mathbb{N}.$$

From (19) we have

$$\psi(p^k) = \sum_{\nu=0}^{k} \frac{p^\nu}{(p^{k-\nu}, p^\nu)} \varphi((p^{k-\nu}, p^\nu))$$

$$= \left\{\sum_{\nu=1}^{k-1} p^\nu \left(1 - \frac{1}{p}\right)\right\} + 1 + p^k = p^{k-1} + p^k = p^k \left(1 + \frac{1}{p}\right),$$

from which the conclusion follows. □

In addition we note that

$$\psi(n) = \mu_0(n) = [\Gamma : \Gamma_0(n)].$$

4. A lemma about pairs of transformations of order n. We now prove the following important

Lemma. *For each pair $M, M' \in \mathcal{M}_n$ there is a pair $S, S' \in \Gamma$ such that*

$$SM = M'S'.$$

Proof. The solvability of this equation for an arbitrarily given pair M and M' follows from the solvability for the special $M_0 = \begin{pmatrix} n & 0 \\ 0 & 1 \end{pmatrix}$ and arbitrary M' of the form $M' = \begin{pmatrix} \alpha & \beta \\ 0 & \delta \end{pmatrix}$. Because

$$S_0 M_0 = M S_1 \quad \text{and} \quad S_0' M_0 = M' S_1'$$

imply

$$M_0 = S_0^{-1} M S_1 = S_0'^{-1} M' S_1',$$

we have

$$S_0' S_0^{-1} M = M' S_1' S_1^{-1}.$$

It follows from Theorem **7**, 1 that M' can be chosen without restriction in the form $\begin{pmatrix} \alpha & \beta \\ 0 & \delta \end{pmatrix}$.

Correspondingly, we show that for every $M' = \begin{pmatrix} \alpha & \beta \\ 0 & \delta \end{pmatrix} \in \mathcal{M}_n$ there is an $S = \begin{pmatrix} a & b \\ c & d \end{pmatrix} \in \Gamma$ such that

$$M'^{-1} S M_0 \in \Gamma. \tag{20}$$

Now

$$M'^{-1} S M_0 = \frac{1}{n} \begin{pmatrix} \delta & -\beta \\ 0 & \alpha \end{pmatrix} \begin{pmatrix} a & b \\ c & d \end{pmatrix} \begin{pmatrix} n & 0 \\ 0 & 1 \end{pmatrix} = \begin{pmatrix} a\alpha - c\beta, & \dfrac{b\delta - d\beta}{n} \\ c\alpha & \dfrac{d\alpha}{n} \end{pmatrix}.$$

These transformations lie in Γ if and only if

$$b\delta - d\beta \equiv 0 \bmod n, \qquad d \equiv 0 \bmod \delta. \tag{21}$$

First we set $d=\delta$ and then we have to solve

$$b-\beta\equiv 0 \bmod \alpha \quad \text{with } (b,\delta)=1 .$$

In order to solve this congruence let

$$P:=\prod_{p\mid\delta,\, p\nmid\alpha} p, \quad \text{so } (P,\alpha)=1 ,$$

and solve

$$\beta+\nu\alpha\equiv 1 \bmod P .$$

Then

$$b=\beta+\nu\alpha, \qquad d=\delta,$$

solves (21). Indeed, $S=\begin{pmatrix} a & b \\ c & d \end{pmatrix}\in\Gamma$ is a solution of (20) for these b and d. This proves our lemma. □

5. Consequences of the Lemma. By multiplying the class ΓM, $M\in\mathcal{M}_n$ on the right by $S\in\Gamma$ we obtain the class $\Gamma M S$. Since

$$\Gamma M S=\Gamma M_1 S \quad \text{implies } \Gamma M=\Gamma M_1 ,$$

multiplication on the right by S leads to a permutation $\sigma(S)$. On the basis of our lemma, in this manner we obtain each equivalence class from every other. Since for fixed M the equation

$$\Gamma M=\Gamma M S, \qquad S\in\Gamma,$$

is satisfied if and only if $S\in\Gamma_M$, we conclude

Theorem 9. *The map*

$$\sigma: S\mapsto\sigma(S)$$

of the elements $S\in\Gamma$ *onto the permutations*

$$\sigma(S):\Gamma M\mapsto\Gamma M S, \qquad M\in\mathcal{M}_n, \tag{22}$$

of the equivalence classes of $\mathcal{M}_n$ *modulo* Γ *is a homomorphism of* Γ *on a transitive permutation group with kernel*

$$K_n=\bigcap_{M\in\mathcal{M}_n}\Gamma_M .$$

At the same time we have shown: for every fixed $M_1\in\mathcal{M}_n$ the transformations $M_1 S$ form a complete system of inequivalent transformations in $\mathcal{M}_n$ provided the cosets $\Gamma_{M_1} S$ form a complete right residue system of $\Gamma \bmod \Gamma_{M_1}$ and conversely.

Equation (15):

$$\Gamma_{SMS'}=S'^{-1}\,\Gamma_M\, S' \quad \text{for all } S, S'\in\Gamma,$$

and our lemma, i. e. the solvability of $M \sim M_1 S$ for $S \in \Gamma$ where M_1 is fixed and $M \in \mathcal{M}_n$ is arbitrary, imply

Theorem 10. *For every pair* $M, M_1 \in \mathcal{M}_n$ *there is an* $S \in \Gamma$ *such that*

$$\Gamma_M = S^{-1} \Gamma_{M_1} S. \tag{23}$$

Thus all Γ_M are conjugate to $\Gamma_{M_0} = \Gamma_0(n)$. Since $\Gamma[n] \subset \Gamma_0(n)$ and since $\Gamma[n]$ is normal in Γ, it follows that

$$\Gamma[n] \subset \Gamma_M \quad \text{for all } M \in \mathcal{M}_n,$$

i.e. the groups Γ_M are congruence groups of level n.

6. The number of different transformation groups of order n. To determine this number we use Theorem **10** with

$$M_1 = M_0 = \begin{pmatrix} n & 0 \\ 0 & 1 \end{pmatrix}.$$

Consequently this number, $\psi_0(n)$, is equal to the index of the normalizer

$$N_0(n) := \{S \mid S \in \Gamma,\ S^{-1} \Gamma_0(n) S = \Gamma_0(n)\} \tag{24}$$

of $\Gamma_0(n)$ in Γ. Since $N_0(n) \supset \Gamma_0(n)$,

$$\psi_0(n) \leqq \mu_0(n) := [\Gamma : \Gamma_0(n)].$$

We already know (cf. Theorem **8**) that $\mu_0(n) = \psi(n)$.

To investigate the group $N_0(n)$ we let

$$S = \begin{pmatrix} a & b \\ c & d \end{pmatrix} \in \Gamma, \qquad V = \begin{pmatrix} u & v \\ x & y \end{pmatrix} \in \Gamma_0(n), \quad \text{i. e. } x \equiv 0 \bmod n.$$

It then follows from

$$S^{-1} V S = \begin{pmatrix} d & -b \\ -c & a \end{pmatrix} \begin{pmatrix} u & v \\ x & y \end{pmatrix} \begin{pmatrix} a & b \\ c & d \end{pmatrix} \equiv \begin{pmatrix} * & * \\ -c(au+cv-ay) & * \end{pmatrix} \bmod n$$

that in order for

$$S^{-1} \Gamma_0(n) S = \Gamma_0(n) \tag{25}$$

to hold, it is necessary and sufficient that the entries of S satisfy the congruences

$$c(au+cv-ay) \equiv 0 \bmod n \quad \text{for all } v \text{ and all } u, y \text{ with } uy \equiv 1 \bmod n. \tag{26}$$

It is clear that (26) is necessary; the sufficiency follows from the lemma of **IV**, **§ 2**. In particular, if we set

$$u = v = y = 1,$$

then it follows that

$$c^2 \equiv 0 \bmod n \tag{27}$$

is a necessary condition on S. Thus necessarily $(a, n) = 1$ and (26) reduces to

$$c(u-y) \equiv 0 \bmod n,$$

or to

$$c(u^2-1) \equiv 0 \bmod n \quad \text{for all } u \text{ with } (u, n) = 1. \tag{28}$$

It is necessary and obviously also sufficient for (25) to hold that the conditions (27) and (28) on the entry c of S be simultaneously satisfied. Let

$$n = p_1^{\alpha_1}, \ldots, p_k^{\alpha_k}, \qquad \alpha_1, \ldots, \alpha_k > 0$$

be the prime factorization of n. Then (28) is equivalent to the system of congruences

$$c(u^2-1) \equiv 0 \bmod p_\nu^{\alpha_\nu} \quad \text{for all } u \text{ with } (u, p_\nu) = 1, \qquad \nu = 1, \ldots, k.$$

These congruences are trivially satisfied by $c \equiv 0 \bmod n$. However, if

$$(c, p_\nu^{\alpha_\nu}) = p_\nu^{\beta_\nu} \quad \text{and} \quad \beta_\nu < \alpha_\nu$$

for some ν, then necessarily for this ν we have

$$u^2 - 1 \equiv 0 \bmod p_\nu^{\alpha_\nu - \beta_\nu} \quad \text{for all } u \text{ with } (u, p_\nu) = 1,$$

which is only possible in the cases

$$\begin{aligned} p_\nu &= 2, \quad \alpha_\nu - \beta_\nu = 1, 2 \text{ or } 3, \\ p_\nu &= 3, \quad \alpha_\nu - \beta_\nu = 1. \end{aligned}$$

Conversely, if these cases occur, then (28) is solvable, and it follows that if

$$n = 2^\alpha 3^{\alpha'} t, \qquad (t, 6) = 1,$$

then the solutions c of the congruence (28) are exactly

$$c \equiv 0 \bmod 2^{\alpha-\varepsilon} 3^{\alpha'-\varepsilon'} t \tag{29}$$

with maximal $\varepsilon, \varepsilon'$ satisfying the inequalities

$$\begin{aligned} 0 \leqq \varepsilon \leqq 3, \quad \varepsilon \leqq \alpha, \\ 0 \leqq \varepsilon' \leqq 1, \quad \varepsilon' \leqq \alpha'. \end{aligned} \tag{30}$$

Those c that satisfy the congruence (28) as well as congruence (27) also satisfy congruence (29) with such maximal $\varepsilon, \varepsilon'$ which, in addition to inequality (30), also satisfy the inequalities

$$\varepsilon \leqq \frac{\alpha}{2}, \quad \varepsilon' \leqq \frac{\alpha'}{2}. \tag{31}$$

Collecting these results we obtain

Theorem 11. *Let*

$$n=2^{\alpha}\cdot 3^{\alpha'}\cdot t, \qquad (t,6)=1\,;$$
$$\varepsilon=\left[\frac{\alpha}{2}\right] \quad \text{for } \alpha<6, \quad \textit{and } \varepsilon=3 \quad \textit{for } \alpha\geqq 6\,;$$
$$\varepsilon'=0 \quad \text{for } \alpha=0,1, \quad \textit{and } \varepsilon'=1 \quad \textit{for } \alpha'\geqq 2\,.$$

Then the normalizer of $\Gamma_{M_0}=\Gamma_0(n)$ *is*

$$N_0(n)=\Gamma_0\left(\frac{n}{2^{\varepsilon}3^{\varepsilon'}}\right),$$

and thus the number $\psi_0(n)$ *of different transformation groups of order n is*

$$\psi_0(n)=\psi\left(\frac{n}{2^{\varepsilon}3^{\varepsilon'}}\right)=\frac{1}{2^{\varepsilon}3^{\varepsilon'}}\,\psi(n)\,.$$

§ 4. Transformation Fields of Order n

We now turn to the fields of modular functions associated with the transformation group Γ_M, that is, to the *transformation fields* K_{Γ_M} *of order* n.

1. Transforms of order n**.** Let f be a non-constant modular function for the full modular group. Then we define the function f_M by

$$f_M(\tau):=f(M(\tau)), \qquad M\in\mathcal{M}_n, \qquad \tau\in\mathcal{H}^*, \tag{32}$$

and call it a *transform of order n of* f. The transform f_M depends only on the equivalence class of M. Indeed, if $S\in\Gamma$ then

$$f_{SM}(\tau)=f(S\,M(\tau))=f(M(\tau))=f_M(\tau)\,.$$

One thus obtains all transforms of f of order n by considering

$$f_{M_\nu},\ M_\nu=\begin{pmatrix}\alpha_\nu & \beta_\nu\\ 0 & \delta_\nu\end{pmatrix}, \qquad \nu=1,\ldots,\psi(n)\,, \tag{33}$$

with the conditions (17) on the entries $\alpha_\nu, \beta_\nu, \delta_\nu$. They are obviously meromorphic in $\mathcal{H}$, and the substitution $S\in\Gamma$ induces the permutation

$$f_{M_\nu}\mapsto f_{M_\nu S}=f_{M_{\nu'}}, \qquad M_\nu S\sim M_{\nu'}\,. \tag{34}$$

In particular, for each $M \in \mathcal{M}_n$ we have

$$f_{MS} = f_M \quad \text{if } S \in \Gamma_M. \tag{35}$$

If f has the expansion

$$f(\tau) = \sum_{\nu \geqq \nu_0} c_\nu e^{2\pi i \tau \nu}, \quad \text{for } \operatorname{Im} \tau > A \quad \text{for some } A > 0,$$

at $i\infty$, then for arbitrary $S \in \Gamma$, f_M has the expansion

$$f_M(\tau) = \sum_{\nu \geqq \nu_0} c_\nu e^{2\pi i M' S(\tau) \nu}, \quad \text{for } \operatorname{Im} S(\tau) > A' \quad \text{for some } A' > 0$$

with some $M' \in \mathcal{M}_n$, $M \sim M'S$ at $S^{-1}(i\infty)$. This, taken with (35), finally implies

Theorem 12. *If f is in K_Γ, then f_M is in K_{Γ_M}.*

Since Γ_M contains the principal congruence group $\Gamma[n]$, f_M belongs to $K_{\Gamma[n]}$ (cf. **V**, **§ 1**, **2**).

2. Proof and applications of a lemma on a extension of Γ.

Lemma. *If Γ_* is a group extension of Γ containing a matrix $M \in \mathcal{M}_n$, $n > 1$, then a fundamental region $\mathcal{F}$ of Γ contains a set of infinitely many Γ_*-equivalent points which has a limit point in $\mathcal{F}$.*

Proof. By **§ 3**, **4**, associated with M are two matrices $S, S_1 \in \Gamma$ such that

$$S M S_1 = M_0 = \begin{pmatrix} n & 0 \\ 0 & 1 \end{pmatrix}.$$

This implies that, if $M \in \Gamma_*$, then so are M_0 and all of its powers. Thus the points $in^k, k \in \mathbb{Z}$, are Γ_*-equivalent. This set has $i\infty$ as a limit point. ▯

As a direct consequence of the lemma we obtain:
If $f \in K_\Gamma$ and if

$$f_M = f, \quad M \in \mathcal{M}_n, \quad n > 1,$$

then f is constant.

Theorem 13. *If f is a non-constant function in K_Γ and if M runs through the $\psi(n)$ different equivalence classes of $\mathcal{M}_n$ modulo Γ, then the functions f_M are pairwise distinct.*

Proof. Suppose $f \in K_\Gamma$ is non-constant and $f_M = f_{M_1}$, $M_1 \in \mathcal{M}_n$, i.e.

$$f(M(\tau)) = f(M_1(\tau)), \quad \text{or } f(M M_1^{-1}(\tau)) = f(\tau), \qquad M, M_1 \in \mathcal{M}_n. \tag{36}$$

Then, by the lemma,

$$r M M_1^{-1} = S \in \Gamma \quad \text{with some } r \in \mathbb{Q}.$$

By taking the determinant we see that $r=1$, and so

$$M M_1^{-1} \in \Gamma, \quad \text{i.e. } M \sim M_1. \quad \square$$

We infer another result from the lemma:
The invariance group of f_M is the transformation group Γ_M,

$$I_{f_M} = \Gamma_M.$$

It follows from

$$f_M(S(\tau)) = f_M(\tau), \qquad f(M S(\tau)) = f(M(\tau)), \qquad S \in \Gamma, \tag{37}$$

that

$$f(M S M^{-1}(\tau)) = f(\tau).$$

Hence, by our lemma,

$$r M S M^{-1} \in \Gamma, \qquad r \in \mathbb{Q},$$

and thus similarly $r=1$; we conclude that

$$S \in \Gamma_M.$$

3. The transformation equation of order n. We now prove

Theorem 14. *The polynomial*

$$Q_n(x, j) := \prod_{(M)} (x - f_M),$$

where the product is taken over a complete system of inequivalent $M \in \mathcal{M}_n$, *is irreducible in* $K_\Gamma[x] = \mathbb{C}(j)[x]$, *and of degree* $\psi(n)$.

Proof. By the Corollary to Theorem **9**, for fixed $M_1 \in \mathcal{M}_n$, $M_1 S$ runs through a complete system (M) of non-equivalent matrices of $\mathcal{M}_n$ when S runs through a complete system (S) of right inequivalent matrices of Γ modulo Γ_{M_1}. Thus

$$\prod_{(M)} (x - f_M) = \prod_{(S)} (x - f_{M_1 S}). \tag{38}$$

By the proof of **V**, Theorem **1** the product on the right is a polynomial in x over K_Γ of degree $\psi(n)$. Since all the functions f_M in (38), and thus also all the functions $f_{M_1 S}$, are pairwise distinct, by Theorem **3** this polynomial is irreducible in K_Γ. $\square$

We call the equation

$$Q_n(x,j)=0 \tag{39}$$

the *transformation equation of order n*. It is satisfied by all the f_M.

The irreducibility of $Q_n(x,j)$ in K_Γ and the first part of Theorem **1** imply that K_{Γ_M} is obtained by adjoining f_M to K_Γ and that it is of degree $\psi(n)$ over K_Γ.

The isomorphism between K_{Γ_M} and $K_{\Gamma_{M_1}}$ is given by

$$f_M \mapsto f_{M_1} \quad \text{with } f_{M_1}(\tau)=f_M(S(\tau)) \quad \text{if } MS\sim M_1 .$$

Thus

$$f_M=f_{M_1} \quad \text{if and only if } M\sim M_1 ,$$

$$\Gamma_M=\Gamma_{M_1} \quad \text{if and only if } MS\sim M_1 \quad \text{with } S\in N_M ,$$

where N_M is the normalizer of Γ_M (cf. Theorem **11**).

§ 5. The Modular Equation of Order n

1. A special case of the transformation equation. We now continue the investigations of the last section for the special case

$$f=j,$$

where j is the absolute invariant. Equation (38) becomes

$$\prod_{(M)}(x-j_M)=\prod_{(S)}(x-j_{M_1S})=:P_n(x,j). \tag{40}$$

$P_n(x,j)$ is again a polynomial of degree $\psi(n)$ in x with coefficients from K_Γ which are polynomials in j, because they are holomorphic in $\mathscr{H}$. Thus,

$$P_n(x,j)=\sum_{\mu=0}^{\psi(n)} p_\mu(j)\,x^\mu, \tag{41}$$

where

$$p_\mu(j)=\sum_{\nu=0}^{\nu_\mu} a_{\mu\nu}j^\nu \quad \text{for some } \nu_\mu,$$

or

$$P_n(x,j)=\sum_{\mu,\nu} a_{\mu\nu}x^\mu j^\nu,$$

with uniquely determined coefficients $a_{\mu\nu}$. The equation

$$P_n(x,j)=0 \tag{42}$$

is called the *modular equation of order n*.

2. The polynomial $P_n(x,y)$. Instead of $P_n(x,j)$ we consider the polynomial

$$P_n(x,y) = \sum_{\mu,\nu} a_{\mu,\nu} x^\mu y^\nu$$

with indeterminates x and y. This polynomial has some remarkable properties which we now derive. First we prove the following

Lemma. *The polynomial* $P_n(x,y)$ *has degree* $\psi(n)$ *in* y.

For the proof we write $P_n(x,j)$ in the form

$$P_n(x,j) = \prod_{M=\left(\begin{smallmatrix}\alpha & \beta\\ 0 & \delta\end{smallmatrix}\right)} (x - j_M) = \sum_{\mu=0}^{\psi(n)} p_\mu(j) x^\mu \tag{43}$$

with the conditions (17)

$$\alpha > 0, \quad \alpha\delta = n, \quad \beta \bmod \delta, \quad (\alpha,\beta,\delta) = 1$$

on the product. The coefficient $p_0(j)$, for which

$$p_0(j(\tau)) = \prod_{\alpha,\beta,\delta} j\left(\frac{\alpha\tau+\beta}{\delta}\right),$$

is the coefficient with the highest order pole at $i\infty$. Its order is

$$\sum_{\alpha\delta=n} \frac{\alpha}{\delta}\,\varphi((\alpha,\delta))\,\frac{\delta}{(\alpha,\delta)} = \sum_{\alpha\mid n} \frac{\alpha}{\left(\alpha,\dfrac{n}{\alpha}\right)}\,\varphi\left(\left(\alpha,\frac{n}{\alpha}\right)\right) = \psi(n),$$

where the last equality follows from (19). Thus the degree of $P_n(x,y)$ in y is also $\psi(n)$. Here we used the fact that $\mathbb{C}(j)$ is transcendental over $\mathbb{C}$.

3. Symmetry of $P_n(x,y)$. With this lemma we prove

Theorem 15. *The polynomial* $P_n(x,y)$ *is symmetric in* x *and* y.

Proof. Let $M^0 := \begin{pmatrix}1 & 0\\ 0 & n\end{pmatrix}$, $M_0 := \begin{pmatrix}n & 0\\ 0 & 1\end{pmatrix}$. The function j_{M^0} is a root of the modular equation:

$$P_n(j_{M^0}, j) = 0.$$

By replacing j by j_{M_0}, and therefore j_{M^0} by j, we have

$$P_n(j, j_{M_0}) = 0.$$

In addition j_{M_0} is a root of the modular equation:

$$P_n(j_{M_0}, j) = 0.$$

From the last two equations we deduce: the two polynomials

$$P_n(x,j), \quad P_n(j,x) \tag{44}$$

over K_Γ have the common root $x = j_{M_0}$. Since the polynomial $P_n(x,j)$ is irreducible over K_Γ and since by our lemma both polynomials have the same degree $\psi(n)$ in x, we have

$$P_n(j,x) = P_n(x,j) \cdot r(j)$$

with a rational function r of j. From this it follows that the equation

$$P_n(y,x) = P_n(x,y) \cdot r(y)$$

holds in the indeterminates x and y, however, since

$$P_n(x,y) = P_n(y,x) \cdot r(x),$$

the first equation can hold only for $r(y) = \pm 1$. Here the case $r(y) = -1$ is excluded. Otherwise $P_n(x,x) = 0$ hence $P_n(x,j)$ has the root j and is thus reducible over K_Γ which, however, is not the case. This proves Theorem **15.** □

4. The coefficients of $P_n(x,y)$. The polynomial $P_n(x,y)$ has an important arithmetical property, more precisely, a property important in the so-called theory of complex multiplication. We express this in

Theorem 16. *The coefficients $a_{\mu\nu}$ of the polynomial*

$$P_n(x,y) = \sum_{\mu,\nu=0}^{\psi(n)} a_{\mu\nu} x^\mu y^\nu$$

are rational integers.

For the proof of this theorem we need the following

Lemma. *If the function $f \in K_\Gamma$ has an expansion*

$$f(\tau) = \sum_{\nu=-k}^{\infty} \alpha_\nu q^\nu, \quad k \geqq 1, \quad \tau \in \mathscr{H}, \tag{45}$$

in powers of $q = e^{2\pi i \tau}$ with rational integral coefficients α_ν, then f has an expansion

$$f = \sum_{\nu=0}^{k} A_\nu j^\nu, \tag{46}$$

in powers of j with rational integral coefficients A_ν.

The lemma is correct for $k=1$. Indeed, by Theorem **24** of **II, § 6** the function

$$\sum_{\nu=-1}^{\infty} \alpha_\nu q^\nu - \alpha_{-1} j(\tau)$$

is holomorphic in $\mathscr{H}$ and at $i\infty$, and is thus a constant; in fact, it is a rational integer. For $k>1$ we have

$$\sum_{\nu=-k}^{\infty} \alpha_\nu q^\nu - \alpha_{-k} j^k(\tau) = \sum_{\nu=-(k-1)}^{\infty} \alpha'_\nu q^\nu$$

with rational integral α'_ν. The proof of the lemma is completed by induction.

One can easily verify the following generalization of this lemma: If the coefficients α_ν in (45) belong to a number ring over $\mathbb{Z}$, then the coefficients A_ν of (46) belong to the same ring.

This generalization of our lemma is called the *q-expansion principle*

Proof of Theorem **16.** With the notation

$$j(\tau) = \sum_{\nu=-1}^{\infty} a_\nu e^{2\pi i\tau\nu}, \qquad a_\nu \text{ rational integral, } a_{-1}=1,$$

we have by (43) that

$$P_n(x, j(\tau)) = \prod_{\alpha,\beta,\delta} \left(x - \sum_{\nu=-1}^{\infty} a_\nu e^{\frac{2\pi i\alpha\tau}{\delta}\nu} e^{\frac{2\pi i\beta}{\delta}\nu} \right) \tag{47}$$

with the conditions

$$\alpha>0, \quad \alpha\delta=n, \quad \beta \bmod \delta, \quad (\alpha,\beta,\delta)=1.$$

By expanding the product as a polynomial in x on the we obtain one hand coefficients that are polynomials in j as in (43). On the other hand a direct calculation of the product yields an expansion of the form

$$\sum_{\mu=0}^{\psi(n)} \left(\sum_{\nu=-\psi(n)}^{\infty} A_{\mu\nu} e^{2\pi i\tau\nu} \right) x^\mu,$$

if one notes that fractional exponents of $e^{2\pi i\tau}$ cannot occur. The $A_{\mu\nu}$ are numbers of the form

$$\sum_{k=0}^{n-1} c_k e^{\frac{2\pi i}{n}k}$$

with rational integral c_k, and thus are integers in the cyclotomic field $\mathbb{Q}\left(e^{\frac{2\pi i}{n}}\right)$. Obviously equation (47) remains valid if in it one replaces β by $\beta\kappa=\beta_1$, where n and κ are relatively prime. Thus the coefficients $A_{\mu\nu}$ also remain unchanged. Such substitutions subject the coefficients to the automorphism

$$e^{\frac{2\pi i}{n}} \to e^{\frac{2\pi i}{n}\kappa}, \quad (\kappa,n)=1,$$

of the cyclotomic field $\mathbb{Q}\left(e^{\frac{2\pi i}{n}}\right)$. Thus the $A_{\mu\nu}$ are rationals, but as integers of $\mathbb{Q}\left(e^{\frac{2\pi i}{n}}\right)$ they are rational integers. By the lemma the coefficients of the polynomial $P_n(x,j)$ are polynomials in j with rational integral coefficients. This establishes our claim. □

§ 6. The Galois Group of the Modular Equation

1. Introduction to the invariance group of all j_M. We know that the group Γ_M is the invariance group I_{j_M} of the transform j_M of j of order n:

$$I_{j_M} = \Gamma_M .$$

This follows from the Lemma of **§ 4, 2**. The group $\Gamma^*(n)$ under which all j_M, $M \in \mathcal{M}_n$, are invariant, is then obviously the intersection of all Γ_M:

$$\Gamma^*(n) = \bigcap_{M \in \mathcal{M}_n} \Gamma_M , \tag{48}$$

where naturally only finitely many $M \in \mathcal{M}_n$ are necessary. We know $\Gamma^*(n)$ is the kernel of the map σ of (22). As such $\Gamma^*(n)$ is normal in Γ, and since $\Gamma[n] \subset \Gamma^*(n)$, it is also of finite index in Γ. By Theorem **4** $K_{\Gamma^*(n)}$ is a Galois extension of K_Γ with Galois group

$$G \cong \Gamma/\Gamma^*(n).$$

Since by **§ 4, 3** the fields K_{Γ_M} are obtained by adjoining j_M to K_Γ, the splitting field of the modular equation is $K_{\Gamma^*(n)}$, because it is the smallest extension of K_Γ containing all roots of the modular equation.

2. Investigation of the invariance group. The group $\Gamma^*(n)$ is a congruence group of level n. We want to specify its elements by congruences mod n. Let $S \in \Gamma$. By (14′) S is in $\Gamma^*(n)$ if and only if

$$M S M^{-1} \in \Gamma \quad \text{for all } M \in \mathcal{M}_n .$$

We set $S = \begin{pmatrix} \alpha & \beta \\ \gamma & \delta \end{pmatrix}$ and, without restriction, take M of the form

$$M = \begin{pmatrix} a & b \\ 0 & d \end{pmatrix}, \qquad a > 0, \qquad ad = n .$$

Then

$$M S M^{-1} = \frac{1}{n} \begin{pmatrix} a & b \\ 0 & d \end{pmatrix} \begin{pmatrix} \alpha & \beta \\ \gamma & \delta \end{pmatrix} \begin{pmatrix} d & -b \\ 0 & a \end{pmatrix} \tag{49}$$

$$= \frac{1}{n} \begin{pmatrix} n\alpha + bd\gamma, & -ab\alpha - b^2\gamma + a^2\beta + ab\delta \\ d^2\gamma, & n\delta - bd\gamma \end{pmatrix};$$

now it is necessary to determine the entries $\alpha, \beta, \gamma, \delta$ so that $MSM^{-1} \in \Gamma$ for each M of the given form.

In particular, if we set $M = \begin{pmatrix} n & b \\ 0 & 1 \end{pmatrix}$, then as necessary conditions for $S \in \Gamma^*(n)$ we obtain the congruence

I $$\gamma \equiv 0 \bmod n$$

and the congruences

$$ab(\alpha - \delta) - a^2 \beta \equiv 0 \bmod n$$

for all divisors a of n and for all b. Setting $a=1$ and $b=0$, we have

II $$\beta \equiv 0 \bmod n,$$

and setting $a=b=1$, we have

III $$\alpha \equiv \delta \bmod n.$$

As we see from (49) these three congruences are also sufficient for S to be in $\Gamma^*(n)$.

Therefore we have

Theorem 17. *The invariance group for all transforms j_M of j of order n is the group*

$$\Gamma^*(n) = \left\{ S \in \Gamma \,|\, S = \begin{pmatrix} \alpha & \beta \\ \gamma & \delta \end{pmatrix} \equiv \begin{pmatrix} \alpha & 0 \\ 0 & \alpha \end{pmatrix} \bmod n \right\}. \tag{50}$$

From this we see that the index of the principal congruence group $\Gamma(n)$ in $\Gamma^*(n)$ is equal to the number z of incongruent solutions modulo n of the congruence

$$x^2 \equiv 1 \bmod n.$$

If n has the prime factorization

$$n = 2^a p_1^{\nu_1} \dots p_r^{\nu_r} \quad \text{with } 2 < p_1 < \dots < p_r,\ a \geqq 0,\ r \geqq 0,$$

then z is known to be

$$z = \begin{cases} 2^r & \text{if } a=0 \text{ or } 1, \\ 2^{r+1} & \text{if } a=2, \text{ and} \\ 2^{r+2} & \text{if } a>2. \end{cases} \tag{51}$$

Thus the index of $\Gamma^*(n)$ in Γ, or what is the same thing, the order of the Galois group of $K_{\Gamma^*(n)}$ over K_Γ, is equal to

$$[\Gamma : \Gamma^*(n)] = \frac{\mu(n)}{z}$$

with $\mu(n) = [\Gamma : \Gamma(n)]$.

We note especially that

$$z=\begin{cases}1 & \text{if } n=1 \quad \text{or } n=2, \text{ and}\\ 2 & \text{if } n=p_1^{\nu_1} \quad \text{or } n=2p_1^{\nu_1}.\end{cases}$$

In these cases

$$\Gamma^*(n)=\Gamma[n], \quad \text{and } K_{\Gamma^*(n)}=K_{\Gamma[n]}.$$

These results, together with the equation (3) for isomorphisms, yield

Theorem 18. *The splitting field of the modular equation of degree n over K_Γ is the field $K_{\Gamma^*(n)}$. Its Galois group*

$$G\cong\Gamma/\Gamma^*(n)$$

has order $\frac{\mu(n)}{z}$, where $\mu(n)$ is the index $[\Gamma:\Gamma(n)]$ and z is the number of incongruent solutions modulo n of the congruence

$$x^2\equiv 1 \bmod n.$$

Each $S\in\Gamma$ induces an automorphism A_S of $K_{\Gamma^*(n)}$ over K_Γ:

$$A_S: f\mapsto f|S \quad \textit{for all } f\in K_{\Gamma^*(n)},$$

and all automorphisms of $K_{\Gamma^*(n)}$ over K_Γ are obtained in this way. Thus for two elements $S, S_1\in\Gamma$,

$$A_S=A_{S_1}$$

if and only if

$$SS_1^{-1}\equiv\begin{pmatrix}\alpha & 0\\ 0 & \alpha\end{pmatrix} \bmod n$$

for some α.

§ 7. Transformations of Order n for Modular Forms

We now construct modular forms of higher level from modular forms of level one by transformations of order n. The construction parallels the corresponding construction of modular functions.

1. Transforms of modular forms. Let g be a homogeneous modular form of dimension $-k$ for the group Γ in the variables ω_1, ω_2. We write this in the form $g\left(\begin{pmatrix}\omega_1\\ \omega_2\end{pmatrix}\right)$ and consider $\begin{pmatrix}\omega_1\\ \omega_2\end{pmatrix}$ as a matrix. We define the function g_M by

$$g_M\left(\begin{pmatrix}\omega_1\\ \omega_2\end{pmatrix}\right):=g\left(M\begin{pmatrix}\omega_1\\ \omega_2\end{pmatrix}\right), \quad M\in\mathscr{M}_n, \quad \operatorname{Im}\left(\frac{\omega_1}{\omega_2}\right)>0$$

and call it a *transform of g of order n*. It obviously satisfies the equations

$$g_M\left(\lambda\begin{pmatrix}\omega_1\\ \omega_2\end{pmatrix}\right)=\lambda^{-k}g_M\left(\begin{pmatrix}\omega_1\\ \omega_2\end{pmatrix}\right) \quad \text{for } \lambda\in\mathbb{C},\ \lambda\neq 0, \tag{52$_1$}$$

$$g_M\left(S\begin{pmatrix}\omega_1\\ \omega_2\end{pmatrix}\right)=g_M\left(\begin{pmatrix}\omega_1\\ \omega_2\end{pmatrix}\right) \quad \text{for } S\in\Gamma_M,\ \text{and} \tag{52$_2$}$$

$$g_{SM}\left(\begin{pmatrix}\omega_1\\ \omega_2\end{pmatrix}\right)=g_M\left(\begin{pmatrix}\omega_1\\ \omega_2\end{pmatrix}\right) \quad \text{for } S\in\Gamma. \tag{52$_3$}$$

Besides the functions g and g_M we consider the inhomogeneous modular form $\hat{g}$:

$$\hat{g}(\tau):=\omega_2^k\, g\left(\begin{pmatrix}\omega_1\\ \omega_2\end{pmatrix}\right),$$

and the functions $\hat{g}_M$:

$$\hat{g}_M(\tau):=\omega_2^k\, g_M\left(\begin{pmatrix}\omega_1\\ \omega_2\end{pmatrix}\right) \tag{53}$$

with $\tau=\dfrac{\omega_1}{\omega_2}$.

From equations (52) we obtain the equations

$$\hat{g}_M(\tau)=(\gamma\tau+\delta)^{-k}\hat{g}(M\tau) \quad \text{for } M=\begin{pmatrix}\alpha & \beta\\ \gamma & \delta\end{pmatrix}\in\mathcal{M}_n, \tag{54$_1$}$$

$$\hat{g}_M(S(\tau))=(c\tau+d)^k\hat{g}_M(\tau) \quad \text{for } S=\begin{pmatrix}a & b\\ c & d\end{pmatrix}\in\Gamma_M,\ \text{and} \tag{54$_2$}$$

$$\hat{g}_{SM}(\tau)=\hat{g}_M(\tau) \quad \text{for } S\in\Gamma. \tag{54$_3$}$$

We now prove

Theorem 19. *The transforms g_M of a homogeneous modular form g for Γ are homogeneous modular forms for Γ_M.*

The functions $\hat{g}_M$ defined in (53) are thus inhomogeneous modular forms for the transformation group Γ_M.

Proof. Because

$$\omega_2^k g_M\left(\begin{pmatrix}\omega_1\\ \omega_2\end{pmatrix}\right)=\hat{g}_M(\tau)=(\gamma\tau+\delta)^{-k}\hat{g}(M(\tau)),$$

$\hat{g}_M$ is meromorphic in $\mathcal{H}$. In order to prove Theorem **19**, in view of (52$_1$) and (52$_2$), we need only show the existence of the required expansions at the cusps. Let $r=-\frac{d}{c}$, $(c,d)=1$, be a cusp, and $S=\begin{pmatrix} a & b \\ c & d \end{pmatrix}\in\Gamma$. By (53)

$$\hat{g}_{MS^{-1}}(S(\tau))=(c\,\omega_1+d\,\omega_2)^k g_{MS^{-1}}\left(S\begin{pmatrix}\omega_1\\ \omega_2\end{pmatrix}\right),$$

which implies that

$$\hat{g}_{MS^{-1}}(S\,\tau)=(c\,\omega_1+d\,\omega_2)^k g_M\left(\begin{pmatrix}\omega_1\\ \omega_2\end{pmatrix}\right)=(c\,\tau+d)^k\,\hat{g}_M(\tau).$$

We now choose $S_1\in\Gamma$ such that

$$S_1 M S^{-1}=\begin{pmatrix}\alpha' & \beta'\\ 0 & \delta'\end{pmatrix}.$$

Then (54_3) and (54_1) imply that

$$\hat{g}_{MS^{-1}}(S\,\tau)=\hat{g}_{S_1MS^{-1}}(S\,\tau)=\delta'^{-k}\,\hat{g}\left(\frac{\alpha' S(\tau)+\beta'}{\delta'}\right).$$

The assumed expansion for $\hat{g}$ at $i\infty$ in powers of $e^{2\pi i\tau}$ yields the expansion of

$$(c\,\tau+d)^k\,\hat{g}_M(\tau)$$

in powers of $e^{\frac{2\pi i}{\delta'}S(\tau)}$. This proves Theorem **19**. □

2. Modular functions for Γ_M. Since g and g_M are modular forms of dimension $-k$ for Γ_M, we have

$$\frac{g_M\left(\begin{pmatrix}\omega_1\\ \omega_2\end{pmatrix}\right)}{g\left(\begin{pmatrix}\omega_1\\ \omega_2\end{pmatrix}\right)}=\frac{\hat{g}_M(\tau)}{\hat{g}(\tau)},\quad\text{and}\quad \frac{\hat{g}_M}{\hat{g}}\in K_{\Gamma_M}.$$

Now if $\psi:=\psi(n)$ is the index $[\Gamma:\Gamma_M]$ and if $\Gamma=\bigcup_{\nu=1}^{\psi}\Gamma_M S_\nu$, then the product

$$\prod_{\nu=1}^{\psi}\left(x-\frac{\hat{g}_M}{\hat{g}}\bigg|S_\nu\right)=\prod_{\nu=1}^{\psi}\left(x-\frac{\hat{g}_{MS_\nu}}{\hat{g}}\right) \tag{55}$$

$$=\prod_{(M)}\left(x-\frac{\hat{g}_M}{\hat{g}}\right)\quad\text{(by the remark to Theorem 9).}$$

The last product is to be taken over a complete system of inequivalent transformations of $\mathcal{M}_n$.

Theorem 20. *If $\hat{g}$ is a modular form for Γ and if $M \in \mathcal{M}_n$, then the modular function $\frac{\hat{g}_M}{\hat{g}}$ satisfies an irreducible equation over K_Γ of degree $\psi(n)$.*

Proof. By Theorem **3** and equation (55) the theorem will be proved when we have shown that

$$g_M \neq g_{M_1} \quad \text{if } M \nsim M_1.$$

Suppose then that $M, M_1 \in \mathcal{M}_n$ and that in homogeneous form

$$g\left(M\begin{pmatrix}\omega_1\\ \omega_2\end{pmatrix}\right) = g\left(M_1\begin{pmatrix}\omega_1\\ \omega_2\end{pmatrix}\right),$$

or

$$g\left(M M_1^{-1}\begin{pmatrix}\omega_1\\ \omega_2\end{pmatrix}\right) = g\left(\begin{pmatrix}\omega_1\\ \omega_2\end{pmatrix}\right).$$

We have $M M_1^{-1} = \frac{1}{n} A$ with integral matrix A and $\det A = n^2$. By § **3, 4** there exist matrices $S, S_1 \in \Gamma$ such that $S A S_1 = \begin{pmatrix} n^2 & 0\\ 0 & 1\end{pmatrix}$. It follows that

$$g\left(\begin{pmatrix} n & 0\\ 0 & n^{-1}\end{pmatrix}\begin{pmatrix}\omega_1\\ \omega_2\end{pmatrix}\right) = g\left(\begin{pmatrix}\omega_1\\ \omega_2\end{pmatrix}\right),$$

or that

$$n^k \hat{g}(n^2 \tau) = \hat{g}(\tau),$$

which is incompatible with the behavior of $\hat{g}$ at $i\infty$ as measured in $e^{2\pi\tau}$ if $n > 1$. □

This then shows that

$$K_{\Gamma_M} = K_\Gamma\left(\frac{\hat{g}_M}{\hat{g}}\right). \qquad \square$$

3. Multiplicator equation. If for g we choose Δ, the homogeneous discriminant, then we obtain

$$\prod_{(M)}\left(x - \frac{\hat{\Delta}_M}{\hat{\Delta}}\right) = \sum_{\nu=0}^{\psi} R_\nu(j)\, x^\nu$$

with rational functions $R_\nu(j)$ in j. In fact these functions are polynomials, since the function $\frac{\hat{\Delta}_M}{\hat{\Delta}}$ has poles only at the rational cusps. The equation

$$\sum_{\nu=0}^{\psi} R_\nu(j)\, x^\nu = 0 \tag{56}$$

is satisfied by all modular functions $\frac{\hat{\Delta}_M}{\hat{\Delta}}$. We call it *multiplicator equation.*

Let the order n of the transforms be a prime number p, then the transformations

$$\begin{pmatrix} p & 0 \\ 0 & 1 \end{pmatrix}, \quad \begin{pmatrix} 1 & \nu \\ 0 & p \end{pmatrix}, \quad \nu \bmod p,$$

represent a complete system of inequivalent transformations of $\mathscr{M}_p$. For

$$M_0 = \begin{pmatrix} p & 0 \\ 0 & 1 \end{pmatrix}, \quad \text{and } M^0 = \begin{pmatrix} 1 & 0 \\ 0 & p \end{pmatrix}$$

we have

$$\Gamma_{M_0} = \Gamma_0(p) \quad \text{and } \Gamma_{M^0} = \Gamma^0(p),$$

with the groups $\Gamma^0(p)$, $\Gamma_0(p)$ introduced in **IV**, § **3** (7). Moreover,

$$\frac{\hat{\Delta}_{M_0}}{\hat{\Delta}} \in K_{\Gamma_0(p)}, \qquad \frac{\hat{\Delta}_{M^0}}{\hat{\Delta}} \in K_{\Gamma^0(p)}. \tag{57}$$

The fundamental region for the group $\Gamma^0(p)$ is given in subsections **2** and **5** of **IV**, § **5**. From there,

$$t_\infty = e^{2\pi i \frac{\tau}{p}} \quad \text{and } t_0 = e^{-\frac{2\pi i}{\tau}} \tag{58}$$

are the local variables at the points $i\infty$ and 0, respectively.

Since $\hat{\Delta}$ is a form for Γ with a zero of order 1 at all rational cusps including $i\infty$ and is non-vanishing in $\mathscr{H}$, the function $\frac{\hat{\Delta}_{M^0}}{\hat{\Delta}}$,

$$\frac{\hat{\Delta}_{M^0}(\tau)}{\hat{\Delta}(\tau)} = \frac{\hat{\Delta}\left(\frac{\tau}{p}\right)}{\hat{\Delta}(\tau)} \cdot p^{12},$$

has a pole of order $p-1$ at $i\infty$ and consequently a zero of order $p-1$ at 0. Moreover, this function is holomorphic and non-vanishing in $\mathscr{H}$.

Hence there is a function φ_p,

$$\varphi_p(\tau) := \left\{ \frac{\hat{\Delta}\left(\frac{\tau}{p}\right)}{\hat{\Delta}(\tau)} \right\}^{\frac{1}{p-1}}, \tag{59}$$

holomorphic in $\mathscr{H}$, with the expansions

$$\varphi_p(\tau) = t_\infty^{-1} \sum_{\nu=0}^{\infty} c_\nu t_\infty^\nu, \quad c_0 \neq 0,$$

$$\varphi_p(\tau) = t_0 \sum_{\nu=0}^{\infty} d_\nu t_0^\nu, \quad d_0 \neq 0,$$

at $i\infty$ and 0, respectively. This function φ_p is meromorphic in $\mathscr{H}^*$ and the same is locally true (not ramified) for the function it induces on the Riemann surface $\mathfrak{R} = \mathscr{H}^*/\Gamma^0(p)$. Thus this function is one-valued on $\mathfrak{R}$ in case $\mathfrak{R}$, or equivalently, the fundamental region of $\Gamma^0(p)$, has genus 0. Since in this case φ_p has exactly one pole of order one on $\mathfrak{R}$, it generates the field $K_{\Gamma^0(p)}$. From the table following **IV**, Theorem **15** the genus of the fundamental region is zero in the cases

$$p = 2, 3, 5, 7 \quad \text{and} \quad 13\,,$$

and, as is easily seen, only in these cases. This gives

Theorem 21. *For* $p = 2, 3, 5, 7$ *and* 13 *the function* φ_p,

$$\varphi_p(\tau) = \left\{ \frac{\hat{\Delta}\left(\frac{\tau}{p}\right)}{\hat{\Delta}(\tau)} \right\}^{\frac{1}{p-1}},$$

generates the field $K_{\Gamma^0(p)}$ *over* $\mathbb{C}$.

Cf. W. Maak [1].

Chapter VII. Eisenstein Series of Higher Level

In this chapter we carry over to congruence groups of higher level the results of Chapter **III.** Again we shall represent modular forms by analytic expressions, namely by Eisenstein series of higher level. Here, in contrast to Chapter **III**, the integral dimension is not required to be even. This will lead to an important difference between forms for the homogeneous group $\Gamma(N)$ and forms for the homogeneous group $\Gamma[N]$, however, for their fields of automorphic functions one has $K_{\Gamma[N]} = K_{\Gamma(N)}$. As an application we discuss the construction of the field of modular functions for the principal congruence subgroup of level N.

§ 1. The Series in the Case of Absolute Convergence

1. Definition and Fourier expansion. Let N and k be natural numbers with $N \geqq 1$ and $k \geqq 3$. Further suppose that

$$\boldsymbol{m} = \begin{pmatrix} m_1 \\ m_2 \end{pmatrix} \quad \text{and} \quad \boldsymbol{a} = \begin{pmatrix} a_1 \\ a_2 \end{pmatrix}$$

are pairs of numbers of $\mathbb{Z}$ and that

$$\boldsymbol{w} = \begin{pmatrix} \omega_1 \\ \omega_2 \end{pmatrix}$$

is the pair of complex variables ω_1, ω_2. We handle the pairs $\boldsymbol{m}, \boldsymbol{a}$ and $\boldsymbol{w}$ as matrices. In particular let $\boldsymbol{m}' = (m_1, m_2)$ so that $\boldsymbol{m}' \boldsymbol{w} = m_1 \omega_1 + m_2 \omega_2$. Also set $\begin{pmatrix} 0 \\ 0 \end{pmatrix} = \boldsymbol{0}$.

For fixed $\boldsymbol{a}$ we define the *homogeneous Eisenstein series* $G_{N,k,\boldsymbol{a}}$ by the function

$$G_{N,k,\boldsymbol{a}}\colon\ W \to \hat{\mathbb{C}}, \qquad \boldsymbol{w} \mapsto G_{N,k,\boldsymbol{a}}(\boldsymbol{w}) := \sum_{\boldsymbol{m} \equiv \boldsymbol{a} (\mathrm{mod}\, N)}' (\boldsymbol{m}' \boldsymbol{w})^{-k},$$

$$W := \left\{ \boldsymbol{w} \,\middle|\, \boldsymbol{w} = \begin{pmatrix} \omega_1 \\ \omega_2 \end{pmatrix},\ \omega_1, \omega_2 \in \mathbb{C},\ \frac{\omega_1}{\omega_2} \in \mathscr{H} \right\}. \tag{1}$$

The congruence $\boldsymbol{m} \equiv \boldsymbol{a} \pmod N$ is to be understood elementwise. The sum extends over all $\boldsymbol{m}$ that are congruent to $\boldsymbol{a}$ modulo N for fixed $\boldsymbol{a}$. Here, and in the following, the prime on the summation sign indicates that the term corresponding to $\boldsymbol{m}=\boldsymbol{0}$ is to be omitted. This series converges absolutely for all $\boldsymbol{w} \in \boldsymbol{W}$ (cf. the proof of **III**, Theorem **1**).

Besides the homogeneous Eisenstein series in the variables ω_1, ω_2 we also consider the *inhomogeneous Eisenstein series*, also denoted by $G_{N,k,\boldsymbol{a}}$,

$$G_{N,k,\boldsymbol{a}}: \mathscr{H} \to \hat{\mathbb{C}}, \qquad \tau \mapsto G_{N,k,\boldsymbol{a}}(\tau) = \sum_{\substack{m_1 \equiv a_1 \\ m_2 \equiv a_2}\bmod N}{}' (m_1\tau + m_2)^{-k},$$

$$G_{N,k,\boldsymbol{a}}(\tau) = \omega_2^k \, G_{N,k,\boldsymbol{a}}(\boldsymbol{w}), \qquad \tau = \frac{\omega_1}{\omega_2}. \tag{2}$$

This series converges absolutely on $\mathscr{H}$ and uniformly on compact subsets of $\mathscr{H}$. Thus it represents a function holomorphic on $\mathscr{H}$.

For both the homogeneous and inhomogeneous series we have

$$\begin{aligned} G_{N,k,-\boldsymbol{a}} &= (-1)^k G_{N,k,\boldsymbol{a}}, \\ G_{N,k,\boldsymbol{a}} &= G_{N,k,\boldsymbol{a}_1} \quad \text{if } \boldsymbol{a} \equiv \boldsymbol{a}_1 \bmod N \text{ and} \\ G_{nN,k,n\boldsymbol{a}} &= n^{-k} G_{N,k,\boldsymbol{a}} \quad \text{if } n \in \mathbb{Z},\ n \neq 0. \end{aligned} \tag{3}$$

We call the series $G_{N,k,\boldsymbol{a}}$ a *primitive Eisenstein series* if the greatest common divisor $(a_1, a_2, N) = 1$. Each Eisenstein series is equal up to a constant factor to a primitive Eisenstein series of suitable level.

Previously we defined $f|_k S$ for functions f on $\mathscr{H}$ and real matrices $S = \begin{pmatrix} a & b \\ c & d \end{pmatrix}$ with $ad - bc > 0$ by

$$f|_k S: \mathscr{H} \to \hat{\mathbb{C}}, \qquad \tau \mapsto |S|^{\frac{1}{2}} (c\tau + d)^{-k} f(S\tau).$$

Now we want to extend this to functions f on $\boldsymbol{W}$ by

$$f|_k S: \boldsymbol{W} \to \hat{\mathbb{C}}, \qquad \boldsymbol{w} \mapsto f(S\boldsymbol{w}).$$

We retain the k in $f|_k S$ so that we will be able to write certain formulas uniformly for both the homogeneous and inhomogeneous $G_{N,k,\boldsymbol{a}}$.

Theorem 1. *For all* $\boldsymbol{a}$ *the Eisenstein series* $G_{N,k,\boldsymbol{a}}$, $N \geqq 1$, $k \geqq 3$ *are entire modular forms of level N and dimension* $-k$.

Proof. We verify the defining properties given in **V, § 2, 1**. First of all we obviously have the property of homogeneity,

$$G_{N,k,\boldsymbol{a}}(\lambda \boldsymbol{w}) = \lambda^{-k} G_{N,k,\boldsymbol{a}}(\boldsymbol{w}) \quad \text{for } \lambda \in \mathbb{C},\ \lambda \neq 0,$$

where

$$\lambda \boldsymbol{w} := \lambda \begin{pmatrix} \omega_1 \\ \omega_2 \end{pmatrix} = \begin{pmatrix} \lambda\, \omega_1 \\ \lambda\, \omega_2 \end{pmatrix}.$$

Then the invariance condition is satisfied since

$$G_{N,k,\boldsymbol{a}}(A\,\boldsymbol{w}) = \sum_{\boldsymbol{m} \equiv \boldsymbol{a}\,(N)}{}' (\boldsymbol{m}'\,A\,\boldsymbol{w})^{-k} = \sum_{\boldsymbol{m} \equiv A'\boldsymbol{a}\,(N)}{}' (\boldsymbol{m}'\,\boldsymbol{w})^{-k} = G_{N,k,A'\boldsymbol{a}}(\boldsymbol{w}),$$

consequently for inhomogeneous series we have

$$G_{N,k,\boldsymbol{a}}(A\,\tau)\,(c\,\tau+d)^{-k} = G_{N,k,A'\boldsymbol{a}}(\tau).$$

In the above $A = \begin{pmatrix} a & b \\ c & d \end{pmatrix} \in \Gamma$, and $A' = \begin{pmatrix} a & c \\ b & d \end{pmatrix}$ denotes its transpose.

So for homogeneous and inhomogeneous series we have

$$\begin{aligned} G_{N,k,\boldsymbol{a}}|_k A &= G_{N,k,A'\boldsymbol{a}} \quad \text{for all } A \in \Gamma, \text{ and} \\ G_{N,k,\boldsymbol{a}}|_k A &= G_{N,k,\boldsymbol{a}} \quad \text{for all } A \in \Gamma(N). \end{aligned} \tag{4}$$

If $G_{N,k,\boldsymbol{a}}$ is primitive, then so is $G_{N,k,\boldsymbol{a}}|_k A$ for all $A \in \Gamma$.

We already saw that the inhomogeneous Eisenstein series are meromorphic, in fact, are holomorphic in $\mathscr{H}$. We only have to show that they are entire and so we derive their Fourier expansion. Clearly

$$\begin{aligned} G_{N,k,\boldsymbol{a}}(\tau) &= \sum_{\substack{m_1 \equiv a_1 \\ m_2 \equiv a_2}\ \mathrm{mod}\, N}{}' (m_1\,\tau + m_2)^{-k} \\ &= \delta\left(\frac{a_1}{N}\right) \sum_{m_2 \equiv a_2 \bmod N}{}' m_2^{-k} + \sum_{m_1 \equiv a_1 \bmod N}{}' \ \sum_{m_2 \equiv a_2 \bmod N} (m_1\,\tau + m_2)^{-k}, \end{aligned}$$

where $\delta\left(\frac{a_1}{N}\right) = 1$ if $\frac{a_1}{N} \in \mathbb{Z}$, and is 0 otherwise. We set

$$\alpha_0(N,k,\boldsymbol{a}) := \delta\left(\frac{a_1}{N}\right) \sum_{m_2 \equiv a_2 \bmod N}{}' m_2^{-k},$$

and, using the formula **III**, (4):

$$\sum_{n \in \mathbb{Z}} (\tau + n)^{-k} = \sum_{\nu \geqq 1} \frac{(-2\pi i)^k\, \nu^{k-1}}{(k-1)!}\, e^{2\pi i \tau \nu}, \qquad \tau \in \mathscr{H},$$

we obtain the expansion

$$\begin{aligned} G_{N,k,\boldsymbol{a}}(\tau) &= \alpha_0(N,k,\boldsymbol{a}) + \sum_{m_1 \equiv a_1 \bmod N}{}' \ \sum_{\nu \in \mathbb{Z}} \frac{1}{N^k} \left(\frac{m_1\,\tau + a_2}{N} + \nu\right)^{-k} \\ &= \alpha_0(N,k,\boldsymbol{a}) + \frac{(-2\pi i)^k}{N^k (k-1)!} \sum_{m_1 \equiv a_1 \bmod N} \ \sum_{m m_1 > 0} m^{k-1} \cdot \operatorname{sgn} m \cdot e^{\frac{2\pi i}{N}(a_2 m + \tau m m_1)}. \end{aligned}$$

With $\zeta_N = e^{\frac{2\pi i}{N}}$ we have

$$G_{N,k,\boldsymbol{a}}(\tau) = \alpha_0(N,k,\boldsymbol{a}) + \frac{(-2\pi i)^k}{N^k(k-1)!} \sum_{\nu \geqq 1} \Bigg(\sum_{\substack{m \mid \nu \\ \frac{\nu}{m} \equiv a_1 \bmod N}} m^{k-1} \cdot \operatorname{sgn} m \cdot \zeta_N^{a_2 m} \Bigg) e^{\frac{2\pi i}{N}\tau\nu}.$$

Thus we may write

$$\left.\begin{aligned} G_{N,k,\boldsymbol{a}}(\tau) &= \sum_{\nu \geqq 0} \alpha_\nu(N,k,\boldsymbol{a}) e^{\frac{2\pi i}{N}\cdot\tau\nu}, \quad \text{with} \\ \alpha_0(N,k,\boldsymbol{a}) &= \delta\left(\frac{a_1}{N}\right) \sum_{m_2 \equiv a_2 \bmod N}' m_2^{-k}, \quad \text{and} \\ \alpha_\nu(N,k,\boldsymbol{a}) &= \frac{(-2\pi i)^k}{N^k(k-1)!} \sum_{\substack{m \mid \nu \\ \frac{\nu}{m} \equiv a_1 \bmod N}} m^{k-1} \operatorname{sgn} m \, \zeta_N^{a_2 m}, \quad \nu \geqq 1. \end{aligned}\right\} \tag{5}$$

From the transformation formula (4) and the expansion (5) it follows that at the cusp $-\frac{d}{c}$ we have the expansion

$$(c\tau+d)^k G_{N,k,\boldsymbol{a}}(\tau) = \sum_{\nu \geqq 0} c_\nu e^{\frac{2\pi i}{N}A(\tau)\nu}, \qquad A = \begin{pmatrix} a & b \\ c & d \end{pmatrix} \in \Gamma,$$

with certain c_ν. Hence $G_{N,k,\boldsymbol{a}}$ is an entire modular form of dimension $-k$ for $\Gamma(N)$. ▯

For $\boldsymbol{a} \not\equiv \boldsymbol{0} \bmod N$ the Eisenstein series of higher level appear as N^{th} division values of the derivatives of the *Weierstrass* $\wp$*-function.* As is well-known, this function, as a function of z, is defined by

$$\wp(z;\boldsymbol{w}) := z^{-2} + \sum_{\boldsymbol{m} \in \mathbb{Z}^2}' \left[(z+\boldsymbol{m}'\boldsymbol{w})^{-2} - (\boldsymbol{m}'\boldsymbol{w})^{-2}\right], \qquad z \in \mathbb{C},\ z \neq \boldsymbol{m}'\boldsymbol{w},$$

where we assume that $\operatorname{Im}\left(\frac{\omega_1}{\omega_2}\right) > 0$. The series converges uniformly on compact sets of the z-plane containing no lattice point $\boldsymbol{m}'\boldsymbol{w}$.

The derivative of order $k-2$ with respect to z is given by

$$\wp^{(k-2)}(z,\boldsymbol{w}) = (-1)^k (k-1)! \left[z^{-k} + \sum_{\boldsymbol{m} \in \mathbb{Z}^2}' (z+\boldsymbol{m}'\boldsymbol{w})^{-k}\right].$$

Here, if we set $z = \frac{\boldsymbol{a}'\boldsymbol{w}}{N}$ with $\boldsymbol{a} \not\equiv \boldsymbol{0} \pmod N$, then for $k>2$ we obtain

$$\begin{aligned} \wp^{(k-2)}\left(\frac{\boldsymbol{a}'\boldsymbol{w}}{N};\boldsymbol{w}\right) &= (-1)^k (k-1)!\, N^k \sum_{\boldsymbol{m} \in \mathbb{Z}^2} (\boldsymbol{a}'\boldsymbol{w} + N\boldsymbol{m}'\boldsymbol{w})^{-k} \\ &= (-1)^k (k-1)!\, N^k G_{N,k,\boldsymbol{a}}(\boldsymbol{w}). \end{aligned} \tag{6}$$

2. The main theorem for principal congruence groups. Since

$$G_{N,k,\boldsymbol{a}} = G_{N,k,\boldsymbol{a}_1} \quad \text{for } \boldsymbol{a} \equiv \boldsymbol{a}_1 \bmod N, \text{ and}$$
$$G_{N,k,-\boldsymbol{a}} = (-1)^k G_{N,k,\boldsymbol{a}},$$

it follows from **IV** Theorem **8**, that for fixed N and k the number of primitive $G_{N,k,\boldsymbol{a}}$ which differ by more than sign is at most $\sigma_\infty(N)$, the number of $\Gamma(N)$-inequivalent cusps. By **IV**, (10)

$$\sigma_\infty(N) = \begin{cases} \dfrac{\mu(N)}{N} & \text{for } N=1,2 \\[2ex] \dfrac{\mu(N)}{2N} & \text{for } N>2 \end{cases}, \quad \text{where } \mu(N) := [\Gamma : \Gamma(N)]$$

and by **IV**, (3)

$$\mu(N) = N^3 \prod_{\substack{p \mid N \\ p \text{ prime}}} \left(1 - \frac{1}{p^2}\right) \quad \text{for all } N.$$

Obviously,

$$G_{N,k,\boldsymbol{a}} = 0 \quad \text{for } N=1 \text{ or } 2, \text{ and } k \equiv 1 \bmod 2.$$

We now prove the main theorem.

Theorem 2. *For $k>2$, and $k \equiv 0 \bmod 2$ if $N=1$ or $N=2$ there is a linear combination of primitive Eisenstein series of level N and dimension $-k$ that does not vanish at an arbitrarily given cusp of a fundamental region but is 0 at the remaining cusps.*

First, under the assumption $(a_1, a_2, N) = 1$, we define a *reduced Eisenstein series* $G^*_{N,k,\boldsymbol{a}}$ by

$$G^*_{N,k,\boldsymbol{a}}(\boldsymbol{w}) := \sum_{\substack{\boldsymbol{m} \equiv \boldsymbol{a} \bmod N \\ (m_1, m_2) = 1}} (\boldsymbol{m}' \boldsymbol{w})^{-k}, \qquad G^*_{N,k,\boldsymbol{a}}(\tau) := \omega_2^k \, G_{N,k,\boldsymbol{a}}(\boldsymbol{w}). \tag{7}$$

As is seen from the proof of the lemma in **III, § 1**, the assumption $(a_1, a_2) = 1$ is not more restrictive than the assumption $(a_1, a_2, N) = 1$. The equations (3) and (4) hold for the functions $G^*_{N,k,\boldsymbol{a}}$; in particular

$$G^*_{N,k,\boldsymbol{a}}|_k A = G^*_{N,k,A'\boldsymbol{a}} \quad \text{for all } A \in \Gamma.$$

Next, we show that the $G^*_{N,k,\boldsymbol{a}}$ *are linear combinations of the primitive* $G_{N,k,\boldsymbol{a}}$, and then we prove that they have the property required by the theorem.

For this purpose we introduce the Möbius function μ that for natural n is defined by

$\mu(1)=1,$
$\mu(n)=0,\quad$ if n contains a square factor other than 1, and
$\mu(n)=(-1)^r,\quad$ if n is the product of r distinct primes.

As is well-known, μ satisfies

$$\sum_{d\mid m,\, d>0}\mu(d)=\delta_{1,m},\quad \text{where } \delta_{1,m}:=\begin{cases}1 & \text{if } m=1,\\ 0 & \text{if } m\neq 1.\end{cases} \tag{8}$$

Hence

$$\begin{aligned} G^*_{N,k,\boldsymbol{a}}(\tau) &= \sum'_{m_\nu\equiv a_\nu \bmod N}\ \sum_{\substack{d\mid(m_1,m_2)\\ d>0}}\mu(d)\,(m_1\tau+m_2)^{-k},\\ &=\sum_{d>0}\ \sum'_{m_\nu d\equiv a_\nu \bmod N}\mu(d)\,(m_1 d\tau+m_2 d)^{-k},\\ &=\sum_{\substack{t \bmod N\\ (t,N)=1}}\ \sum_{\substack{d\equiv t \bmod N\\ d>0}}\ \sum'_{m_\nu\equiv a_\nu d^{-1} \bmod N}\frac{\mu(d)}{d^k}(m_1\tau+m_2)^{-k},\\ &=\sum_{t^{-1} \bmod N}\ \sum_{\substack{dt\equiv 1 \bmod N\\ d>0}}\frac{\mu(d)}{d^k}\ \sum'_{m_\nu\equiv a_\nu t \bmod N}(m_1\tau+m_2)^{-k}.\end{aligned}$$

Thus $G^*_{N,k,\boldsymbol{a}}$ *is a linear combination of primitive* $G_{N,k,t\boldsymbol{a}}$:

$$G^*_{N,k,\boldsymbol{a}}=\sum_{t \bmod N}\left[\sum_{\substack{dt\equiv 1 \bmod N\\ d>0}}\frac{\mu(d)}{d^k}\right]G_{N,k,t\boldsymbol{a}}. \tag{9}$$

Now for $N>2$ or for $N=1$ or $N=2$ and $k\equiv 0 \bmod 2$ we show:

If $(a_1,a_2)=1$, *then* $G^*_{N,k,\boldsymbol{a}}(\tau)$ *is different from* 0 *at the cusps* $\Gamma(N)$-*equivalent to* $-\dfrac{a_2}{a_1}$, *and is* 0 *at the remaining cusps.*

By (4), if $A:=\begin{pmatrix}a & b\\ c & d\end{pmatrix}\in\Gamma$, then

$$G_{N,k,\boldsymbol{a}}=G_{N,k,A'^{-1}\boldsymbol{a}}|_k A\,.$$

We set

$$A'^{-1}\boldsymbol{a}=\begin{pmatrix}a_1'\\ a_2'\end{pmatrix}.$$

Then the constant term of the Fourier expansion of $(c\tau+d)^k G_{N,k,t\boldsymbol{a}}(\tau)$ at the point $-\dfrac{d}{c}$ has the value

$$\alpha_0(N_1k,tA'^{-1}\boldsymbol{a})=\delta\left(\frac{ta_1'}{N}\right)\sum'_{m_2\equiv ta_2' \bmod N}m_2^{-k}.$$

Hence for the constant term of the Fourier expansion of $(c\tau+d)^k G^*_{N,k,\boldsymbol{a}}(\tau)$ we obtain

$$\begin{aligned}\alpha_0^*(N,k,A'^{-1}\boldsymbol{a}) &= \sum_{t \bmod N} \sum_{\substack{dt\equiv 1 \bmod N\\ d>0}} \frac{\mu(d)}{d^k}\,\delta\left(\frac{t\,a_1'}{N}\right) \sum_{m_2\equiv t a_2' \bmod N}{}^{'} m_2^{-k},\\ &= \delta\left(\frac{a_1'}{N}\right) \sum_{d>0} \sum_{m_2 d\equiv a_2' \bmod N}{}^{'} \frac{\mu(d)}{(d\,m_2)^k}\\ &= \delta\left(\frac{a_1'}{N}\right) \sum_{m\equiv a_2' \bmod N}{}^{'} m^{-k} \sum_{\substack{d\mid m\\ d>0}} \mu(d)\,.\end{aligned}$$

Equation (8) implies that $\alpha_0^*(N,k,A'^{-1}\boldsymbol{a})$ is different from 0 only if c and d satisfy the congruences

$$\begin{aligned} a_1' &= d\,a_1 - c\,a_2 \equiv 0 \bmod N\,,\\ a_2' &= -b\,a_1 + a\,a_2 \equiv \pm 1 \bmod N\,,\end{aligned} \tag{10}$$

that is, if simultaneously

$$\begin{aligned} c &\equiv \pm a_1 \bmod N, \quad \text{and}\\ d &\equiv \pm a_2 \bmod N\,.\end{aligned}$$

Thus in this case $-\dfrac{d}{c}$ must be equivalent to $-\dfrac{a_2}{a_1}$. At this cusp $(c\tau+d)^k G^*_{N,k,\boldsymbol{a}}(\tau)$ has the value $(\pm 1)^k$ if $N \neq 1$ or 2, and $(-1)^k+1$ if $N=1$ or 2—thus, as we already observed, it is 0 for odd k. This completes the proof of Theorem **2**.

As an immediate consequence of this theorem we have:

For each entire modular form f of level N and dimension $-k$ there is a linear combination L of primitive $G_{N,k,\boldsymbol{a}}$ so that the difference $f-L$ vanishes at all cusps.

The $\sigma_\infty(N)$ functions $G^*_{N,k,\boldsymbol{a}}$ where $\boldsymbol{a}$ runs mod N with $(a_1,a_2,N)=1$ and where only one of the two $G^*_{N,k,\pm\boldsymbol{a}}$ appears are linearly independent. Thus the primitive $G_{N,k,\boldsymbol{a}}$ can be expressed linearly in terms of these functions. For primitive functions we have

$$G_{N,k,\boldsymbol{a}} = \sum_{\substack{t \bmod N\\ (t,N)=1}} \left[\sum_{\substack{dt\equiv 1 \bmod N\\ d>0}} \frac{1}{d^k}\right] G^*_{N,k,t\boldsymbol{a}}\,,$$

as one verifies by substitution.

Finally, we show:

The non-primitive Eisenstein series can be expressed as a linear combination of primitive Eisenstein series of the same level.

This follows from the fact that for $N = N_0 N_1, G^*_{N_0,k,\boldsymbol{a}}$ is a linear combination of the $G_{N,k,\boldsymbol{b}}$, which in turn is proved by

$$G^*_{N_0,k,\boldsymbol{a}}(\boldsymbol{w}) = \sum_{\substack{\boldsymbol{m} \equiv \boldsymbol{a} \bmod N_0 \\ (m_1,m_2)=1}} (\boldsymbol{m}'\boldsymbol{w})^{-k} = \sum_{\substack{\boldsymbol{b} \bmod N_0 N_1 \\ \boldsymbol{b} \equiv \boldsymbol{a} \bmod N_0}} \sum_{\substack{\boldsymbol{m} \equiv \boldsymbol{b} \bmod N_0 N_1 \\ (m_1,m_2)=1}} (\boldsymbol{m}'\boldsymbol{w})^{-k} = \sum_{\substack{\boldsymbol{b} \bmod N \\ \boldsymbol{b} \equiv \boldsymbol{a} \bmod N_0}} G^*_{N,k,\boldsymbol{b}}(\boldsymbol{w}). \tag{11}$$

In particular we have shown

Theorem 3. *For $k > 2$, and $k \equiv 0 \bmod 2$ if $N = 1$ or $N = 2$, the maximal number of linearly independent Eisenstein series of level N is $\sigma_\infty(N)$, the number of cusps of a fundamental region of $\Gamma(N)$. A linear combination of Eisenstein series vanishes in $\mathcal{H}$ if it vanishes at all cusps.*

We present an application to the groups $\Gamma(2)$ and $\Gamma(3)$. Their fundamental regions have the genus 0. Hence

$$K_{\Gamma(2)} = \mathbb{C}(j_2), \qquad K_{\Gamma(3)} = \mathbb{C}(j_3),$$

where j_2 and j_3 assume each value exactly once in a fundamental region and for which we also assume that they have a zero at an arbitrarily given cusp of the fundamental region and a pole at a different arbitrary cusp. *These functions can be represented as the quotients* of *reduced Eisenstein series.* Indeed, the number $\frac{\mu k}{12}$, $\mu = [\Gamma : \Gamma[N]]$, of zeros of an entire modular form of level N and dimension $-k$ is $\sigma_\infty(2) - 1$ for $N = 2$ and $k = 4$, and is $\sigma_\infty(3) - 1$ for $N = 3$ and $k = 3$. For this reason and by Theorem **2**, there is a reduced Eisenstein series that has no zeros in $\mathcal{H}$, is different from 0 at a single arbitrarily given cusp of a fundamental region, and has zeros of order one at the remaining cusps. The functions j_2 and j_3 are quotients of such series. Following F. Klein we call the functions j_2 and j_3 Hauptfunktionen (principal functions).

3. The principal theorem for arbitrary congruence groups. If Γ_1 is a congruence group of level N with

$$\Gamma_1 = \bigcup_{\nu=1}^{\mu_1} \Gamma(N) A_\nu, \qquad A_\nu = \begin{pmatrix} \alpha_\nu & \beta_\nu \\ \gamma_\nu & \delta_\nu \end{pmatrix}, \qquad \mu_1 = [\Gamma_1 : \Gamma(N)],$$

then we obtain an entire modular form for Γ_1 by considering

$$G^*_{\Gamma_1,k,\boldsymbol{a}} := \sum_{\nu=1}^{\mu_1} G^*_{N,k,\boldsymbol{a}}|_k A_\nu = \sum_{\nu=1}^{\mu_1} G^*_{N,k,A'_\nu \boldsymbol{a}}, \qquad (a_1, a_2) = 1. \tag{12}$$

Indeed, for $A \in \Gamma_1$,

$$A_\nu A = G_\nu A_{\nu'} \quad \text{with } G_\nu \in \Gamma(N)$$

and with ν' uniquely determined by ν which together with ν runs through the set $\{1, 2, \ldots, \mu_1\}$. Consequently

$$G^*_{\Gamma_1,k,\boldsymbol{a}}|_k A = \sum_{\nu=1}^{\mu_1} G^*_{N,k,\boldsymbol{a}}|_k A_\nu A = \sum_{\nu'=1}^{\mu_1} G^*_{N,k,\boldsymbol{a}}|_k G_\nu A_{\nu'}$$
$$= \sum_{\nu'=1}^{\mu_1} G^*_{N,k,\boldsymbol{a}}|_k A_{\nu'} = G^*_{\Gamma_1,k,\boldsymbol{a}}.$$

Obviously the $G^*_{\Gamma_1,k,\boldsymbol{a}}$ satisfy the remaining defining properties of an entire modular form for Γ_1.

They also satisfy the equations

$$\begin{aligned} &G^*_{\Gamma_1,k,-\boldsymbol{a}} = (-1)^k G^*_{\Gamma_1,k,\boldsymbol{a}} \\ &G^*_{\Gamma_1,k,A'\boldsymbol{a}} = G^*_{\Gamma_1,k,\boldsymbol{a}}, \quad \text{for } A\in\Gamma_1 . \end{aligned} \tag{13}$$

Now we investigate their behavior at the cusps $-\frac{d}{c}$. For this we assume that $N \geqq 2$ and, if $N=2$, that $k \equiv 0 \pmod 2$. By the proof of Theorem **2** the constant term in the expansion of

$$(c\tau+d)^k G^*_{N,k,A'_\nu \boldsymbol{a}}(\tau) \tag{14}$$

in powers of

$$e^{\frac{2\pi i}{N} S(\tau)}, \quad S = \begin{pmatrix} a & b \\ c & d \end{pmatrix} \in \Gamma,$$

is different from 0 if and only if

$$-\frac{a_2^\nu}{a_1^\nu} = V\left(-\frac{d}{c}\right) \quad \text{for some } \nu\in\{1,\ldots,\mu_1\} \quad \text{and } V\in\Gamma(N),$$

where we use the notation

$$\boldsymbol{a} = \begin{pmatrix} a_1 \\ a_2 \end{pmatrix}, \qquad \boldsymbol{a}^\nu = A'_\nu \boldsymbol{a}, \qquad \boldsymbol{a}^\nu = \begin{pmatrix} a_1^\nu \\ a_2^\nu \end{pmatrix}.$$

Since

$$-\frac{a_2^\nu}{a_1^\nu} = -\frac{\beta_\nu a_1 + \delta_\nu a_2}{\alpha_\nu a_1 + \gamma_\nu a_2} = \frac{\delta_\nu\left(-\frac{a_2}{a_1}\right) - \beta_\nu}{-\gamma_\nu\left(-\frac{a_2}{a_1}\right) + \alpha_\nu} = A_\nu^{-1}\left(-\frac{a_2}{a_1}\right),$$

and

$$A_\nu \Gamma(N) = \Gamma(N) A_\nu,$$

this relation is equivalent with

$$-\frac{a_2}{a_1} = V A_\nu\left(-\frac{d}{c}\right), \quad V\in\Gamma(N). \tag{15}$$

If k is even, then $G^*_{\Gamma_1,k,\boldsymbol{a}}$ is different from 0 at exactly those cusps that have the form

$$V\left(-\frac{a_2}{a_1}\right), \quad V \in \Gamma_1 ,$$

because the constant term in the explosion of (14) is non-negative, and if (15) is satisfied is in fact positive. This was shown in subsection **2**.

However, taking (13) into consideration there is essentially only one $G^*_{\Gamma_1,k,\boldsymbol{a}}$ with this property. Since the equality of the sets

$$\left\{V\left(-\frac{a_2}{a_1}\right) : V \in \Gamma_1\right\} = \left\{V\left(-\frac{a_2^0}{a_1^0}\right) : V \in \Gamma_1\right\}$$

implies that

$$-\frac{a_2^0}{a_1^0} = V_0\left(-\frac{a_2}{a_1}\right), \quad V_0 \in \Gamma_1 ,$$

or

$$\boldsymbol{a}_0 = V_0'^{-1}\boldsymbol{a}, \quad \boldsymbol{a}_0 = \begin{pmatrix} a_1^0 \\ a_2^0 \end{pmatrix}, \quad V_0 \in \Gamma_1 ,$$

therefore (13) implies that

$$G^*_{\Gamma_1,k,\boldsymbol{a}_0} = G^*_{\Gamma_1,k,\boldsymbol{a}} .$$

Finally, every modular form for Γ_1 belonging to the vector space generated by Eisenstein series of level N and dimension $-k$ is a linear combination of $G^*_{\Gamma_1,k,\boldsymbol{a}}$. Indeed,

$$G = \sum_{\substack{\boldsymbol{a} \bmod N \\ (a_1,a_2)=1}} c_{\boldsymbol{a}} G^*_{N,k,\boldsymbol{a}} \quad \text{and} \quad G|_k A_\nu = G, \qquad \nu = 1, \dots, \mu_1$$

implies that

$$G = \frac{1}{\mu_1} \sum_{\substack{\nu=1 \\ \boldsymbol{a}}}^{\mu_1} c_{\boldsymbol{a}} G^*_{N,k,\boldsymbol{a}}|_k A_\nu = \frac{1}{\mu_1} \sum_{\boldsymbol{a}} c_{\boldsymbol{a}} G^*_{\Gamma_1,k,\boldsymbol{a}} .$$

This proves

Theorem 4. *If $k \equiv 0 \bmod 2$ and $k > 2$, then for every congruence group Γ_1 of level N there is a linear combination of Eisenstein series $G_{N,k,\boldsymbol{a}}$ which is a form for Γ_1, does not vanish at an arbitrarily given cusp of a fundamental region for Γ_1, and is 0 at the remaining cusps. The $\mathbb{C}$-dimension of the space spanned by these linear combinations of $G_{N,k,\boldsymbol{a}}$ which are forms for Γ_1 is $\sigma_\infty(\Gamma_1)$, where $\sigma_\infty(\Gamma_1)$ denotes the number of cusps of a fundamental region for Γ_1.*

There are no modular forms of odd dimension for congruence groups containing $-I$.

§ 2. The Series of Dimension -1 and -2

1. Definition and formulation of the series expansions. The definition of the Eisenstein series for $k=1$ and $k=2$ cannot be carried over from the last section where $k>2$. Following E. Hecke, analogous to **III, § 2**, we introduce instead the series

$$\phi_{N,k,\boldsymbol{a}}(\tau,s) := \sum_{m_\nu \equiv a_\nu \bmod N}{}' (m_1\tau+m_2)^{-k}|m_1\tau+m_2|^{-s}, \qquad \tau\in\mathscr{H}. \tag{16}$$

If $\tau\in\mathscr{H}$ is fixed, then these series represent holomorphic functions of s in the half-plane $\operatorname{Re}s>2-k$, which for $k>2$ coincide at $s=0$ with the Eisenstein series of the preceding section. For $k=1$ and $k=2$ we shall show that $\phi_{N,k,\boldsymbol{a}}$ can be analytically continued to the entire s-plane as a meromorphic function which is holomorphic at $s=0$. This continuation shall again be denoted by $\phi_{N,k,\boldsymbol{a}}$, and we shall define *inhomogeneous* and *homogeneous Eisenstein series* by

$$\begin{aligned} G_{N,k,\boldsymbol{a}}(\tau) &:= \phi_{N,k,\boldsymbol{a}}(\tau,0), \qquad \text{and} \\ G_{N,k,\boldsymbol{a}}(\boldsymbol{w}) &:= \omega_2^{-k} G_{N,k,\boldsymbol{a}}(\tau), \qquad \tau=\frac{\omega_1}{\omega_2}, \end{aligned} \tag{17}$$

for $k=1,2$.

If we define

$$\alpha_0(s;N,k,\boldsymbol{a}) := \delta\left(\frac{a_1}{N}\right) \sum_{m_2\equiv a_2 \bmod N}{}' m_2^{-k}|m_2|^{-s}, \tag{18}$$

then

$$\begin{aligned} &\phi_{N,k,\boldsymbol{a}}(\tau,s) \\ &= \alpha_0(s;N,k,\boldsymbol{a}) + \sum_{m_1\equiv a_1 \bmod N}{}' \sum_{m_2\equiv a_2 \bmod N} (m_1\tau+m_2)^{-k}|m_1\tau+m_2|^{-s} \\ &= \alpha_0(s;N,k,\boldsymbol{a}) + \sum_{m_1\equiv a_1 \bmod N}{}' \sum_{n=-\infty}^{\infty} N^{-k-s}\left(\frac{m_1\tau+a_2}{N}+n\right)^{-k}\left|\frac{m_1\tau+a_2}{N}+n\right|^{-s}. \end{aligned}$$

We set

$$\tau^* := \frac{m_1\tau+a_2}{N}, \qquad m_1 \neq 0.$$

The series

$$\psi(n,s) := \sum_{n=-\infty}^{\infty} (\tau^*+n+u)^{-k}[(\tau^*+n+u)(\bar\tau^*+n+u)]^{-\frac{s}{2}}, \qquad \operatorname{Re}s>1-k,$$

represents a holomorphic function of u in the strip

$$\operatorname{Im}\left(\frac{\bar\tau}{N}\right) < \operatorname{Im}(u) < \operatorname{Im}\left(\frac{\tau}{N}\right)$$

which is periodic with period 1. Hence it has a Fourier expansion:

$$\psi(u,s)=\sum_{m=-\infty}^{\infty} c_{k,m}(\tau^*,s)\,e^{2\pi imu}$$

with

$$\begin{aligned} c_{k,m}(\tau^*,s) &= \int_0^1 \sum_{n=-\infty}^{\infty} (\tau^*+n+u)^{-k}\left[(\tau^*+n+u)(\bar{\tau}^*+n+u)\right]^{-\frac{s}{2}} e^{-2\pi imu}\,du \\ &= \int_{-\infty}^{\infty} (\tau^*+u)^{-k}|\tau^*+u|^{-s} e^{-2\pi imu}\,du\,. \end{aligned} \tag{19}$$

Since

$$\psi(0,s)=\sum_{m=-\infty}^{\infty} c_{k,m}(\tau^*,s)\,,$$

we have

$$\phi_{N,k,\boldsymbol{a}}(\tau,s)=\alpha_0(s;N,k,\boldsymbol{a})+N^{-k-s}\sum_{m_1\equiv a_1 \bmod N}{}' \sum_{m=-\infty}^{\infty} c_{k,m}\left(\frac{m_1\tau+a_2}{N},s\right) \tag{20}$$

for $\operatorname{Re}s>2-k$. As in **III, § 2**, by bending the path of integration up if $m<0$ and down if $m>0$, we obtain the $c_{k,m}\left(\frac{m_1\tau+a_2}{N},s\right)$, excluding $c_{k,0}$, in a form by which the function defined by

$$\sum_{m_1\equiv a_1 \bmod N}{}' \sum_{m=-\infty}^{\infty} c_{k,m}\left(\frac{m_1\tau+a_2}{N},s\right),\quad \tau\in\mathscr{H},\ \operatorname{Re}s>2-k,$$

is continued to an entire function in the s-plane. As in **III, § 2**, we calculate from this form of $c_{k,m}\left(\frac{m_1\tau+a_2}{N},s\right)$ that

$$c_{k,m}\left(\frac{m_1\tau+a_2}{N},0\right)=\begin{cases} (-2\pi i)^k \dfrac{m^{k-1}\operatorname{sgn} m}{(k-1)!}\,\zeta_N^{a_2 m}\, e^{\frac{2\pi i}{N} m m_1 \tau} & \text{if } mm_1>0,\\[2ex] 0 & \text{if } mm_1<0. \end{cases} \tag{21}$$

We now determine the summands of $\phi_{N,k,\boldsymbol{a}}$ corresponding to $m=0$ in (20) and denote them by

$$\beta_{N,k,\boldsymbol{a}}(\tau,s):=N^{-k-s}\sum_{m_1\equiv a_1 \bmod N}{}' c_{k,0}\left(\frac{m_1\tau+a_2}{N},s\right). \tag{22}$$

As above we obtain

$$\beta_{N,k,\boldsymbol{a}}(\tau,s)=N^{-k-s}\sum_{m_1\equiv a_1 \bmod N}{}' \int_{-\infty}^{\infty}\left(\frac{m_1\tau+a_2}{N}+u\right)^{-k}\left|\frac{m_1\tau+a_2}{N}+u\right|^{-s} du\,.$$

Hence by substituting $\frac{m_1 u - a_2}{N}$ for u, we have

$$\beta_{N,k,\boldsymbol{a}}(\tau,s) = \frac{1}{N} \sum_{m_1 \equiv a_1 \bmod N}' \frac{\operatorname{sgn} m_1}{m_1^{k-1}|m_1|^s} \int_{-\infty}^{\infty} (\tau+u)^{-k}|\tau+u|^{-s} du\,. \quad (23)$$

2. Conclusion of the series expansions for $k=2$. If $k=2$, it follows from **III, § 2** that

$$\int_{-\infty}^{\infty} (\tau+u)^{-k}|\tau+u|^{-s} du = -\frac{\sqrt{\pi}}{(\operatorname{Im}\tau)^{1+s}} \frac{\Gamma\left(\frac{s+1}{2}\right)}{\Gamma\left(\frac{s}{2}+1\right)} \frac{s}{s+2}, \qquad \operatorname{Re} s > 0\,.$$

We introduce the *Hurwitz zeta function* defined by

$$\zeta(s,\alpha) := \sum_{n>-\alpha} (n+\alpha)^{-s}, \qquad \operatorname{Re} s > 1, \qquad \alpha \in \mathbb{R}\,.$$

This function can be continued to the entire s-plane as a meromorphic function which is holomorphic except at $s=1$ where it has a simple pole with residue 1. See E. T. Whittaker and G. N. Watson [1], p. 226. If we introduce

$$\begin{aligned} \sum_{m_1 \equiv a_1 \bmod N}' |m_1|^{-1-s} &= \sum_{\substack{m_1 \equiv a_1 \bmod N \\ m_1 > 0}} m_1^{-1-s} + \sum_{\substack{m_1 \equiv -a_1 \bmod N \\ m_1 > 0}} m_1^{-1-s} \\ &= \frac{1}{N^{1+s}} \left[\zeta\left(1+s, \frac{a_1}{N}\right) + \zeta\left(1+s, -\frac{a_1}{N}\right) \right] \end{aligned}$$

into (23) we have

$$\begin{aligned} \beta_{N,k,\boldsymbol{a}}(\tau,s) = &-\frac{1}{N^{2+s}} \left[\zeta\left(1+s, \frac{a_1}{N}\right) + \zeta\left(1+s, -\frac{a_1}{N}\right) \right] \cdot \\ &\cdot \frac{\sqrt{\pi}}{(\operatorname{Im}(\tau))^{1+s}} \frac{\Gamma\left(\frac{s+1}{2}\right)}{\Gamma\left(\frac{s}{2}+1\right)} \frac{s}{s+2}\,. \end{aligned} \quad (24)$$

The functions $\phi_{N,2,\boldsymbol{a}}$ can be continued as meromorphic functions to the entire s-plane. This follows from the continuability of the double series (20) without the terms with $m=0$, of the series of (18)—which up to elementary functions is the sum of two Hurwitz zeta functions—and finally of the functions of (24).

As $s \to 0$ in (24) we obtain the limit

$$\beta_{N,2,\boldsymbol{a}}(\tau,0) = -\frac{1}{N^2} \frac{\pi}{\operatorname{Im}(\tau)}\,.$$

Hence for the Eisenstein series of dimension -2 we have the series expansion

$$G_{N,2,\boldsymbol{a}}(\tau) = -\frac{2\pi i}{N^2(\tau-\bar{\tau})} + \sum_{\nu \geqq 0} \alpha_\nu(N,\boldsymbol{a})\, e^{\frac{2\pi i}{N}\cdot \nu\tau}, \quad \text{where}$$

$$\alpha_0(N,\boldsymbol{a}) = \delta\left(\frac{a_1}{N}\right) \sum_{m_2 \equiv a_2 \bmod N}' m_2^{-2}, \quad \text{and} \tag{25}$$

$$\alpha_\nu(N,\boldsymbol{a}) = -\frac{4\pi^2}{N^2} \sum_{\substack{m \mid \nu \\ \frac{\nu}{m} \equiv a_1 \bmod N}} |m|\, \zeta_N^{a_2 m}, \qquad \nu \geqq 1.$$

This formally corresponds to the expansion of the series of higher dimension up to its non-analytic part $-\dfrac{2\pi i}{N^2(\tau-\bar{\tau})}$ (cf. (5))—which, by the way, is independent of $\boldsymbol{a}$.

3. Conclusion of the series expansion for $k=1$. If $k=1$ and $\operatorname{Re} s > 1$, then by (23) the sum of the terms of $\phi_{N,1,\boldsymbol{a}}$ with $m=0$ in (20) is

$$\beta_{N,1,\boldsymbol{a}}(\tau,s) = \frac{1}{N} \sum_{m_1 \equiv a_1 \bmod N}' \frac{\operatorname{sgn} m_1}{|m_1|^s} \int_{-\infty}^{\infty} (\tau+u)^{-1} |\tau+u|^{-s}\, du. \tag{26}$$

Here

$$\sum_{m_1 \equiv a_1 \bmod N}' \frac{\operatorname{sgn} m_1}{|m_1|^s} = \sum_{\substack{m_1 \equiv a_1 \bmod N \\ m_1 > 0}} \frac{1}{m_1^s} - \sum_{\substack{m_1 \equiv -a_1 \bmod N \\ m_1 > 0}} \frac{1}{m_1^s}$$

$$= N^{-s}\left[\zeta\left(s, \frac{a_1}{N}\right) - \zeta\left(s, -\frac{a_1}{N}\right)\right]$$

can be continued to the s-plane as an entire function. Furthermore, if $\tau = x+iy$ then

$$\int_{-\infty}^{\infty} (\tau+u)^{-1} |\tau+u|^{-s}\, du = \int_{-\infty}^{\infty} (iy+u)^{-1} |iy+u|^{-s}\, du$$

$$= \int_0^{\infty} \frac{2iy}{(y^2+u^2)^{1+\frac{s}{2}}}\, du = \frac{i\sqrt{\pi}}{y^s} \frac{\Gamma\left(\frac{s+1}{2}\right)}{\Gamma\left(\frac{s}{2}+1\right)}.$$

These two properties imply that $\beta_{N,1,\boldsymbol{a}}(\tau,s)$ can be continued to the entire s-plane and is holomorphic at $s=0$. Thus $\phi_{N,1,\boldsymbol{a}}$ is a meromorphic function of s and is holomorphic at $s=0$.

For the expansion of the Eisenstein series we obtain

$$
\begin{aligned}
G_{N,1,\boldsymbol{a}}(\tau) &= \sum_{\nu \geqq 0} \alpha_\nu(N,\boldsymbol{a})\, e^{\frac{2\pi i}{N}\cdot \nu\tau}, \quad \text{where} \\
\alpha_0(N,\boldsymbol{a}) &= \frac{1}{N}\,\delta\left(\frac{a_1}{N}\right) \lim_{s\to 0}\left[\zeta\left(1+s,\frac{a_2}{N}\right) - \zeta\left(1+s,-\frac{a_2}{N}\right)\right] \\
&\quad -\frac{\pi i}{N}\left[\zeta\left(0,\frac{a_1}{N}\right) - \zeta\left(0,-\frac{a_1}{N}\right)\right], \quad \text{and} \\
\alpha_\nu(N,\boldsymbol{a}) &= -\frac{2\pi i}{N} \sum_{\substack{m\mid\nu \\ \frac{\nu}{m}\equiv a_1 \bmod N}} \operatorname{sgn} m\, \zeta_N^{a_2 m}, \quad \text{for } \nu \geqq 1.
\end{aligned}
\tag{27}
$$

These series expansions for $G_{N,1,\boldsymbol{a}}$ differ formally from the corresponding expansions (5) of the series of higher dimension by the addition of the imaginary term in $\alpha_0(N,\boldsymbol{a})$.

§ 3. Properties of the Series of Dimension -1 and -2. Applications

1. Their properties as modular forms. The properties (3) of the $G_{N,k,\boldsymbol{a}}$ and the representability of the non-primitive $G_{N,k,\boldsymbol{a}}$, $(a_1,a_2,N)>1$, by the primitive ones, $(a_1,a_2,N)=1$, obviously also hold when $k=1$ and $k=2$. If

$$
A=\begin{pmatrix} a & b \\ c & d\end{pmatrix}\in\Gamma, \qquad \boldsymbol{a}=\begin{pmatrix} a_1 \\ a_2\end{pmatrix}, \qquad A'\boldsymbol{a}=\begin{pmatrix} a_1' \\ a_2'\end{pmatrix}, \qquad k=1,2
$$

then

$$
\begin{aligned}
\phi_{N,k,\boldsymbol{a}}(A(\tau),s) &= \sum_{m_\nu\equiv a_\nu \bmod N}{}' (m_1 A(\tau)+m_2)^{-k}\,|m_1 A(\tau)+m_2|^{-s} \\
&= (c\tau+d)^k\,|c\tau+d|^s \sum_{m_\nu\equiv a_\nu' \bmod N}{}' (m_1\tau+m_2)^{-k}\,|m_1\tau+m_2|^{-s} \\
&= (c\tau+d)^k\,|c\tau+d|^s\,\phi_{N,k,A'\boldsymbol{a}}(\tau,s).
\end{aligned}
$$

Thus by the definition (17) for the Eisenstein series we have:

$$
\begin{aligned}
G_{N,k,\boldsymbol{a}}|_k A &= G_{N,k,A'\boldsymbol{a}} \quad \text{for all } A\in\Gamma, \\
G_{N,k,\boldsymbol{a}}|_k A &= G_{N,k,\boldsymbol{a}} \quad \text{for all } A\in\Gamma(N).
\end{aligned}
\tag{28}
$$

The Eisenstein series $G_{N,1,\boldsymbol{a}}$ are holomorphic in $\mathcal{H}$. By **§ 2** they have the required expansion at $i\infty$ for modular forms, and hence in view of the above transformation formula, they have similar expansions at the other cusps.

Theorem 5. *The Eisenstein series $G_{N,1,\boldsymbol{a}}$ are entire modular forms of dimension -1.*

This statement is not correct for the series $G_{N,2,\boldsymbol{a}}$ because of their non-analytic part. However, since this part is independent of $\boldsymbol{a}$, we can state

Theorem 6. *The difference between two distinct series $G_{N,2,\boldsymbol{a}}$ is an entire modular form of dimension -2.*

2. A connection between the Weierstrass $\wp$-function

$$\wp(z;\boldsymbol{w}) = z^{-2} + \sum_{\boldsymbol{m}}' \{(z+\boldsymbol{m}'\boldsymbol{w})^{-2} - (\boldsymbol{m}'\boldsymbol{w})^{-2}\}$$

and the Eisenstein series of dimension -2.

For $N>1$, $\boldsymbol{a}\not\equiv\boldsymbol{0} \bmod N$ and $z=\dfrac{\boldsymbol{a}'\boldsymbol{w}}{N}$ we have the Nth *division values* $\wp_{N,\boldsymbol{a}}$ of $\wp$:

$$\wp_{N,\boldsymbol{a}}(\boldsymbol{w}) := \wp\left(\frac{\boldsymbol{a}'\boldsymbol{w}}{N};\boldsymbol{w}\right) = \left(\frac{\boldsymbol{a}'\boldsymbol{w}}{N}\right)^{-2} + \sum_{\boldsymbol{m}}' \left\{\left(\frac{\boldsymbol{a}'\boldsymbol{w}}{N} + \boldsymbol{m}'\boldsymbol{w}\right)^{-2} - (\boldsymbol{m}'\boldsymbol{w})^{-2}\right\}. \tag{29}$$

The periodicity of $\wp(z;\boldsymbol{w})$ implies that

$$\wp_{N,\boldsymbol{a}_1} = \wp_{N,\boldsymbol{a}} \quad \text{for } \boldsymbol{a}_1 \equiv \boldsymbol{a} \bmod N.$$

Multiplying by ω_2^2 and separating the terms with $m_1=0$ we obtain the equation

$$\wp_{N,\boldsymbol{a}}(\tau) := \omega_2^2\,\wp\left(\frac{\boldsymbol{a}'\boldsymbol{w}}{N};\boldsymbol{w}\right) = -\sum_{m_2=-\infty}^{\infty}{}' m_2^{-2} + \sum_{m_2=-\infty}^{\infty}\left(\frac{a_1\tau+a_2}{N}+m_2\right)^{-2}$$
$$+\sum_{m_1=-\infty}^{\infty}{}' \sum_{m_2=-\infty}^{\infty}\left\{\left(\frac{(a_1+Nm_1)\tau+a}{N}+m_2\right)^{-2} - (m_1\tau+m_2)^{-2}\right\}.$$

We write the inner sum as a difference of sums and develop the right side in a Fourier expansion under the tentative assumption $a_1\neq 0$:

$$\wp_{N,\boldsymbol{a}}(\tau) = -\frac{\pi^2}{3} - 4\pi^2\sum_{\nu\geq 1}\nu e^{2\pi i\frac{a_1\tau+a_2}{N}(\operatorname{sgn}a_1)\nu} + \delta\left(\frac{a_1}{N}\right)N^2\sum_{m_2\equiv a_2 \bmod N}{}' m_2^{-2}$$
$$-4\pi^2\sum_{m_1=-\infty}^{\infty}\left(\sum_{\substack{\nu\geq 1\\(a_1+Nm_1\neq 0)}}\nu e^{2\pi i\frac{(a_1+Nm_1)\tau+a_2}{N}(\operatorname{sgn}(a_1+Nm_1))\nu} - \sum_{\nu\geq 1}\nu e^{2\pi i|m_1|\tau\nu}\right).$$

The first inner sum is to be replaced by 0, if $a_1+Nm_1=0$.

The expression just obtained is independent of the order of summation, thus

$$\wp_{N,\boldsymbol{a}}(\tau)=\delta\left(\frac{a_1}{N}\right)N^2\sum_{m_2\equiv a_2(N)}{}' m_2^{-2}-4\pi^2\sum_{\substack{\nu\geqq 1\\ m_1\equiv a_1(N)\\ m_1\neq 0}}\nu e^{2\pi i\frac{m_1\tau+a_2}{N}(\operatorname{sgn} m_1)\nu}$$

$$-\frac{\pi^2}{3}+4\pi^2\sum_{\substack{\nu\geqq 1\\ m_1\neq 0}}\nu e^{2\pi i|m_1|\tau\nu}$$

$$=N^2\left[G_{N,2,\boldsymbol{a}}(\tau)-G_{N,2,\boldsymbol{0}}(\tau)\right],$$

the last equality following from equation (25). We collect these results in

Theorem 7. *The Nth division values of $\wp$ are entire modular forms of dimension -2 for the principal congruence group $\Gamma(N)$ and can be represented in the form*

$$\wp_{N,\boldsymbol{a}}=N^2(G_{N,2,\boldsymbol{a}}-G_{N,2,\boldsymbol{0}}). \tag{30}$$

Since both sides of this equation only depend on $\boldsymbol{a}\bmod N$, (30) is also correct when $a_1=0$. Obviously

$$\wp_{N,\boldsymbol{a}}|_2 A=\wp_{N,A'\boldsymbol{a}} \quad \text{for all } A\in\Gamma.$$

3. The main theorem for division values of $\wp$. We now carry the rest of the first section over to the series $G_{N,2,\boldsymbol{a}}$. There are at most $\sigma_\infty(N)$ linearly independent $G_{N,2,\boldsymbol{a}}$. Since they are not holomorphic in $\mathscr{H}$ and therefore cannot be modular forms, we must choose a method different from the one used in **§ 1, 2**. Beside the primitive $G_{N,2,\boldsymbol{a}}$ we consider the *primitive $\wp$-division values*, i.e. those $\wp_{N,\boldsymbol{a}}$ for which g.c.d. $(a_1,a_2,N)=1$. Without restriction we assume that $(a_1,a_2)=1$. Analogous to the system (9) of $G^*_{N,k,\boldsymbol{a}}$, $k>2$, we form the system of $\sigma_\infty(N)$ series

$$\wp^*_{N,\boldsymbol{a}}:=\sum_{t\bmod N}\sum_{\substack{d>0\\ dt\equiv 1\bmod N}}\frac{\mu(d)}{d^2}\wp_{N,t\boldsymbol{a}}, \tag{31}$$

where $\boldsymbol{a}$ runs $\bmod N$ with g.c.d. $(a_1,a_2)=1$ and where only one of the two series $\wp^*_{N,\pm\boldsymbol{a}}$ is included.

We want to determine the maximal number of linearly independent such series. For this purpose we use equation (30), valid for $\boldsymbol{a}\not\equiv\boldsymbol{0}\bmod N$, and the representation (25):

$$G_{N,2,\boldsymbol{a}}=-\frac{2\pi i}{\tau-\bar\tau}+\sum_{\nu\geqq 0}\alpha_\nu(N,\boldsymbol{a})e^{\frac{2\pi i\tau}{N}\cdot\nu}$$

valid for all $\boldsymbol{a}$. If $N>2$, according to the calculations in the proof of Theorem **2**, which are also valid for $k=2$, the constant term of the Fourier expansion of

$$(c\tau+d)^2\,\wp^*_{N,\boldsymbol{a}}(\tau)$$

at the cusp $-\frac{d}{c}$, $(c,d)=1$ has the value

$$\begin{aligned} &N^2+\omega, \quad \text{if } -\frac{d}{c} \text{ is } \Gamma[N]\text{-equivalent to } -\frac{a_2}{a_1}, \quad \text{and} \\ &\omega \qquad \text{otherwise}, \end{aligned} \tag{32}$$

where ω stems from $G_{N,2,\mathbf{0}}$ and is the same for all cusps. If $N=2$, here and in the following we must replace N^2 by $2N^2$. The sum

$$\sum_{(\boldsymbol{a})} \wp^*_{N,\boldsymbol{a}}$$

of the functions of the system (31) is an entire modular form of dimension -2 for the full modular group Γ and as such is 0 (cf. Theorem **12**). Hence

$$\omega=-N^2/\sigma_\infty(N).$$

By suitably enumerating the forms and cusps, the scheme (32) of constant terms can be put in the form of the matrix

$$S=(\alpha_{i,k}), \qquad \alpha_{i,k}=N^2\,\delta_{i,k}-N^2/\sigma_\infty(N); \qquad i,k=1,2,\ldots,\sigma_\infty(N). \tag{33}$$

As one can easily convince oneself, the rank of S is $\sigma_\infty(N)-1$. Hence the maximal number of linearly independent functions (31) is at least $\sigma_\infty(N)-1$.

Suppose $\mathscr{R}$, $\mathscr{R}_1$ and $\mathscr{R}_2$ are the vector spaces generated by the $G_{N,2,\boldsymbol{a}}$, $\wp^*_{N,\boldsymbol{a}}$ and the $\wp_{N,\boldsymbol{a}}$, respectively; let $\mathscr{R}_3$ be the vector space of all modular forms contained in $\mathscr{R}$. Then

$$\mathscr{R}_1\subset\mathscr{R}_2\subset\mathscr{R}_3\subsetneqq\mathscr{R}.$$

The inequalities

$$\dim\mathscr{R}\leqq\sigma_\infty(N), \quad \text{and} \quad \dim\mathscr{R}_1\geqq\sigma_\infty(N)-1$$

imply that

$$\dim\mathscr{R}=\sigma_\infty(N), \quad \text{and} \quad \dim\mathscr{R}_1=\sigma_\infty(N)-1,$$

and hence that

$$\mathscr{R}_1=\mathscr{R}_2=\mathscr{R}_3.$$

Collecting these results we have shown

Theorem 8. *The vector space $\mathscr{R}$ generated by the $G_{N,2,\boldsymbol{a}}$ has $\mathbb{C}$-dimension $\sigma_\infty(N)$ and the subspace $\mathscr{R}_1$ of all modular forms in $\mathscr{R}$ has $\mathbb{C}$-dimension $\sigma_\infty(N)-1$. Every set of $\sigma_\infty(N)-1$ distinct primitive division values of $\wp$ is a basis for $\mathscr{R}_1$.*

An entire modular form f of dimension -2 for $\Gamma(N)$, or equivalently for $\Gamma[N]$, has an expansion.

$$(c_\mu\tau+d_\mu)^2 f(\tau)=\sum_{\nu\geqq 0} c_\nu^{(\mu)} e^{\frac{2\pi i}{N}A_\mu(\tau)\nu}, \quad A_\mu=\begin{pmatrix} a_\mu & b_\mu \\ c_\mu & d_\mu \end{pmatrix}\in\Gamma,$$

at the $\sigma_\infty(N)$ inequivalent cusps $r_\mu=\dfrac{-d_\mu}{c_\mu}$ of a fundamental region. We show that

$$\sum_{\mu=1}^{\sigma_\infty(N)} c_0^{(\mu)}=0. \tag{34}$$

By **V, § 3**, $f=\dfrac{dF}{d\tau}$ for some integral F for $\Gamma(N)$, and moreover

$$(c_\mu\tau+d_\mu)^2\frac{dF}{d\tau}=\frac{2\pi i}{N}t_{r_\mu}\frac{dF}{dt_{r_\mu}}, \qquad t_{r_\mu}=e^{\frac{2\pi i}{N}A_\mu(\tau)}.$$

Except at the points $\langle r_\mu\rangle$, $\mu\in\{1,\dots,\sigma_\infty(N)\}$, the differential dF is obviously holomorphic on the Riemann surface $\mathfrak{R}$ associated with $\Gamma(N)$. At these exceptional points it has the residue $\dfrac{N}{2\pi i}c_0^{(\mu)}$. However, on a compact Riemann surface the sum of the residues of a meromorphic differential is 0, thus

$$\sum_{\mu=1}^{\sigma_\infty(N)}\frac{N}{2\pi i}c_0^{(\mu)}=0,$$

which was to be shown.

Alternatively one can prove (34) by the method of proof of Theorem **2** or Theorem **13** of Chapter **II**. After these preparations we prove the main theorem concerning the division values of $\wp$.

Theorem 9. *To each entire modular form f of dimension -2 and of level N corresponds a linear combination of primitive $\wp$-division values of level N such that their difference is an entire cusp form of dimension -2.*

Proof. For fixed $\boldsymbol{g}=\begin{pmatrix} g_1 \\ g_2 \end{pmatrix}$,

$$G^*_{N,2,\boldsymbol{a}}-G^*_{N,2,\boldsymbol{g}}$$

is a linear combination of $\wp$-partial values. If $\boldsymbol{a}$ and $\boldsymbol{g}$ are chosen with relatively prime components and $\boldsymbol{a}$ is not $\Gamma(N)$-equivalent to $\boldsymbol{g}$ then the above function is non-vanishing at the cusps $\Gamma(N)$-equivalent to either $-\dfrac{a_2}{a_1}$ or $-\dfrac{g_2}{g_1}$, and is 0 at the remaining ones. If one subtracts a suitable linear combination of these functions from f one obtains a function that vanishes at all cusps except possibly at those $\Gamma(N)$-equivalent to $-\frac{g_2}{g_1}$. However, in view of (34) this difference is 0 at these cusps as well. □

Obviously, Theorem **8** holds for arbitrary congruence groups $\Gamma_1(N)$ when $\sigma_\infty(N)$ is replaced by the number of $\Gamma_1(N)$-inequivalent cusps (cf. Theorem **4**).

4. The series $G_{N,1,\boldsymbol{a}}$. As opposed to the series $G_{N,2,\boldsymbol{a}}$, the Eisenstein series $G_{N,1,\boldsymbol{a}}$ are modular forms. They differ formally from the $G_{N,k,\boldsymbol{a}}$, $k \geqq 3$, by the replacement of the constant term

$$\delta\left(\frac{a_1}{N}\right) \sum_{m_2 \equiv a_2 \bmod N}' m_2^{-k}, \quad k \geqq 3,$$

of their Fourier expansion by the real value at $s=0$ of the entire function, which for $\operatorname{Re} s > 0$ is defined by

$$\delta\left(\frac{a_1}{N}\right) \sum_{m_2 \equiv a_2 \bmod N}' m_2^{-1} |m_2|^{-s},$$

and by an additional purely imaginary term (cf. (27)).

We shall now construct a linear combination $G^*_{N,1,\boldsymbol{a}}$ of the $G_{N,1,\boldsymbol{a}}$ the real part of whose constant term in its expansion at exactly one cusp and its equivalents under $\Gamma(N)$ is different from 0. At the remaining cusps its real part is 0.

For this purpose we introduce some facts from the theory of Dirichlet's L-series. Let χ be a *residue character modulo* N, i. e. a function on $\mathbb{Z}$ with the properties

$$\chi(n) \neq 0 \quad \text{if } (n, N) = 1, \qquad \chi(n) = 0 \quad \text{if } (n, N) > 1,$$
$$\chi(n') = \chi(n) \quad \text{if } n' \equiv n \bmod N, \quad \text{and } \chi(n n') = \chi(n)\chi(n').$$

We denote by χ_0 the character that has the value 1 for all n relatively prime to N. Then

$$\sum_\chi \chi(a) = \begin{cases} 0 & \text{if } a \not\equiv 1 \bmod N, \quad \text{and} \\ \varphi(N) & \text{if } a \equiv 1 \bmod N, \end{cases}$$

where the sum runs over all distinct residue characters $\bmod N$ and where φ denotes Euler's φ-function and

$$\sum_{a \bmod N} \chi(a) = \begin{cases} 0 & \text{if } \chi \neq \chi_0, \quad \text{and} \\ \varphi(N) & \text{if } \chi = \chi_0. \end{cases}$$

The *Dirichlet L-series* L_χ are defined by

$$L_\chi(s) = \sum_{n=1}^{\infty} \chi(n) n^{-s} \begin{cases} \text{for } \operatorname{Re} s > 0 & \text{if } \chi \neq \chi_0, \quad \text{or} \\ \text{for } \operatorname{Re} s > 1 & \text{if } \chi = \chi_0 . \end{cases}$$

For $\chi \neq \chi_0$ these functions can be continued to the entire s-plane as entire functions, and for $\chi = \chi_0$ the continuation yields a meromorphic function in the s-plane with a single simple pole at $s=1$. This follows from the representation of L_χ by the Hurwitz zeta function. The reciprocals $(L_\chi)^{-1}$ are thus meromorphic functions of s. If μ denotes the Möbius function, one finds that

$$(L_\chi(s))^{-1} = \sum_{n=1}^{\infty} \mu(n) \chi(n) n^{-s} \quad \text{for } \operatorname{Re}(s) > 1. \tag{35}$$

A famous, deep and often proved assertion is the inequality

$$L_\chi(1) \neq 0 \quad \text{for all } \chi .$$

For a proof of this inequality we refer to E. Hecke [1], p. 169 or H. Davenport [1], p. 33 ff. This proof implies that L_χ^{-1} is holomorphic at $s=1$. By summing (35) over all distinct characters one obtains

$$\begin{aligned} \sum_{\chi} \chi(a) (L_\chi(s))^{-1} &= \sum_{\substack{n=1 \\ \chi}}^{\infty} \chi(a) \mu(n) \chi(n) n^{-s} \\ &= \varphi(N) \sum_{\substack{n>0 \\ an \equiv 1 \bmod N}} \mu(n) n^{-s} \quad \text{for } \operatorname{Re} s > 1, \end{aligned}$$

and fixed $a \in \mathbb{Z}$. This yields a function which is meromorphic in the s-plane, holomorphic at $s=1$, and which has the representation

$$\sum_{\substack{n>0 \\ an \equiv 1 \bmod N}} \mu(n) n^{-s} \quad \text{for } \operatorname{Re} s > 1 .$$

The value of such a function at $s = s_0$ is denoted by adjoining $|_{s=s_0}$ to the sum.

Analogous to (9), we now define

$$G^*_{N,1,\boldsymbol{a}} := \sum_{t \bmod N} \left(\tfrac{1}{2} \sum_{dt \equiv 1 (\bmod N)} \operatorname{sgn} d \, \mu(d) |d|^{-1-s} \right) \Bigg|_{s=0} G_{N,1,t\boldsymbol{a}}, \qquad (a_1, a_2) = 1 ;$$

with the inversion (36)

$$G_{N,1,\boldsymbol{a}} = \sum_{\substack{t \bmod N \\ (t,N)=1}} \left(\tfrac{1}{2} \sum_{dt \equiv 1 \bmod N} \operatorname{sgn} d \, |d|^{-1-s} \right) \Bigg|_{s=0} G^*_{N,1,t\boldsymbol{a}}, \qquad (a_1, a_2) = 1 .$$

The constant term in the expansion of

$$(c\tau + d) G^*_{N,1,\boldsymbol{a}}(\tau), \qquad (c,d) = 1, \qquad A = \begin{pmatrix} a & b \\ c & d \end{pmatrix} \in \Gamma_1 ,$$

in powers of $e^{\frac{2\pi i}{N}A(\tau)}$ has a non-zero real part if and only if the cusp $-\frac{d}{c}$ is $\Gamma(N)$-equivalent to the cusp $-\frac{a_2}{a_1}$. One sees this immediately if one carries out the proof of Theorem **2**—in particular the proof of congruence (10)—for the functions $\phi_{N,1,\boldsymbol{a}}$ of τ and s defined in (16) for $\operatorname{Re} s > 1$ and if one then observes (17).

Hence for every entire modular form G_1 of level N and dimension -1 there is a linear combination L_1 of Eisenstein series $G_{N,1,\boldsymbol{a}}$ such that the constant term in the expansion of

$$(c\tau+d)(G_1(\tau)-L_1(\tau))$$

in powers of $e^{2\pi i A(\tau)/N}$, $A=\begin{pmatrix} a & b \\ c & d \end{pmatrix}\in\Gamma$, has vanishing real part for all cusps $-\frac{d}{c}$. The square $(G_1-L_1)^2$, an entire modular form of dimension -2, thus has non-positive constant terms at all cusps which because of (34) are actually 0. This shows

Theorem 10. *The $G_{N,1,\boldsymbol{a}}$ generate a vector space of dimension $\sigma_\infty(N)$ over the field $\mathbb{R}$ of real numbers. To each entire modular form of dimension -1 of level N there is a linear combination of the $G_{N,1,\boldsymbol{a}}$ such that their difference is an entire cusp form of dimension -1.*

In the case of prime level we prove

Theorem 10*. *If $p>2$ is a prime, then the $G_{p,1,\boldsymbol{a}}$ generate a vector space over $\mathbb{C}$ of dimension $r=\frac{1}{2}\sigma_\infty(p)$.*

Proof. Let p be a prime, $p>2$. Theorem **10** implies that $r \geqq \frac{1}{2}\sigma_\infty(p)$. We show that $r \leqq \frac{1}{2}\sigma_\infty(p)$. To do this we write equation (27) for arbitrary natural N in the form

$$G_{N,1,\boldsymbol{a}}(\tau)=\alpha_0(N,\boldsymbol{a})-\frac{2\pi i}{N}\sum_{\substack{mm_1>0\\ m_1\equiv a_1 \bmod N}} \operatorname{sgn} m\, \zeta_N^{a_2 m}\, e^{\frac{2\pi i\tau}{N}mm_1}.$$

We next replace the family of $G_{N,1,\boldsymbol{a}}$, $\boldsymbol{a} \bmod N$, by the linearly equivalent family

$$H_{\boldsymbol{b}}:=\alpha_0(\boldsymbol{b})-2\pi i\sum_{\substack{m_1m_2>0\\ \boldsymbol{m}\equiv \boldsymbol{b} \bmod N}} \operatorname{sgn} m_1\, e^{\frac{2\pi i\tau}{N}m_1m_2}, \qquad \boldsymbol{b}=\begin{pmatrix} b_1\\ b_2\end{pmatrix},\ b_1,b_2\in\mathbb{Z},\ \boldsymbol{b} \bmod N.$$

This equivalence is a consequence of

$$H_{\boldsymbol{b}}=\sum_{a_2 \bmod N}\zeta_N^{-a_2b_2}\,G_{N,1,\boldsymbol{a}} \quad \text{with } b_1=a_1, \quad \text{and}$$

$$\alpha_0(\boldsymbol{b})=\sum_{a_2 \bmod N}\zeta_N^{-a_2b_2}\,\alpha_0(N,\boldsymbol{a}).$$

Clearly the functions $H_{\boldsymbol{b}}$ satisfy the relations:

1) $H_{\boldsymbol{b}^*}=\pm H_{\boldsymbol{b}}$, if $\boldsymbol{b}^*\equiv\pm\boldsymbol{b}\bmod N$,

2) $H_{\boldsymbol{b}^*}=H_{\boldsymbol{b}}$, if $\boldsymbol{b}=\begin{pmatrix}b_1\\b_2\end{pmatrix}$ and $\boldsymbol{b}^*=\begin{pmatrix}b_2\\b_1\end{pmatrix}$.

Among the functions $H_{\boldsymbol{b}}$ we choose the functions

$$\begin{aligned}&H_{\boldsymbol{b}} \quad \text{with } 1\leqq b_1\leqq b_2\leqq N, \quad b_1+b_2<N-1, \quad \text{and}\\&H_{\boldsymbol{b}} \quad \text{with } 1\leqq b_1\leqq\left[\frac{N-1}{2}\right], \quad b_2=N.\end{aligned}\tag{$*$}$$

It follows from 1) and 2) that $H_{\boldsymbol{b}}=0$ if $b_1+b_2\equiv 0\bmod N$. Hence the remaining $H_{\boldsymbol{b}}$ coincide with the $H_{\boldsymbol{b}}$ of the system (*) or with their negatives. The number A of functions in (*) satisfies

$$A\geqq r.$$

If $N=p>2$ is prime, one easily sees that

$$A=\tfrac{1}{4}(p^2-1)=\tfrac{1}{2}\sigma_\infty(p),$$

and that consequently

$$r\leqq\tfrac{1}{2}\sigma_\infty(p). \quad \square$$

Theorem **10*** is actually true when the prime p is replaced by an arbitrary $N>2$. For a proof see H. Petersson [2].

5. Applications. For $\Gamma(N)$, $N\in\{2,3,4,5\}$, the number $\sigma_\infty(N)$ of cusps and the number $\nu_0(N)$ of zeros of an entire modular form of dimension -2 are given in:

N	2	3	4	5
$\sigma_\infty(N)$	3	4	6	12
$\nu_0(N)$	1	2	4	10

Let $p_{\mu,\nu}^{(N)}$ be linear combinations of the division values of $\wp$ that are different from zero at the cusps s_μ and s_ν of a fundamental region and are equal to zero at the remaining cusps. It follows that these zeros have order 1 and that $p_{\mu,\nu}^{(N)}(\tau)\neq 0$ if $\tau\in\mathcal{H}$. If $\lambda\neq\mu$, then $\dfrac{p_{\mu,\nu}^{(N)}}{p_{\lambda,\nu}^{(N)}}$ are Hauptfunktionen for $\Gamma(N)$, $N\in\{2,3,4,5\}$ (cf. § **1, 2**), which have a pole of order 1 at s_μ and a zero of order 1 at s_λ.

$(p_{1,2}^{(2)}p_{2,3}^{(2)}p_{3,1}^{(2)})^2$ is a cusp form of dimension -12 and level 2 which has zeros of order 2 at the cusps. As such it is a constant multiple of Δ. Thus this leads to the representation

$$\sqrt{\Delta}=c_2\,p_{1,2}^{(2)}p_{2,3}^{(2)}p_{3,1}^{(2)} \quad \text{with a constant } c_2.$$

Correspondingly, for $N=3$ we have

$$\sqrt[3]{\Delta}=c_3\, p_{1,2}^{(3)}\, p_{3,4}^{(3)}\,.$$

By forming the quotient

$$\sqrt[6]{\Delta}=\sqrt{\Delta}\,\sqrt[3]{\Delta}^{-1}$$

we obtain a modular form of dimension -2 for $\Gamma(6)$.

Let $p_{(4)}^{\mu,\nu}$ be a non-trivial linear combination of $G_{4,1,\boldsymbol{a}}$ which vanishes at the cusps s_μ and s_ν. Its existence can be shown by computation. Then $p_{\mu,\nu}^{(4)}\, p_{(4)}^{\mu,\nu}$ with $\mu \neq \nu$ is a cusp form of dimension -3 for $\Gamma(4)$, however not for $\Gamma[4]$, and

$$\sqrt[4]{\Delta}=c_4\, p_{\nu,\mu}^{(4)}\, p_{(4)}^{\nu,\mu}=c_4^1\, p_{(4)}^{1,2}\, p_{(4)}^{3,4}\, p_{(4)}^{5,6}\,.$$

The quotient

$$\sqrt[12]{\Delta}=\sqrt[3]{\Delta}\,\sqrt[4]{\Delta}^{-1}$$

is thus a modular cusp form of dimension -1 for $\Gamma(12)$, but not for $\Gamma[12]$.

The fundamental region $\mathscr{F}_p$ of $\Gamma_0(p)$ for prime $p>2$ has two cusps of fan width p and 1, respectively. It follows that up to a multiplicative constant there is exactly one modular form of dimension -2 for $\Gamma_0(p)$ in the space of Eisenstein series.

From **III, § 9** we take the function G_2^* with

$$G_2^*(\tau)=\frac{\pi^2}{3}-8\pi^2\sum_{m,m_1\geqq 1} m\, e^{2\pi i m m_1 \tau},$$

holomorphic in $\mathscr{H}$, which for $A=\begin{pmatrix} a & b\\ c & d\end{pmatrix}\in\Gamma$, transforms as follows:

$$G_2^*(A(\tau))=(c\tau+d)^2\, G_2^*(\tau)-2\pi i c(c\tau+d)\,.$$

Hence for natural $N\geqq 2$,

$$E(\tau;N):=N\,G_2^*(N\tau)-G_2^*(\tau)=N\sum_{a_2=1}^{N-1}\wp_{N,\binom{0}{a_2}}(\tau) \tag{37}$$

is also holomorphic in $\mathscr{H}$, has the expansion

$$E(\tau;N)=\frac{N-1}{3}\pi^2-8\pi^2\sum_{n\geqq 1}\Bigg(\sum_{\substack{d\mid n\\ d>0\\ N\nmid d}} d\Bigg)e^{2\pi i n\tau}$$

and is an entire modular form of dimension -2 for $\Gamma(N)$. Obviously for

$$A=\begin{pmatrix} a & b\\ c & d\end{pmatrix}\in\Gamma_0(N),\qquad \tilde{A}:=\begin{pmatrix} a & Nb\\ \dfrac{c}{N} & d\end{pmatrix}\in\Gamma\,,$$

we have

$$E(A(\tau); N) = N\, G_2^*(\tilde{A}(N\tau)) - G_2^*(A(\tau))$$
$$= (c\tau+d)^2\, G_2^*(N\tau) - (c\tau+d)^2\, G_2^*(\tau) = (c\tau+d)^2\, E(\tau; N).$$

Thus $E(\tau; N)$ is an entire modular form for $\Gamma_0(N)$.

With the notations of **III, § 3**,

$$\int E(\tau; N)\, d\tau = -4\pi i(\psi(N\tau) - \psi(\tau)) + C$$
$$= -4\pi i \log\frac{\eta(N\tau)}{\eta(\tau)} + C = -\frac{\pi i}{6}\log\frac{\Delta(N\tau)}{\Delta(\tau)} + C.$$

is an integral of the 3rd kind for $\Gamma_0(N)$.

§ 4. Division Equation

The content of this section belongs to the realm of ideas of Chapter **VI**, however for its treatment we need the analytic investigations of this chapter. For $N \geqq 2$ we seek to construct the field $K_{\Gamma[N]}$ by means of the Nth division values of $\wp$. Although everything can easily be formulated in terms of inhomogeneous series, in what follows the series $G_{N,k,\boldsymbol{a}}$, $k \geqq 1$, are always to be considered as homogeneous series. In this section we assume that $N \geqq 2$.

1. The distinct $G_{N,k,\boldsymbol{a}}$. For fixed natural N and k we need a maximal system of distinct primitive $G_{N,k,\boldsymbol{a}}$. As an immediate consequence of our earlier results we prove

Theorem 11. *Two primitive series $G_{N,k,\boldsymbol{a}}$ and $G_{N,k,\boldsymbol{a}_1}$ are equal if and only if*

$$\boldsymbol{a} \equiv \pm \boldsymbol{a}_1 \bmod N \quad \text{for } k \equiv 0 \bmod 2, \quad \text{and}$$

$$\boldsymbol{a} \equiv \boldsymbol{a}_1 \bmod N \quad \text{for } k \equiv 1 \bmod 2.$$

The number of distinct primitive $G_{N,k,\boldsymbol{a}}$ is

$$\sigma_\infty(N), \quad \text{if } k \equiv 0 \bmod 2,$$

$$2\sigma_\infty(N), \quad \text{if } k \equiv 1 \bmod 2 \quad \text{and } N > 2, \quad \text{or}$$

$$0, \quad \text{if } k \equiv 1 \bmod 2 \quad \text{and } N = 2.$$

Proof. Let N and k be fixed. Then by **IV** Theorem **8** the number of primitive $G_{N,k,\boldsymbol{a}}$, $\boldsymbol{a} \bmod N$, is $2\sigma_\infty(N)$ for $N > 2$, and is $\sigma_\infty(2)$ if $N = 2$.

By § **1, 2** and § **3, 3** of this chapter, for $k \geqq 2$ the primitive $G_{N,k,\boldsymbol{a}}$ generate a vector space over $\mathbb{C}$ of dimension $\sigma_\infty(N)$. The claim for $k \geqq 2$ then follows from

$$\begin{gathered}\boldsymbol{a} \not\equiv -\boldsymbol{a} \bmod N \quad \text{if } N>2, \qquad \boldsymbol{a} \equiv -\boldsymbol{a} \bmod 2, \quad \text{and} \\ G_{N,k,-\boldsymbol{a}} = (-1) G_{N,k,\boldsymbol{a}} \neq 0\end{gathered} \tag{38}$$

For $k=1$ the $G_{N,1,\boldsymbol{a}}$ generate a vector space over $\mathbb{R}$, the field of real numbers, of dimension $\sigma_\infty(N)$ (cf. Theorem **10**), and the assertion of the theorem again follows from (38). □

We refer to the proof of the lemma in **IV**, § **2**, **1** to show that the condition $(a_1, a_2, N)=1$ for $G_{N,k,\boldsymbol{a}}$, $\boldsymbol{a} = \begin{pmatrix} a_1 \\ a_2 \end{pmatrix}$, can be replaced by $(a_1, a_2)=1$.

2. The action of Γ on the $G_{N,k,\boldsymbol{a}}$. Invariance groups. For fixed N and k let $\mathcal{M}$ denote the set of distinct primitive $G_{N,k,\boldsymbol{a}}$. Then we have

Theorem 12.
a) *Γ acts on $\mathcal{M}$ by permuting its elements,*
b) *Γ is transitive on $\mathcal{M}$.*

Proof. By (4)

$$G_{N,k,\boldsymbol{a}}|_k S = G_{N,k,S'\boldsymbol{a}}, \qquad S \in \Gamma,$$

and if $G_{N,k,\boldsymbol{a}}$ is primitive, so is $G_{N,k,S'\boldsymbol{a}}$. Moreover,

$$S'\boldsymbol{a} \equiv \pm S'\boldsymbol{a}_1 \bmod N \quad \text{if and only if } \boldsymbol{a} \equiv \pm \boldsymbol{a}_1 \bmod N.$$

This implies a).
Since

$$\begin{pmatrix} a_1 \\ a_2 \end{pmatrix} = S' \begin{pmatrix} 1 \\ 0 \end{pmatrix}, \qquad S \in \Gamma,$$

is solvable for every pair of relatively prime numbers a_1, a_2, it follows that

$$\begin{pmatrix} a_1 \\ a_2 \end{pmatrix} = S' \begin{pmatrix} b_1 \\ b_2 \end{pmatrix}, \qquad S \in \Gamma,$$

is also solvable for given relatively prime pairs a_1, a_2 and b_1, b_2. This proves b). □

Theorem 13. *All of the $G_{N,k,\boldsymbol{a}}$ in $\mathcal{M}$ are invariant under $S \in \Gamma$ if and only if*

$$\begin{aligned} &S \in \Gamma(N) \quad \textit{for } k \equiv 1 \bmod 2, \textit{ and} \\ &S \in \Gamma[N] \quad \textit{for } k \equiv 0 \bmod 2. \end{aligned}$$

Proof. We make the special choices $\boldsymbol{a}=\begin{pmatrix}1\\0\end{pmatrix}$ and $\boldsymbol{a}=\begin{pmatrix}0\\1\end{pmatrix}$. Then for these we have

$$S'\begin{pmatrix}1\\0\end{pmatrix}\equiv\pm\begin{pmatrix}1\\0\end{pmatrix}\bmod N \quad \text{if and only if } S\equiv\pm\begin{pmatrix}1&0\\c&1\end{pmatrix}\bmod N, \text{ and}$$

$$S'\begin{pmatrix}0\\1\end{pmatrix}\equiv\pm\begin{pmatrix}0\\1\end{pmatrix}\bmod N \quad \text{if and only if } S\equiv\pm\begin{pmatrix}1&b\\0&1\end{pmatrix}\bmod N$$

with arbitrary b and c. Furthermore, since

$$G_{N,k,-\boldsymbol{a}}=(-1)^k G_{N,k,\boldsymbol{a}},$$

it follows that the intersection of the invariance groups for $G_{N,k,\boldsymbol{a}}$ with $\boldsymbol{a}=\begin{pmatrix}1\\0\end{pmatrix}$ and $\boldsymbol{a}=\begin{pmatrix}0\\1\end{pmatrix}$ is $\Gamma(N)$ for odd k, and is $\Gamma[N]$ for even k. Of course all $G_{N,k,\boldsymbol{a}}$ are invariant under these groups. □

If Γ^* is the invariance group of the Eisenstein series $G_{N,k,\boldsymbol{a}}$ then $A^{-1}\Gamma^*A$ is the invariance group of $G_{N,k,A'\boldsymbol{a}}$, $A\in\Gamma$, because $S'\boldsymbol{a}\equiv\pm\boldsymbol{a}\bmod N$ implies

$$(A^{-1}SA)'A'\boldsymbol{a}=A'S'A'^{-1}A'\boldsymbol{a}\equiv\pm A'\boldsymbol{a}\bmod N$$

and conversely.

3. The division equation. Here this equation takes the place of the transformation equation of Chapter **VI**.

Theorem 14. *The primitive Eisenstein series* $G_{N,k,\boldsymbol{a}}$, $k\neq 2$, *satisfy an algebraic equation*

$$P(x):=\sum_{\nu=0}^{\sigma_k(N)} P_\nu x^\nu=0,$$

where the coefficients P_ν *belong to the ring* $\mathbb{C}[G_4,G_6]$ *of polynomials in the Eisenstein series* G_4 *and* G_6 *of level* 1. *The degree of* P *is*

$$\sigma_k(N)=\begin{cases}\sigma_\infty(N), & \text{if } k\equiv 0\bmod 2, \text{ and}\\ 2\sigma_\infty(N), & \text{if } k\equiv 1\bmod 2.\end{cases}$$

P *is irreducible over* $\mathbb{C}[G_4,G_6]$.

Proof. We set

$$P(x):=\prod_{(\boldsymbol{a})}(x-G_{N,k,\boldsymbol{a}}) \tag{39}$$

and extend the product over a maximal system of such $\boldsymbol{a}$ that lead to pairwise distinct primitive $G_{N,k,\boldsymbol{a}}$. Here, and in what follows, this is what the parentheses under a product symbol are to signify. Obviously, $P(G_{N,k,\boldsymbol{a}})=0$ for all primitive $G_{N,k,\boldsymbol{a}}$. The degree of the polynomial P

is $\sigma_\infty(N)$ for even k and is $2\sigma_\infty(N)$ for odd k. By multiplying the factors of the product one obtains a polynomial whose coefficients, as symmetric functions of the distinct $G_{N,k\boldsymbol{a}}$, are entire modular forms for Γ by Theorem **12**, and thus belong to the polynomial ring $\mathbb{C}[G_4, G_6]$.

The irreducibility of $P(x)$ over $\mathbb{C}[G_4, G_6]$ follows because Γ acts transitively on the simple roots $G_{N,k,\boldsymbol{a}}$ of $P(x)$. □

In the case $k=2$ the $G_{N,2,\boldsymbol{a}}$ are not holomorphic, so we consider the division values of $\wp$:

$$\wp_{N,\boldsymbol{a}} = N^2(G_{N,2,\boldsymbol{a}} - G_{N,2,\boldsymbol{0}}). \tag{40}$$

Analogous to Theorem **14** we have

Theorem 15. *The primitive $\wp$-division values $\wp_{N,\boldsymbol{a}}$ satisfy an algebraic equation*

$$P(x) := \sum_{\nu=0}^{\sigma(N)} P_\nu x^\nu = 0$$

with coefficients P_ν in $\mathbb{C}[G_4, G_6]$. P is irreducible over the ring $\mathbb{C}[G_4, G_6]$.

Proof. By (40) the $\wp_{N,\boldsymbol{a}}$ run through a system of pairwise distinct functions if the $G_{N,2,\boldsymbol{a}}$ do. None of them is identically zero, and

$$\wp_{N,\boldsymbol{a}}|_2 S = \wp_{N,S'\boldsymbol{a}}.$$

The conclusion is drawn as in the proof of Theorem **14**. □

The equation

$$P(x) := \prod_{(\boldsymbol{a})} (x - \wp_{N,\boldsymbol{a}}) = 0 \tag{41}$$

is called the *division equation.*

4. Generation of the field $K_{\Gamma[N]}$ of modular functions for the principal congruence group $\Gamma[N]$. The functions $(G_4/G_6)\wp_{N,\boldsymbol{a}}$ are modular functions of level N by Theorem **7**. We prove

Theorem 16. *The $\sigma(N)$ distinct modular functions*

$$\frac{G_4}{G_6}\wp_{N,\boldsymbol{a}}, \quad \textit{for primitive } \wp_{N,\boldsymbol{a}}, \tag{42}$$

satisfy an algebraic equation

$$P(x) := \sum_{\nu=0}^{\sigma(N)} R_\nu x^\nu = 0$$

with coefficients R_ν in $\mathbb{C}(j) = K_\Gamma$ that is irreducible over $\mathbb{C}(j)$. The field $K_{\Gamma[N]}$ is obtained by adjoining the functions (42) *to $\mathbb{C}(j)$.*

Proof. Recalling the proof of Theorem **14**, one easily sees that the required polynomial is of the form

$$P(x)=\prod_{(a)}\left(x-\frac{G_4}{G_6}\wp_{N,a}\right). \tag{43}$$

Obviously $\Gamma[N]$ is the invariance group of the set of functions $\frac{G_4}{G_6}\wp_{N,a}$ (cf. Theorem **3**). This implies that $K_{\Gamma[N]}$ is the splitting field of the polynomial $P(x)$ over $\mathbb{C}(j)$, which proves the second part of the theorem. □

Theorem **16** remains valid if in (42) one replaces the functions $(G_4/G_6)\wp_{N,a}$ by the functions $(G_4 G_6/\Delta)\wp_{N,a}$; if $k>2$ is even, by the functions

$$\frac{G_4^\alpha G_6^\beta}{\Delta^{k_0}}\cdot G_{N,k,a}$$

with non-negative α,β satisfying

$$k+4\alpha+6\beta\equiv 0 \bmod 12 \text{ and } k_0=\frac{k+4\alpha+6\beta}{12};$$

and if $k>0$ is odd, by the functions

$$\frac{G_4^\alpha G_6^\beta}{\Delta^{k_0}}\cdot G_{N,k,a}^2$$

with non-negative α,β satisfying

$$2k+4\alpha+6\beta\equiv 0 \bmod 12 \text{ and } k_0=\frac{2k+4\alpha+6\beta}{12}.$$

In this connection we note that

$$G_4=c\sum_{(a)}\wp_{N,a}^2,\qquad G_6=c'\sum_{(a)}\wp_{N,a}'^2, \tag{44}$$

where $\wp'_{N,a}$ is the Nth division value of the derivative of the $\wp$-function (cf. (6)). This follows because the right hand sides of equations (44) are entire modular forms for Γ of dimension -4 and -6, respectively, which do not vanish identically since the constant terms in the expansion (5) and in those of § **3**, **2** are real and not all zero. We conclude with the determination of the discriminant of the division equation.

Theorem 17. *The discriminant of the division equation is*

$$\prod_{\substack{(a),(a_1)\\ a\not\equiv\pm a_1 \bmod N}}(\wp_{N,a}-\wp_{N,a_1})=C\Delta^{k_0}$$

with a constant $C\neq 0$ and suitable natural k_0.

Proof. Obviously, the product is an entire modular form for Γ. It has no zeros in $\mathscr{H}$ since, as is known, the values of the Weierstrass $\wp$-function at z_1 and z_2 are only equal if

$$z_1\equiv\pm z_2 \bmod \mathscr{G},$$

where $\mathscr{G}$ is the period lattice of the $\wp$-function. Thus the product has integral total zero order, and consequently its dimension $-k=-12k_0$, $k_0 \in \mathbb{N}$. Therefore it follows that the product is $c\Delta^{k_0}$, where $c \neq 0$. Clearly $k_0 = \sigma(N)(\sigma(N)-1)\frac{1}{6}$. □

Chapter VIII. The Integrals of $\wp$-Division Values

In this chapter we determine the behavior of the integrals of $\wp$-division values under arbitrary modular substitutions. We make a transition from Eisenstein series of dimension -2 and fixed level N to a linearly equivalent system. This system contains the non-analytic function G_2 introduced in **III, § 2** that leads to the Dedekind η-function. In addition, this system contains functions holomorphic in $\mathscr{H}$ whose integrals turn out to be logarithms of the generalized Dedekind η-function. We will study the behavior of these integrals as well as the behavior of the function η under the transformation $T=\begin{pmatrix} 0 & -1 \\ 1 & 0 \end{pmatrix}$ by two methods. The first method began with B. Riemann and R. Dedekind (see B. Riemann [1] pp. 438—447), the other method with C. L. Siegel ([1] V. 3 p. 188). We then investigate the behavior of our functions under arbitrary modular transformations applying an often used method due to E. Hecke [2], p. 224 or [3], p. 442. In the course of study we encounter the so-called Dedekind sums.

For the functions under investigation the notation used in this chapter will omit the explicit dependence on the level N.

§ 1. The Space of $\wp$-Division Values. Integrals

1. Transition to a new basis. We introduce a new system of generators for the space of Eisenstein series of dimension -2 and level $N \geqq 1$ by

$$\phi_{\boldsymbol{g}} := \sum_{\boldsymbol{a} \bmod N} \zeta_N^{a_1 h - a_2 g} G_{N,2,\boldsymbol{a}}, \qquad \boldsymbol{g} \bmod N, \tag{1}$$

where $\boldsymbol{g} = \begin{pmatrix} g \\ h \end{pmatrix}$, $\boldsymbol{a} = \begin{pmatrix} a_1 \\ a_2 \end{pmatrix}$ and $\zeta_N := e^{2\pi i/N}$. Inversely we have

$$G_{N,2,\boldsymbol{a}} = N^{-2} \sum_{\boldsymbol{g} \bmod N} \zeta_N^{g a_2 - h a_1} \phi_{\boldsymbol{g}}.$$

For $\boldsymbol{g} \not\equiv \boldsymbol{0} \bmod N$, $N \geqq 2$, the series $\phi_{\boldsymbol{g}}$ represent functions holomorphic in $\mathscr{H}$ since

$$\phi_{\boldsymbol{g}}(\tau) = \sum_{\boldsymbol{a} \bmod N} \zeta_N^{a_1 h - a_2 g}(G_{N,2,\boldsymbol{a}}(\tau) - G_{N,2,\boldsymbol{0}}(\tau))$$
$$= N^{-2} \sum_{\boldsymbol{a} \bmod N}' \zeta_N^{a_1 h - a_2 g} \wp_{N,\boldsymbol{a}}(\tau).$$

On the other hand $\phi_{\boldsymbol{0}}$ is the non-analytic function of **III, § 2**, (25) with

$$\phi_{\boldsymbol{0}}(\tau) = G_2(\tau) = \frac{\pi^2}{3} - 8\pi^2 \sum_{mk \geqq 1} m e^{2\pi i \tau m k} - \frac{2\pi i}{\tau - \bar{\tau}}. \tag{2}$$

If $\boldsymbol{g} \not\equiv \boldsymbol{0} \bmod N$ we obtain the following expansion

$$\phi_{\boldsymbol{g}}(\tau) = \sum_{\nu \geqq 0} \left\{ \sum_{\boldsymbol{a} \bmod N} \zeta_N^{a_1 h - a_2 g} \alpha_\nu(N, \boldsymbol{a}) \right\} e^{\frac{2\pi i \tau}{N} \nu}$$

by replacing $G_{N,2,\boldsymbol{a}}$ by the series given in **VII, § 2**, (25). Thus

$$\phi_{\boldsymbol{g}}(\tau) = \sum_{\nu \geqq 0} \beta_\nu(\boldsymbol{g}) e^{\frac{2\pi i \tau}{N} \cdot \nu},$$

where

$$\beta_0(\boldsymbol{g}) = \sum_{\boldsymbol{a} \bmod N} \zeta_N^{a_1 h - a_2 g} \delta\left(\frac{a_1}{N}\right) \sum_{m_2 \equiv a_2 \bmod N}' m_2^{-2}$$
$$= \sum_{m=-\infty}^{\infty}{}' m^{-2} \zeta_N^{-mg} = 2 \sum_{m \geqq 1} m^{-2} \cos\left(\frac{2\pi}{N} m g\right).$$

As is well-known and easy to verify,

$$\frac{1}{\pi^2} \sum_{m \geqq 1} m^{-2} \cos(2\pi m x), \qquad x \in \mathbb{R},$$

is the Fourier expansion of the *second Bernoulli polynomial*

$$P_2(x) := (x - [x])^2 - (x - [x]) + \tfrac{1}{6}, \qquad x \in \mathbb{R}.$$

Thus

$$\beta_0(\boldsymbol{g}) = 2\pi^2 P_2\left(\frac{g}{N}\right).$$

Moreover, for $\nu \geqq 1$,

$$\beta_\nu(\boldsymbol{g}) = -\frac{4\pi^2}{N^2} \sum_{\boldsymbol{a} \bmod N} \zeta_N^{a_1 h - a_2 g} \sum_{\substack{m \mid \nu \\ \frac{\nu}{m} \equiv a_1 \bmod N}} |m| \zeta_N^{a_2 m},$$
$$= -\frac{4\pi^2}{N} \sum_{a_1 \bmod N} \sum_{\substack{m \mid \nu \\ \frac{\nu}{m} \equiv a_1 \bmod N}} |m| \zeta_N^{a_1 h} \delta\left(\frac{m-g}{N}\right),$$
$$= -\frac{4\pi^2}{N} \sum_{\substack{m \mid \nu \\ \frac{\nu}{m} \equiv g \bmod N}} \left|\frac{\nu}{N}\right| \zeta_N^{mh}.$$

For the series ϕ_g, $g \not\equiv \mathbf{0} \bmod N$, we have found the expansion

$$\begin{aligned}
\phi_g(\tau) &= \sum_{\nu \geqq 0} \beta_\nu(g)\, e^{\frac{2\pi i \tau}{N}\nu}, \quad \text{with} \\
\beta_0(g) &= 2\pi^2 P_2\left(\frac{g}{N}\right), \quad \text{and} \\
\beta_\nu(g) &= -\frac{4\pi^2}{N} \sum_{\substack{m \mid \nu \\ \frac{\nu}{m} \equiv g \bmod N}} \left|\frac{\nu}{m}\right| \zeta_N^{mh}, \qquad \nu \geqq 1 .
\end{aligned} \tag{3}$$

2. The integrals of the ℘-division values. For all $g \bmod N$ we define

$$\psi_g(\tau) := -\frac{1}{2\pi i} \int \left\{\phi_g(\tau) + \frac{2\pi i}{\tau - \bar{\tau}}\,\delta_g\right\} d\tau \tag{4}$$

with

$$\delta_g = \begin{cases} 1 & \text{if } g \equiv \mathbf{0} \bmod N, \quad \text{and} \\ 0 & \text{otherwise}, \end{cases}$$

where in view of the transformation formulas (23) and (24) the constant of integration will be fixed by setting the constant term in the series expansion of ψ_g equal to

$$\gamma_0(g) := \begin{cases} \pi i P_1\left(\frac{h}{N}\right) - \sum_{m \geqq 1} m^{-1} \zeta_N^{-hm}, & \text{if } g \equiv \mathbf{0} \text{ and } h \not\equiv \mathbf{0} \bmod N, \text{ and} \\ 0 & \text{otherwise}. \end{cases} \tag{5}$$

In the formula above,

$$P_1(x) := x - [x] - \tfrac{1}{2}, \qquad x \in \mathbb{R},$$

is the *first Bernoulli polynomial.*

From the definition (4) for $\psi_g(\tau)$ it follows that

$$\psi_g(\tau) = \gamma_0(g) + \pi i P_2\left(\frac{g}{N}\right)\tau + \frac{N}{4\pi^2} \sum_{\nu \geqq 1} \frac{\beta_\nu(g)}{\nu} e^{\frac{2\pi i \tau}{N}\cdot\nu} \tag{6}$$

for all $g \bmod N$ if the coefficient $\beta_\nu(\mathbf{0})$ is also defined by (3). This is consistent with equation (2). If $g \not\equiv \mathbf{0} \bmod N$, then by **V**, Theorem **11**, ψ_g is an integral of the 3rd kind for $\Gamma[N]$.

We rewrite the series expansion (6) in the following way:

$$\begin{aligned}\psi_{\boldsymbol{g}}(\tau) &= \gamma_0(\boldsymbol{g}) + \pi i P_2\left(\frac{g}{N}\right)\tau - \sum_{\substack{m, m_1 > 0 \\ m_1 \equiv g \bmod N}} |m|^{-1} \zeta_N^{mh} e^{\frac{2\pi i \tau}{N} m m_1} \\ &= \gamma_0(\boldsymbol{g}) + \pi i P_2\left(\frac{g}{N}\right)\tau - \sum_{\substack{m, m_1 > 0 \\ m_1 \equiv g \bmod N}} m^{-1} \left[\zeta_N^h e^{\frac{2\pi i \tau}{N} \cdot m_1}\right]^m \\ &\quad - \sum_{\substack{m, m_1 > 0 \\ m_1 \equiv -g \bmod N}} m^{-1} \left[\zeta_N^{-h} e^{\frac{2\pi i \tau}{N} m_1}\right]^m .\end{aligned} \tag{7}$$

3. The Dedekind functions. The functions

$$\eta_{\boldsymbol{g}}(\tau) := e^{\psi_{\boldsymbol{g}}(\tau)}, \qquad \boldsymbol{g} \bmod N,$$

have the product expansions

$$\eta_{\boldsymbol{g}}(\tau) = \alpha_0(\boldsymbol{g}) e^{\pi i P_2\left(\frac{g}{N}\right)\tau} \prod_{\substack{m > 0 \\ m \equiv g \bmod N}} \left(1 - \zeta_N^h e^{\frac{2\pi i \tau}{N} m}\right) \prod_{\substack{m > 0 \\ m \equiv -g \bmod N}} \left(1 - \zeta_N^{-h} e^{\frac{2\pi i \tau}{N} m}\right) \tag{8}$$

with

$$\alpha_0(\boldsymbol{g}) := \begin{Bmatrix} (1 - \zeta_N^{-h}) e^{\pi i P_1\left(\frac{h}{N}\right)} & \text{if } \boldsymbol{g} \equiv \boldsymbol{0} \text{ and } \boldsymbol{h} \not\equiv \boldsymbol{0} \bmod N, \\ 1 & \text{otherwise} \end{Bmatrix} = e^{\gamma_0(\boldsymbol{g})}.$$

The expansion (8) is a trivial consequence of (7). Since

$$\eta_{\boldsymbol{0}}(\tau) = \eta^2(\tau),$$

the functions $\eta_{\boldsymbol{g}}$ are generalizations of the square of Dedekind's η-function:

$$\eta(\tau) = e^{\frac{\pi i \tau}{12}} \prod_{m \geqq 1} (1 - e^{2\pi i \tau m}) \qquad (\text{cf. } \mathbf{III}, \S\, \mathbf{3}).$$

We also call them Dedekind functions. Naturally,

$$\psi_{\boldsymbol{g}}(\tau) = \log \eta_{\boldsymbol{g}}(\tau), \qquad \boldsymbol{g} \bmod N, \tag{9}$$

for a suitable branch of the logarithm.

From

$$\phi_{\boldsymbol{0}}(\tau) + \frac{2\pi i}{\tau - \bar{\tau}} = G_2^*(\tau)$$

and the transformation formula for G_2^* (cf. **III**, Theorem **7**) it follows that for $A = \begin{pmatrix} a & b \\ c & d \end{pmatrix} \in \Gamma$,

$$\psi_{\boldsymbol{0}}(A(\tau)) - \psi_{\boldsymbol{0}}(\tau) - \log(c\tau + d) =: \pi_{\boldsymbol{0}}(A) \tag{10}$$

with a constant $\pi_{\boldsymbol{0}}(A)$ depending upon the choice of the logarithm.

By applying

$$\phi_{\boldsymbol{g}}(A(\tau))=(c\tau+d)^2\,\phi_{A'\boldsymbol{g}}(\tau)$$

to the definition of $\psi_{\boldsymbol{g}}(\tau)$ for $\boldsymbol{g}\not\equiv\boldsymbol{0} \bmod N$ we obtain the relation

$$\psi_{\boldsymbol{g}}(A\tau)-\psi_{A'\boldsymbol{g}}(\tau)=:\pi_{\boldsymbol{g}}(A)\,. \tag{11}$$

The constants $\pi_{\boldsymbol{g}}(A)$, $A\in\Gamma[N]$, are periods of the integral $\psi_{\boldsymbol{g}}$. The next section is devoted to the computation of the values $\pi_{\boldsymbol{g}}(A)$, $A\in\Gamma$.

§ 2. An Asymptotic Formula and the Behavior of the Integrals under the Transformation T

In this section we choose a systematic way to evaluate $\pi_{\boldsymbol{g}}(T)$, $T=\begin{pmatrix}0 & -1\\ 1 & 0\end{pmatrix}$. In the next section we give a shorter and more elegant derivation.

1. An asymptotic expansion. As usual let $\zeta(s,\alpha)$ denote the Hurwitz zeta function defined by

$$\zeta(s,\alpha)=\sum_{n>-\alpha}\frac{1}{(n+\alpha)^s}\quad\text{for }\operatorname{Re}(s)>1,\qquad \alpha\in\mathbb{R}\,,$$

which can be continued to the entire s-plane (cf. **VII, § 2**). For an arbitrary root of unity ξ, let $Z(s;\xi)$ denote the analytic continuation to the entire s-plane of the function defined by

$$Z(s;\xi):=\sum_{n>0}\frac{\xi^n}{n^s}\quad\text{for }\operatorname{Re}(s)>1\,.$$

The function U, defined for $x>0$ by the series

$$U(x;\alpha,\xi):=\sum_{\substack{n>0\\ m>-\alpha}}\frac{\xi^n}{n}e^{-(m+\alpha)nx},\qquad \alpha\in\mathbb{Q}\,, \tag{12}$$

has the asymptotic expansions

$$U(x;\alpha,\xi)=\begin{cases}\dfrac{1}{x}Z(2;\xi)+Z(1;\xi)\zeta(0,\alpha)+\mathrm{o}(x) & \text{for }\xi\neq1,\\[2ex] \dfrac{1}{x}Z(2;1)+\log\dfrac{\Gamma(\alpha^*)}{\sqrt{2\pi}}-\zeta(0,\alpha)\log x+\mathrm{o}(x) & \text{for }\xi=1\end{cases} \tag{13}$$

as $x>0$ tends to 0, where

$$\alpha^*:=\alpha-[\alpha]+\delta(\alpha),\quad\text{and}$$

$$\delta(\alpha)=\begin{cases}1 & \text{for }\alpha\in\mathbb{Z},\\ 0 & \text{for }\alpha\notin\mathbb{Z}.\end{cases}$$

Here and in what follows, o denotes a function f of $x>0$ such that $\lim_{x\to 0+} f(x)=0$. For positive values we always choose the real logarithm.

Proof of the asymptotic formula. Let $s=\sigma+it$. E.T. Whittaker and G.N. Watson [1] give the following estimates:

(a) $\Gamma(s)=\mathrm{O}\left(e^{-\frac{\pi}{2}|t|}|t|^{\sigma-\frac{1}{2}}\right)$, (p. 279), and

(b) $\zeta(s,\alpha)=\mathrm{O}(|t|^C)$ for suitable $C>0$, (p. 276),

which are uniform in every finite interval $\sigma_1\leqq\sigma\leqq\sigma_2$ as $|t|\to\infty$. Clearly, the last approximation also holds for $Z(s;\xi)$.

First we apply the Mellin inversion formula (cf. E.C. Titchmarsh [2])

$$e^{-z}=\frac{1}{2\pi i}\int_{\delta-i\infty}^{\delta+i\infty} z^{-s}\Gamma(s)\,ds,\qquad \operatorname{Re}z>0,\qquad \delta>0,$$

to $U(x;\alpha,\xi)$. The integral is to be taken over the path $s=\delta+it$, $-\infty<t<\infty$, with fixed δ; this will also be the case in the analogous situations below. Now we have

$$U(x,\alpha,\xi)=\frac{1}{2\pi i}\sum_{n>0}\frac{\xi^n}{n}\sum_{m>-\alpha}\int_{\delta-i\infty}^{\delta+i\infty}\frac{\Gamma(s)}{((m+\alpha)nx)^s}\,ds,\quad \text{where } \delta>1.$$

For $\sigma=\delta>1$, the series $\sum_{n>0}\frac{\xi^n}{n^{s+1}}$ and $\sum_{m>-\alpha}(m+\alpha)^{-s}$ are absolutely and uniformly convergent. In addition, the series are also bounded. Hence by (*a*), summation and integration can be interchanged and we have

$$\begin{aligned} U(x;\alpha,\xi)&=\frac{1}{2\pi i}\int_{\delta-i\infty}^{\delta+i\infty}\sum_{n>0}\frac{\xi^n}{n^{s+1}}\cdot\frac{\Gamma(s)}{x^s}\cdot\zeta(s,\alpha)\,ds\\ &=\frac{1}{2\pi i}\int_{\delta-i\infty}^{\delta+i\infty}x^{-s}\Gamma(s)Z(s+1;\xi)\zeta(s,\alpha)\,ds.\end{aligned}$$

Because of (*a*) and (*b*) this integral equals

$$\begin{aligned} U(x;\alpha,\xi)=&\frac{1}{2\pi i}\int_{\eta-i\infty}^{\eta+i\infty}x^{-s}\Gamma(s)Z(s+1;\xi)\zeta(s,\alpha)\,ds\\ &+\operatorname*{Res}_{s=0,1}\{x^{-s}\Gamma(s)Z(s+1;\xi)\zeta(s,\alpha)\},\end{aligned}$$

where η satisfies the condition $-1<\eta<0$. Since $\eta<0$, (a) and (b) imply that

$$\int_{\eta-i\infty}^{\eta+i\infty} x^{-s}\Gamma(s)Z(s+1;\xi)\zeta(s,\alpha)\,ds=o(x).$$

Without difficulty we obtain

$$\begin{aligned}&\operatorname*{Res}_{s=1}\{x^{-s}\Gamma(s)Z(s+1,\xi)\zeta(s,\alpha)\}=\frac{1}{x}Z(2,\xi) \qquad \text{for all } \xi, \text{ and}\\ &\operatorname*{Res}_{s=0}\{x^{-s}\Gamma(s)Z(s+1,\xi)\zeta(s,\alpha)\}=Z(1;\xi)\zeta(0,\alpha) \quad \text{if } \xi\neq 1.\end{aligned} \tag{14}$$

To calculate the residue at $s=0$ when $\xi=1$, following E.T. Whittaker and G.N. Watson [1], we consider the Laurent expansions

$$\Gamma(s)=\frac{1}{s}-\gamma+\cdots \qquad \text{(p. 236)},$$

$$Z(s+1,1)=\zeta(s+1)=\frac{1}{s}+\gamma+\cdots \qquad \text{(p. 271)},$$

$$\zeta(s,\alpha)=\zeta(0,\alpha)+\log\left(\frac{\Gamma(\alpha^*)}{\sqrt{2\pi}}\right)s+\cdots \qquad \text{(p. 271)} \quad \text{and}$$

$$x^{-s}=1-s\log x+\cdots.$$

Here γ denotes Euler's constant. From these expansions we find that

$$\operatorname*{Res}_{s=0}\{x^{-s}\Gamma(s)Z(s+1;1)\zeta(s,\alpha)\}=\frac{\log\Gamma(\alpha^*)}{\sqrt{2\pi}}-\zeta(0,\alpha)\log x. \tag{15}$$

This completes the proof of the asymptotic expansion. □

We note that

$$\zeta(0,\alpha)=\tfrac{1}{2}-\alpha^* \qquad \text{(p. 271)}.$$

2. Calculation of $\pi_g(T)$. In order to determine the value of

$$\pi_g(T)=\psi_g(T(\tau))-\psi_{T'g}(\tau)-\delta_g\log\tau, \tag{16}$$

which by the way is independent of τ, we set $\tau=\frac{i}{x}$, $x>0$, and apply the asymptotic formula of the previous subsection to the calculation of the limit of the right hand side of (16) as $x\to 0+$. We always choose the principal value of the logarithm. Then $\pi_g(T)$ is uniquely determined (cf. (10)).

In **§ 1**, (7) we found the representation

$$\psi_{g}(\tau)=\gamma_0(g)+\pi i P_2\left(\frac{g}{N}\right)\tau-\chi_{g}(\tau)-\chi_{-g}(\tau), \tag{17}$$

where

$$\chi_{g}(\tau):=\sum_{\substack{m,m_1>0\\ m_1\equiv g \bmod N}} m^{-1}\left(\zeta_N^h e^{\frac{2\pi i\tau}{N}\cdot m_1}\right)^m.$$

If we set $\tau=\frac{i}{x}$ in (16) and let $x\to 0+$, we obtain

$$\begin{aligned}
\psi_{g}\left(-\frac{1}{\tau}\right)&=\gamma_0(g)-\chi_{g}(ix)-\chi_{-g}(ix)+o(x),\\
\psi_{T'g}(\tau)&=\gamma_0(T'g)-\pi P_2\left(\frac{h}{N}\right)\cdot\frac{1}{x}+o(x),\\
\log\tau&=\frac{\pi i}{2}-\log x
\end{aligned} \tag{18}$$

for the terms on the right in (16). Further, if $\tau=\frac{i}{x}$, then

$$\chi_{g}\left(-\frac{1}{\tau}\right)=\sum_{\substack{m>0\\ \mu>-g/N}}\frac{\zeta_N^{hm}}{m}e^{-2\pi x\left(\mu+\frac{g}{N}\right)m}.$$

Our asymptotic formula (13) now implies that

$$\chi_{g}\left(-\frac{1}{\tau}\right)=\begin{cases}\dfrac{1}{2\pi x}Z(2;\zeta_N^h)+Z(1;\zeta_N^h)\zeta\left(0,\dfrac{g}{N}\right)+o(x)\\ \text{for } h\not\equiv 0 \bmod N, \quad \text{and}\\ \dfrac{1}{2\pi x}Z(2;1)+\log\dfrac{\Gamma\left(\left(\frac{g}{N}\right)^*\right)}{\sqrt{2\pi}}-\zeta\left(0,\dfrac{g}{N}\right)\log 2\pi x+o(x)\\ \text{for } h\equiv 0 \bmod N.\end{cases} \tag{19}$$

An analogous expression holds for $\chi_{-g}\left(-\frac{1}{\tau}\right)$. Since $\pi_g(T)$ is a constant, the terms $o(x)$, $C\cdot\frac{1}{x}$ and $C\log x$ must cancel in the asymptotic expansion of the right hand side of (16):

$$\pi_g(T)=\psi_g\left(-\frac{1}{\tau}\right)-\psi_{T'g}(\tau)-\delta_g\log\tau.$$

Therefore in the expression

$$\chi_g\left(-\frac{1}{\tau}\right)+\chi_{-g}\left(-\frac{1}{\tau}\right)$$

we consider only the terms that do not contain such factors.

We first treat the case $h \not\equiv 0 \bmod N$. These terms yield the sum

$$X_g := Z(1;\zeta_N^h)\,\zeta\left(0,\frac{g}{N}\right)+Z(1;\zeta_N^{-h})\,\zeta\left(0,\frac{-g}{N}\right), \qquad (h \not\equiv 0 \bmod N). \tag{20}$$

Since

$$\zeta(0,\alpha)=\tfrac{1}{2}-\alpha^* = -P_1(\alpha)-\delta(\alpha), \qquad \zeta(0,-\alpha)=P_1(\alpha),$$

we derive upon substitution that

$$\begin{aligned} X_g &= -\left\{P_1\left(\frac{g}{N}\right)+\delta\left(\frac{g}{N}\right)\right\} Z(1,\zeta_N^h)+P_1\left(\frac{g}{N}\right)Z(1;\zeta_N^{-h}) \\ &= -P_1\left(\frac{g}{N}\right)\{Z(1;\zeta_N^h)-Z(1,\zeta_N^{-h})\}-\delta\left(\frac{g}{N}\right)Z(1;\zeta_N^h). \end{aligned}$$

So if $h \not\equiv 0 \bmod N$, then

$$\begin{aligned} Z(1;\zeta_N^h)-Z(1;\zeta_N^{-h}) &= \sum_{m\geqq 1}\frac{\zeta_N^{mh}-\zeta_N^{-mh}}{m} = \sum_{m\geqq 1} m^{-1}\sin\left(\frac{2\pi}{N}mh\right) \\ &= -2\pi i P_1\left(\frac{h}{N}\right) \end{aligned}$$

because, as is known,

$$-\frac{1}{\pi}\sum_{m\geqq 1} m^{-1}\sin(2\pi m x), \qquad x\in\mathbb{R},$$

is the Fourier expansion of

$$((x)) := x-[x]-\frac{1}{2}+\frac{\delta(x)}{2}=P_1(x)+\frac{\delta(x)}{2}.$$

Thus

$$X_g = 2\pi i P_1\left(\frac{g}{N}\right)P_1\left(\frac{h}{N}\right)-\delta\left(\frac{g}{N}\right)Z(1;\zeta_N^h) \quad \text{for } h \not\equiv 0 \bmod N. \tag{20_1}$$

Analogously, for the case $h \equiv 0 \bmod N$ we find that

$$\begin{aligned} X_g &:= \log\frac{\Gamma\left(\left(\frac{g}{N}\right)^*\right)}{\sqrt{2\pi}}+\log\frac{\Gamma\left(\left(-\frac{g}{N}\right)^*\right)}{\sqrt{2\pi}} \\ &\quad -\left\{\zeta\left(0,\frac{g}{N}\right)+\zeta\left(0,-\frac{g}{N}\right)\right\}\log(2\pi). \end{aligned}$$

Now if $g \not\equiv 0 \bmod N$, then by the display following (20)

$$X_g = \log \frac{\Gamma\left(\left(\frac{g}{N}\right)^*\right)\Gamma\left(1-\left(\frac{g}{N}\right)^*\right)}{2\pi} = -\log\left\{2\sin\left(\pi\left(\frac{g}{N}\right)^*\right)\right\}.$$

It follows from

$$(\sin\pi x)^2 = \tfrac{1}{4}(1-e^{2\pi i x})(1-e^{-2\pi i x})$$

that

$$\log\left\{2\sin\left(\pi\left(\frac{g}{N}\right)^*\right)\right\} = -\frac{1}{2}\{Z(1;\zeta_N^g)+Z(1;\zeta_N^{-g})\},$$

and thus that

$$X_g = \pi i P_1\left(-\frac{g}{N}\right) - Z(1;\zeta_N^g) \quad \text{if } h\equiv 0,\ g\not\equiv 0 \bmod N. \tag{20_2}$$

Finally,

$$X_{\boldsymbol{g}} = 0 \quad \text{for } \boldsymbol{g}\equiv \boldsymbol{0} \bmod N. \tag{20_3}$$

We have now completed all the necessary calculations for the determination of $\pi_{\boldsymbol{g}}(T)$. Combining these calculations we obtain the result

$$\pi_{\boldsymbol{g}}(T) = \gamma_0(\boldsymbol{g}) - \gamma_0(T'\boldsymbol{g}) - X_{\boldsymbol{g}} - \tfrac{1}{2}\pi i \delta_{\boldsymbol{g}}, \tag{21}$$

where $\gamma_0(\boldsymbol{g})$ and $\gamma_0(T'\boldsymbol{g})$ are given in (5), and where $X_{\boldsymbol{g}}$ is determined in $(20_{1,2,3})$. Using the function

$$((x)) = P_1(x) + \frac{\delta(x)}{2}$$

introduced above, we get

$$\pi_{\boldsymbol{g}}(T) = -2\pi i\left(\left(\frac{g}{N}\right)\right)\left(\left(\frac{h}{N}\right)\right) - \frac{1}{2}\pi i \delta_{\boldsymbol{g}}, \quad \boldsymbol{g} = \begin{pmatrix} g \\ h \end{pmatrix}. \tag{22}$$

One verifies the equality of the right hand sides of (21) and (22) by considering separately the four cases:

$$g\not\equiv 0, h\not\equiv 0; \quad g\equiv 0, h\not\equiv 0; \quad g\not\equiv 0, h\equiv 0; \quad g\equiv 0, h\equiv 0 \bmod N.$$

Equation (16) now takes the form

$$\psi_{\boldsymbol{g}}(T\tau) - \psi_{T'\boldsymbol{g}}(\tau) - \left(\log\tau - \frac{\pi i}{2}\right)\delta_{\boldsymbol{g}} = -2\pi i\left(\left(\frac{g}{N}\right)\right)\left(\left(\frac{h}{N}\right)\right). \tag{23}$$

§ 3. A Second Look at the Behavior of the Integrals under the Transformation T. The General Transformation Formula

1. A proof of the transformation formula (23) that uses the residue theorem. In what follows we use the abbreviations

$$q_N := e^{\frac{2\pi i \tau}{N}}, \quad q'_N := e^{-\frac{2\pi i}{N\tau}}.$$

With this notation equation (7) implies that

$$\psi_{\boldsymbol{g}}(\tau+1) = \gamma_0(\boldsymbol{g}) + \pi i P_2\left(\frac{g}{N}\right)(\tau+1) - \sum_{\substack{m m_1 > 0 \\ m_1 \equiv g \bmod N}} |m|^{-1} (\zeta_N^{h+g} q_N^{m_1})^m .$$

From equation (5) we see that

$$\gamma_0(\boldsymbol{g}) = \gamma_0(U'\boldsymbol{g})$$

and that hence we may also write

$$\psi_{\boldsymbol{g}}(U\tau) - \psi_{U'\boldsymbol{g}}(\tau) = \pi i P_2\left(\frac{g}{N}\right) = \pi_{\boldsymbol{g}}(U). \tag{24}$$

Now we compute $\pi_{\boldsymbol{g}}(T)$ by the second method mentioned above which does not use the asymptotic formulas of **§ 2**. For this we choose $0 < g \leqq N$. Then by (7) we find that

$$\begin{aligned}\psi_{\boldsymbol{g}}(\tau) &= \gamma_0(\boldsymbol{g}) + \pi i P_2\left(\frac{g}{N}\right)\tau - \sum_{\substack{m, m_1 > 0 \\ m_1 \equiv g \bmod N}} \frac{(\zeta_N^h q_N^{m_1})^m}{m} - \sum_{\substack{m, m_1 > 0 \\ m_1 \equiv -g \bmod N}} \frac{(\zeta_N^{-h} q_N^{m_1})^m}{m} \\ &= \gamma_0(\boldsymbol{g}) + \pi i P_2\left(\frac{g}{N}\right)\tau - \sum_{m>0} \frac{(\zeta_N^h q_N^g)^m}{m(1-q_1^m)} - \sum_{m<0} \frac{\zeta_N^{hm}}{m}\left\{\frac{q_N^{gm}}{1-q_1^m} + \delta\left(\frac{g}{N}\right)\right\}.\end{aligned}$$

Cf. B. Schoeneberg [2].

Replacing $\gamma_0(\boldsymbol{g})$ by its value in (5) we obtain

$$\begin{aligned}\psi_{\boldsymbol{g}}(\tau) = \pi i P_1\left(\frac{h}{N}\right)\alpha_{\boldsymbol{g}} + \pi i P_2\left(\frac{g}{N}\right)\tau - \sum_{m>0} \frac{1}{m} \frac{(\zeta_N^h q_N^g)^m}{1-q_1^m} \\ - \sum_{m<0} \frac{1}{m}\left\{\frac{(\zeta_N^h q_N^g)^m}{1-q_1^m} - \delta_{\boldsymbol{g}}\right\}\end{aligned} \tag{25}$$

with

$$\alpha_{\boldsymbol{g}} = \begin{cases} 1, & \text{if } g \equiv 0,\ h \not\equiv 0 \bmod N, \text{ and} \\ 0, & \text{otherwise.} \end{cases}$$

We determine the transformation behavior of these functions by applying the residue theorem to the sequence of functions

$$f_n(z; \boldsymbol{g}) := 2 \frac{e^{-\frac{2n+1}{N}\pi i h z} \cdot e^{\frac{2n+1}{N}\pi i g \frac{z}{\tau}}}{z\left(1 - e^{-(2n+1)\pi i z}\right)\left(1 - e^{(2n+1)\pi i \frac{z}{\tau}}\right)}, \quad n \in \mathbb{N}. \tag{26}$$

For $g, h \in \mathbb{Z}$ and $\tau \in \mathscr{H}$ these functions are meromorphic functions of z with simple poles at the points $z = \dfrac{2m}{2n+1}$, $z = \dfrac{2m}{2n+1}\tau$, $m \in \mathbb{Z} \setminus \{0\}$; and a pole of order 3 at $z=0$. For the residues we obtain

$$\operatorname*{Res}_{z=\frac{2m}{2n+1}} f_n(z;\boldsymbol{g}) = \lim_{z \to \frac{2m}{2n+1}} \left(z - \frac{2m}{2n+1}\right) f_n(z;\boldsymbol{g})$$

$$= \frac{2n+1}{m} \cdot \frac{e^{-\frac{2\pi i}{N} hm} e^{\frac{2\pi i}{N\tau} gm}}{1 - e^{\frac{2\pi i}{\tau} m}} \cdot \frac{1}{(2n+1) i\pi} = \frac{1}{m\pi i} \frac{(\zeta_N^h q_N'^g)^{-m}}{1 - q_1'^{-m}},$$

and

$$\operatorname*{Res}_{z=\frac{2m}{2n+1}\tau} f_n(z;\boldsymbol{g}) = \lim_{z \to \frac{2m}{2n+1}\tau} \left(z - \frac{2m}{2n+1}\tau\right) f_n(z;\boldsymbol{g})$$

$$= \frac{2n+1}{m\tau} \cdot \frac{e^{-\frac{2\pi i \tau}{N} hm} e^{\frac{2\pi i}{N} gm}}{1 - e^{-2\pi i \tau m}} \cdot \frac{-\tau}{(2n+1) i\pi} = -\frac{1}{m\pi i} \frac{(\zeta_N^{-g} q_N^h)^{-m}}{1 - q_1'^{-m}}.$$

The Laurent expansion of $f_n(z;\boldsymbol{g})$ at $z=0$ begins with

$$f_n(z;\boldsymbol{g}) = \frac{2\tau}{[(2n+1)\pi]^2} z^{-3} + \frac{2i}{(2n+1)\pi}\left[\frac{g}{N} - \frac{1}{2}\tau\left(\frac{h}{N} - \frac{1}{2}\right)\right] z^{-2} +$$

$$\left\{2\left(\frac{g}{N} - \frac{1}{2}\right)\left(\frac{h}{N} - \frac{1}{2}\right) - \frac{1}{\tau}\left[\left(\frac{g}{N}\right)^2 - \frac{g}{N} + \frac{1}{6}\right] - \tau\left[\left(\frac{h}{N}\right)^2 - \frac{h}{N} + \frac{1}{6}\right]\right\} z^{-1} + \cdots.$$

Hence for $0 < g, h \leq N$ and with

$$((x)) = P_1(x) + \tfrac{1}{2}\delta(x) \quad \text{for } x \in \mathbb{R},$$

we have

$$\begin{aligned} \operatorname*{Res}_{z=0} f_n(z;\boldsymbol{g}) = {} & 2\left(\left(\frac{g}{N}\right)\right)\left(\left(\frac{h}{N}\right)\right) + P_1\left(\frac{h}{N}\right)\alpha_g + P_1\left(\frac{g}{N}\right)\alpha_{Tg} \\ & + \frac{1}{2}\delta_{\boldsymbol{g}} - \frac{1}{\tau} P_2\left(\frac{g}{N}\right) - \tau P_2\left(\frac{h}{N}\right). \end{aligned} \tag{27}$$

We now integrate each f_n along the boundary $\mathscr{C}$ of the parallelogram with vertices $1, \tau, -1$ and $-\tau$. By the residue theorem

$$\frac{1}{2\pi i}\oint_{\mathscr{C}} f_n(z;\boldsymbol{g})\,dz = -\frac{1}{\pi i} \sum_{|m| \leq n}' \frac{1}{n}\left\{\frac{(\zeta_N^h q_N'^g)^m}{1 - q_1'^m} - \frac{(\zeta_N^{-g} q_N^h)^m}{1 - q_1^m}\right\} + \operatorname*{Res}_{z=0} f_n(z;\boldsymbol{g}).$$

If we denote the sum of the terms of the expansion (25) of $\psi_{\boldsymbol{g}}$ with $|m|\leqq n$ by $\psi_{\boldsymbol{g}}^n$, and if we substitute for the residue its value (27), we obtain

$$\frac{1}{2\pi i}\oint_{\mathscr{C}} f_n(z;\boldsymbol{g})\,dz=\frac{1}{\pi i}\{\psi_{\boldsymbol{g}}^n(T(\tau))-\psi_{T'\boldsymbol{g}}^n(\tau)\}+2\left(\left(\frac{g}{N}\right)\right)\left(\left(\frac{h}{N}\right)\right)+\frac{1}{2}\delta_{\boldsymbol{g}}.$$

The sequence f_n converges on the open edges of the parallelogram; indeed, if $(g,h)=(N,N)$, it converges to $\frac{2}{z}$ on the open side $(1,\tau)$ and to 0 on the other open sides, and if $\boldsymbol{g}\neq\boldsymbol{0}\bmod N$, it converges to 0 on all open sides. Moreover, the sequence is uniformly bounded on $\mathscr{C}$. To convince oneself of the convergence one only need consider the signs of the real parts $\operatorname{Re}(iz)$ and $\operatorname{Re}\left(\frac{iz}{\tau}\right)$ in (26). The uniform boundedness of the sequence on $\mathscr{C}$ follows from

$$\left|\frac{e^{-\frac{2n+1}{N}\pi ihz}}{1-e^{-(2n+1)\pi iz}}\right|\leqq\left|\frac{1}{1-e^{\pm(2n+1)\pi iz}}\right|,$$

where the upper sign is taken if $\operatorname{Re}(iz)\leqq 0$, and the lower sign is taken if $\operatorname{Re}(iz)\geqq 0$. To verify the boundedness we note that

$$e^{\pm(2n+1)\pi i(\pm 1+\lambda(\pm\tau\mp 1))}$$

does not come arbitrarily close to 1 for $n\in\mathbf{N}$ and $0\leqq\lambda\leqq 1$ under any distribution of the signs, for this is the case only if

$$-e^{i\Lambda(\pm\tau\pm 1)},\qquad -\infty<\Lambda<+\infty,$$

comes arbitrarily close to 1, which does not happen since $\operatorname{Im}\tau\neq 0$. The corresponding properties hold when z is replaced by $-\frac{z}{\tau}$. The sequence $f_n(z;\boldsymbol{g})$ is therefore boundedly convergent, and from Lebesgues convergence theorem (cf. E.C. Titchmarsh [1] p. 40, we find that

$$\lim_{n\to\infty}\oint_{\mathscr{C}} f_n(z;\boldsymbol{g})\,dz=2\delta_{\boldsymbol{g}}\int_1^{\tau} z^{-1}\,dz=2\delta_{\boldsymbol{g}}\log\tau$$

and then that

$$\psi_{\boldsymbol{g}}(T\tau)-\psi_{T'\boldsymbol{g}}(\tau)-\left(\log\tau-\frac{\pi i}{2}\right)\delta_{\boldsymbol{g}}=-2\pi i\left(\left(\frac{g}{N}\right)\right)\left(\left(\frac{h}{N}\right)\right),\qquad(23)$$

obviously without restriction on $\boldsymbol{g}$.

2. Proof of an auxiliary formula. We know the transformation behavior of the $\psi_{\boldsymbol{g}}$ under $U=\begin{pmatrix}1&1\\0&1\end{pmatrix}$ and $T=\begin{pmatrix}0&-1\\1&0\end{pmatrix}$. To determine the general transformation formula we need an auxiliary formula. This formula contains forms of level N as well as nN. Therefore in the proof of this formula and in its applications we must indicate the dependence on the respective levels. Hence we write $\psi_{g,h}(\tau;N)$ in place of $\psi_{\boldsymbol{g}}(\tau)$, $\gamma_0(g,h;N)$ in place of $\gamma_0(\boldsymbol{g})$ and $\delta^N_{g,h}$ instead of $\delta_{\boldsymbol{g}}$.

The formula to be proved is:

$$\psi_{g,h}(\tau;N)=\sum_{v \bmod n}\psi_{g+vN,nh}(n\tau;nN),\qquad n\in\mathbb{N}. \tag{28}$$

Proof. First of all

$$\sum_{v \bmod n}P_2\left(\frac{\alpha+v}{n}\right)=\sum_{v \bmod n}\left\{\frac{1}{6}-\left(\frac{\alpha+v}{n}-\left[\frac{\alpha+v}{n}\right]\right)+\left(\frac{\alpha+v}{n}-\left[\frac{\alpha+v}{n}\right]\right)^2\right\}$$

$$=\frac{n}{6}-\left\{\frac{\alpha-[\alpha]}{n}\cdot n+\frac{n-1}{2}\right\}+\left\{\left(\frac{\alpha-[\alpha]}{n}\right)^2 n+\frac{\alpha-[\alpha]}{n}(n-1)+\frac{(n-1)(2n-1)}{6n}\right\}$$

$$=\frac{1}{n}\left\{(\alpha-[\alpha])^2-(\alpha-[\alpha])+\frac{1}{6}\right\}=\frac{1}{n}P_2(\alpha),$$

and thus

$$\sum_{v \bmod n}\psi_{g+vN,nh}(n\tau;nN)$$

$$=\sum_{v \bmod n}\left[\gamma_0(g+vN,nh;nN)+\pi i P_2\left(\frac{g+vN}{nN}\right)\tau n-\sum_{\substack{mm_1>0\\ m_1\equiv g+vN \bmod nN}}|m|^{-1}\zeta_N^{mh}q_N^{mm_1}\right]$$

$$=\gamma_0(g,h;N)+\pi i P_2\left(\frac{g}{N}\right)\tau-\sum_{\substack{mm_1>0\\ m_1\equiv g \bmod N}}|m|^{-1}\zeta_N^{mh}q_N^{mm_1}=\psi_{g,h}(\tau;N).\quad\square$$

3. Proof of the general transformation formula. If $A=\begin{pmatrix}a&b\\0&d\end{pmatrix}\in\Gamma$, the transformation behavior of $\psi_{\boldsymbol{g}}(\tau)$ is given by

$$\psi_{\boldsymbol{g}}(A(\tau))=\psi_{\boldsymbol{g}}\left(\tau+\frac{b}{d};N\right)=\psi_{A'\boldsymbol{g}}(\tau)+\frac{b}{d}\pi i P_2\left(\frac{g}{N}\right),$$

as can be read off from (24). If $A=\begin{pmatrix}a&b\\c&d\end{pmatrix}\in\Gamma$ with $c\neq 0$, and if for the present $c>0$, then with the notation

$$\begin{pmatrix}a&b\\c&d\end{pmatrix}'\begin{pmatrix}g\\h\end{pmatrix}=\begin{pmatrix}g'\\h'\end{pmatrix},$$

we obtain one after another

$$\begin{aligned}
\psi_{g,h}(A(\tau);N) &= \psi_{g,h}\left(\frac{a}{c}-\frac{1}{c(c\tau+d)};N\right)\\
&= \sum_{\nu \bmod c} \psi_{g+\nu N,ch}\left(a-\frac{1}{c\tau+d};cN\right) && \text{(by (28))}\\
&= \sum_{\nu \bmod c} \left\{\psi_{g+\nu N,g'+a\nu N}\left(\frac{-1}{c\tau+d};cN\right)+a\pi i P_2\left(\frac{g+\nu N}{cN}\right)\right\} && \text{(by (24))}\\
&= \sum_{\nu \bmod c} \left\{\psi_{g'+a\nu N,-g-\nu N}(c\tau+d;cN)-2\pi i\left(\left(\frac{g+\nu N}{cN}\right)\right)\left(\left(\frac{g'+a\nu N}{cN}\right)\right)\right.\\
&\quad \left.+\left[\log(c\tau+d)-\frac{\pi i}{2}\right]\delta^{cN}_{g+\nu N,g'+a\nu N}\right\}+\pi i\frac{a}{c}P_2\left(\frac{g}{N}\right)\\
&&& \text{(by (23) and the proof of (28))}\\
&= \sum_{\nu \bmod c} \left\{\psi_{g'+a\nu N,h'c}(c\tau;cN)+d\pi i P_2\left(\frac{g'+a\nu N}{cN}\right)\right\}-2\pi i s^N_{g,h}(a,c)\\
&\quad +\left(\log(c\tau+d)-\frac{\pi i}{2}\right)\delta^N_{g,h}+\pi i\frac{a}{c}P_2\left(\frac{g}{N}\right) && \text{(by (24))}.
\end{aligned}$$

Here $s^N_{g,h}(a,c)$ is the *generalized Dedekind sum*

$$s^N_{g,h}(a,c):=\sum_{\nu \bmod c}\left(\left(\frac{g+\nu N}{cN}\right)\right)\left(\left(\frac{g'+a\nu N}{cN}\right)\right), \tag{29}$$

which for $c<0$ is also defined by (29). Applying (28) we further have

$$\psi_{g,h}(A(\tau);N)=\psi_{g',h'}(\tau;N)+\pi i\frac{d}{c}P_2\left(\frac{g'}{N}\right)+\pi i\frac{a}{c}P_2\left(\frac{g}{N}\right)-2\pi i s^N_{g,h}(a,c)$$
$$+\left(\log(c\tau+d)-\frac{\pi i}{2}\right)\delta^N_{g,h}.$$

Using our earlier notation we get

$$\psi_{\boldsymbol{g}}(A(\tau))-\psi_{A'\boldsymbol{g}}(\tau)-\left(\log(c\tau+d)-\frac{\pi i}{2}\right)\delta_{\boldsymbol{g}}$$
$$=\pi i\left(\frac{a}{c}P_2\left(\frac{g}{N}\right)+\frac{d}{c}P_2\left(\frac{g'}{N}\right)-2s^N_{\boldsymbol{g}}(a,c)\right).$$

Observing that $s^N_{\boldsymbol{g}}(-\alpha,-\beta)=-s^N_{\boldsymbol{g}}(\alpha,\beta)$ and $P_2(-\alpha)=P_2(\alpha)$, we now formulate the result for arbitrary c:

$$\psi_{\boldsymbol{g}}(A(\tau))-\psi_{A'\boldsymbol{g}}(\tau)-\operatorname{sgn} c\left(\operatorname{sgn} c\cdot\log(c\tau+d)-\frac{\pi i}{2}\right)\delta_{\boldsymbol{g}}$$

$$=\begin{cases}\pi i\left\{\frac{a}{c}P_2\left(\frac{g}{N}\right)+\frac{d}{c}P_2\left(\frac{g'}{N}\right)\right\}-2\operatorname{sgn} c\cdot s_{\boldsymbol{g}}^N(a,c) & \text{for } c\neq 0, \text{ and} \\ \pi i\frac{b}{d}P_2\left(\frac{g}{N}\right) & \text{if } c=0.\end{cases} \tag{30}$$

Here

$$A=\begin{pmatrix} a & b \\ c & d\end{pmatrix}\in\Gamma,\qquad A'\boldsymbol{g}=\begin{pmatrix} g' \\ h'\end{pmatrix}.$$

Cf. B. Schoeneberg [3].

§ 4. Consequences of the Transformation Formula

In this section we always suppose that $\boldsymbol{g}\not\equiv\mathbf{0} \bmod N$ so that $\pi_{\boldsymbol{g}}(A)$, $A\in\Gamma[N]$, represent periods of integrals for $\Gamma[N]$.

1. Dedekind's functions. Let $A\in\Gamma$. Since $(-A)(\tau)=A(\tau)$ and $\psi_{-\boldsymbol{g}}=\psi_{\boldsymbol{g}}$, it follows that

$$\pi_{\boldsymbol{g}}(-A)=\pi_{\boldsymbol{g}}(A)=\pi_{-\boldsymbol{g}}(A).$$

If $\mathbb{M}$ is the set of linear maps

$$M(A):\begin{cases}\psi_{\boldsymbol{g}}(\tau)\mapsto\psi_{\boldsymbol{g}}(A(\tau)) \\ 1\mapsto 1\end{cases},\quad \boldsymbol{g} \bmod N, \tag{31}$$

of the $\mathbb{C}$-vector space

$$\Psi:=\left\{\Psi(\tau):=\sum_{\substack{\boldsymbol{g} \bmod N \\ \boldsymbol{g}\not\equiv 0 \bmod N}} x_{\boldsymbol{g}}\psi_{\boldsymbol{g}}(\tau)+x_0\,\middle|\, x_0,x_{\boldsymbol{g}}\in\mathbb{C}\right\}$$

into itself, then the correspondence

$$\Gamma\to\mathbb{M},\qquad A\mapsto M(A) \tag{32}$$

is a homomorphism of Γ on $\mathbb{M}$. First, by (11), for arbitrary $A,B\in\Gamma$,

$$\psi_{\boldsymbol{g}}(AB(\tau))=\psi_{(AB)'\boldsymbol{g}}(\tau)+\pi_{\boldsymbol{g}}(AB).$$

On the other hand

$$\psi_{\boldsymbol{g}}(AB(\tau))=\psi_{A'\boldsymbol{g}}(B(\tau))+\pi_{\boldsymbol{g}}(A)=\psi_{(AB)'\boldsymbol{g}}(\tau)+\pi_{A'\boldsymbol{g}}(B)+\pi_{\boldsymbol{g}}(A),$$

and thus

$$\pi_{\boldsymbol{g}}(AB)=\pi_{\boldsymbol{g}}(A)+\pi_{A'\boldsymbol{g}}(B).$$

If in particular $A \in \Gamma[N]$, then the last formula becomes

$$\pi_g(AB) = \pi_g(A) + \pi_g(B). \tag{33}$$

Thus *the* π_g *are additive characters* of $\Gamma[N]$ or $\bar{\Gamma}(N)$, respectively, and the map

$$A \mapsto e^{\pi_g(A)}$$

is a homomorphism of $\Gamma[N]/(\Gamma[N])'$, the factor group $\Gamma[N]$ by its commutator subgroup $(\Gamma[N])'$, in the multiplicative group $\mathbb{E} = \{\zeta \in \mathbb{C} \mid |\zeta| = 1\}$. As we have seen

$$\pi_g(U) = \pi i P_2\left(\frac{g}{N}\right) \quad \text{and} \quad \pi_g(T) = -2\pi i\left(\left(\frac{g}{N}\right)\right)\left(\left(\frac{h}{N}\right)\right).$$

Hence $\frac{1}{2\pi i}\pi_g(U)$ and $\frac{1}{2\pi i}\pi_g(T)$ are rational numbers whose denominators devide $N_1 := \frac{12N^2}{(6,N)}$. Since U and T generate Γ, we deduce that

$$\frac{N_1}{2\pi i}\pi_g(A) \in \mathbb{Z} \quad \text{for all } A \in \Gamma \text{ with } N_1 = \frac{12N^2}{(6,N)}. \tag{34}$$

We conclude: *The Dedekind function* η_g (cf. (8)) *satisfies the equation*

$$\eta_g^{N_1}(A(\tau)) = \eta_{A'g}^{N_1}(\tau) \tag{35}$$

for all $A \in \Gamma$*; in particular* $\eta_g^{N_1}$ *is a modular function for* $\Gamma[N]$.

More precise investigations show that if $N_2 := \frac{12N}{(6, N)}$ and $A \in \Gamma[N]$, then

$$\eta_g^{N_2}(A(\tau)) = \eta_g^{N_2}(\tau).$$

For $r \in \mathbb{Q}$,

$$\begin{aligned} K_r(g) &:= \text{kernel of } (\Gamma[N] \to \mathbb{E},\ A \mapsto e^{r \cdot \pi_g(A)}) \\ &= \left\{ A \in \Gamma[N] \,\middle|\, \frac{r}{2\pi i}\pi_g(A) \in \mathbb{Z} \right\} \end{aligned} \tag{36}$$

is obviously a normal subgroup of finite index in $\Gamma[N]$. For the functions η_g^r this implies that

$$\eta_g^r(A(\tau)) = \eta_g^r(\tau) \quad \text{for } A \in K_r(g).$$

2. The space of characters on $\Gamma[N]$. Now we investigate the $\mathbb{C}$-vector space Π spanned by $\pi_g(A)$, $A \in \Gamma[N]$. The functions

$$\Psi(\tau) := \sum_{\substack{g \bmod N \\ g \not\equiv 0}} x_g \psi_g(\tau) + x_0; \qquad x_0, x_g \in \mathbb{C},$$

satisfy the transformation formula

$$\Psi(A(\tau)) = \Psi(\tau) + \Pi(A), \qquad A \in \Gamma[N],$$

where

$$\Pi(A) = \sum_{\substack{\boldsymbol{g} \bmod N \\ \boldsymbol{g} \not\equiv 0}} x_{\boldsymbol{g}} \pi_{\boldsymbol{g}}(A).$$

For $A, B \in \Gamma[N]$ corresponding to (33) we have

$$\Pi(AB) = \Pi(A) + \Pi(B).$$

The derivatives $\frac{d}{d\tau} \psi_{\boldsymbol{g}}(\tau)$ span the space of $\wp$-division values of level N. As we know from **VII, § 3, 3**, this space contains series of the type

$$\Phi_\lambda := C_\lambda(\wp^*_{N, \boldsymbol{a}_\lambda} - \wp^*_{N, \boldsymbol{a}_\sigma}).$$

In $\mathbb{Z}^2$ we choose

$$\boldsymbol{a}_\lambda = \begin{pmatrix} c_\lambda \\ d_\lambda \end{pmatrix}, \quad \lambda = 1, \ldots, \sigma_\infty(N) - 1; \quad \boldsymbol{a}_\sigma = \begin{pmatrix} c_\sigma \\ d_\sigma \end{pmatrix}, \quad \sigma = \sigma_\infty(N),$$

so that $r_\lambda = -\frac{d_\lambda}{c_\lambda}$, $\lambda = 1, \ldots, \sigma_\infty(N)$, are $\Gamma[N]$-inequivalent. The series Φ_λ have the expansions

$$(c_\mu \tau + d_\mu)^2 \Phi_\lambda(\tau) = \sum_{\nu \geqq 0} c_{\lambda,\nu}^{(\mu)} e^{\frac{2\pi i}{N} A_\mu(\tau) \cdot \nu}, \qquad A_\mu = \begin{pmatrix} a_\mu & b_\mu \\ c_\mu & d_\mu \end{pmatrix} \in \Gamma,$$

at the cusps r_λ.

We can choose C_λ so that the coefficients $c_{\lambda,0}^{(\mu)}$ satisfy the conditions

$$c_{\lambda,0}^{\mu} = \begin{cases} \frac{1}{N} \delta_{\lambda,\mu} & \text{if } 1 \leqq \lambda, \ \mu < \sigma_\infty(N), \text{ and} \\[2ex] -\frac{1}{N} & \text{if } 1 \leqq \lambda < \sigma_\infty(N), \ \mu = \sigma_\infty(N). \end{cases}$$

The integrals

$$\Psi_\lambda(\tau) := \int \Phi_\lambda(\tau) \, d\tau$$

lie in the vector space Ψ. Let A_μ be a generator of the group of transformations in $\Gamma[N]$ with fixed point r_μ. Then the periods

$$\Pi_\lambda(A) := \Psi_\lambda(A(\tau)) - \Psi_\lambda(\tau)$$

satisfy the equation

$$\Pi_\lambda(A_\mu) = \int_{\tau_0}^{A_\mu(\tau_0)} \Phi_\lambda(\tau) \, d\tau = \begin{cases} \delta_{\lambda,\mu} & \text{if } 1 \leqq \lambda, \ \mu < \sigma_\infty(N), \\[2ex] -1 & \text{if } 1 \leqq \lambda < \sigma_\infty(N), \ \mu = \sigma_\infty(N), \end{cases} \tag{37}$$

for arbitrary $\tau_0 \in \mathscr{H}$. Thus the Π_λ are linearly independent. Obviously:

The Π_λ, $1 \leqq \lambda < \sigma_\infty(N)$, *form a basis of* Π *and*

$$\Pi(A) = \sum_{\lambda=1}^{\sigma_\infty(N)-1} \Pi(A_\lambda)\,\Pi_\lambda(A), \quad \text{if } A \in \Gamma[N].$$

For $N \geqq 2$ the dimension of the space of characters of the transformation group $\bar{\Gamma}(N)$ is

$$\sigma_\infty(N) - 1 + 2g,$$

where g is the genus of the fundamental region for $\bar{\Gamma}(N)$. This is an immediate consequence of the following theorem:

For $N \geqq 2$ *the transformation group* $\bar{\Gamma}(N)$ *is a free group of rank* $\sigma_\infty(N) - 1 + 2g$.

For a proof see Lehner [1], p. 362.

In the case $g = 0$, that is, for $2 \leqq N \leqq 5$, the Π_λ, $\lambda \in \{1, 2, \ldots, \sigma_\infty(N) - 1\}$, form a basis of all characters for $\bar{\Gamma}(N)$.

Chapter IX. Theta Series

In this chapter we construct entire modular forms of higher level and arbitrary dimension by means of positive definite quadratic forms. We assume that the quadratic form has an even number of variables. This is an unnatural assumption from a number-theoretic point of view, however, only in this case do we obtain modular forms as we have defined them previously. In the case of an odd number of variables we get a new type of function, namely a modular form of half-integral dimension. An even more general theory of modular forms has been developed. In particular, the work initiated by H. Petersson [1] allows the results of this chapter to be carried over to definite quadratic forms in an odd number of variables. For this see W. Pfetzer [1].

§ 1. General Theta Series. An Operator

1. Introduction to theta series. Let A be a real symmetric matrix of order $f \geqq 1$:

$$A=(a_{\mu,\nu}); \quad a_{\mu,\nu}\in\mathbb{R}; \quad a_{\mu,\nu}=a_{\nu,\mu}, \quad 1\leqq\mu,\nu\leqq f.$$

Let $\boldsymbol{x}$ be a column vector with the f complex variable components $x_1,\ldots,x_f$ and $\boldsymbol{n}$ a column vector with the f integral components $n_1,\ldots,n_f$. We denote their respective transposes by $\boldsymbol{x}'$ and $\boldsymbol{n}'$. By the *norm* of x we understand the non-negative number

$$|\boldsymbol{x}| = \left\{\sum_{\nu=1}^{f} |x_\nu|^2\right\}^{\frac{1}{2}}.$$

Concerning the matrix A, we assume that the associated quadratic form

$$\boldsymbol{x}'A\boldsymbol{x} = \sum_{\mu,\nu=1}^{f} a_{\mu,\nu}x_\mu x_\nu$$

is *positive definite*, i. e.

$$\boldsymbol{x}'A\boldsymbol{x}>0 \quad \text{for all real } \boldsymbol{x}\neq\boldsymbol{0},$$

and that therefore the determinant of A is positive, $|A|>0$. With these preparations we consider the series

$$\theta_{A,x} := \sum_{n \in \mathbb{Z}^f} e^{-(n+x)' A (n+x)}. \tag{1}$$

Lemma. *The series $\theta_{A,x}$ converges absolutely and uniformly on the set*

$$\{x \mid x \in \mathbb{C}^f, |x| \leqq C\}$$

for arbitrary $C>0$.

Proof. Let $C>0$ be given. Since $x' A x$ for $x \in \mathbb{R}^f$ assumes a positive minimum c on $|x|=1$, we have

$$x' A x \geqq c|x|^2 \quad \text{for all } x \text{ in } \mathbb{R}^f.$$

Using $A'=A$, we have

$$(n+x)' A(n+x) = n' A n + 2n' A x + x' A x,$$

and hence for the real part of the left handside we have the inequality

$$\operatorname{Re}\{(n+x)' A(n+x)\} \geqq \tfrac{1}{2} n' A n \quad \text{for } |x| \leqq C \quad \text{and } |n| \geqq N_0(C),$$

where $N_0(C)>0$ depends only on C. Thus by the first inequality

$$\operatorname{Re}\{(n+x)' A(n+x) \geqq \frac{c}{2}|n|^2.$$

The lemma follows from this. Moreover we also see:

The series $\theta_{A,x}$ represents a holomorphic function of each x_ν in the finite plane.

Since $\theta_{A,x}$ is periodic with period 1 in each of its variables, it possesses the Fourier expansion

$$\theta_{A,x} = \sum_{m \in \mathbb{Z}^f} a_m e^{2\pi i m' x}, \qquad x \in \mathbb{C}^f \tag{2}$$

with the Fourier coefficients

$$a_m = \int_0^1 \cdots \int \sum_{n \in \mathbb{Z}^f} e^{-(n+x)' A(n+x)} e^{-2\pi i m' x} dx_1 \ldots dx_f,$$

where $x_1, \ldots, x_f$ are real integration variables. We write dX instead of $dx_1 \ldots dx_f$ and get

$$a_m = \int_{-\infty}^{\infty} \cdots \int e^{-x' A x - 2\pi i m' x} dX. \tag{3}$$

2. Proof of a transformation formula. We now calculate the Fourier coefficients a_m. To that end we set

$$x = A^{-1} y, \quad \text{so } x' A x = y' A^{-1} y.$$

By (3) we obtain the representation

$$a_m = \frac{1}{|A|} \int_{-\infty}^{+\infty} \cdots \int e^{-(y' A^{-1} y + 2\pi i m' A^{-1} y)} dY.$$

Furthermore,

$$a_m = \frac{1}{|A|} e^{-\pi^2 m' A m} \int_{-\infty}^{+\infty} \cdots \int e^{-(y+\pi i m)' A^{-1} (y+\pi i m)} dY$$

$$= \frac{1}{|A|} e^{-\pi^2 m' A m} \int_{-\infty}^{+\infty} \cdots \int e^{-y' A^{-1} y} dY.$$

To calculate this last integral we transform the exponent to a sum of squares. For this we set $y = Lz$ with a real matrix L so that

$$y' A^{-1} y = |z|^2.$$

Then $|L|^2 = |A|$. Hence

$$\int_{-\infty}^{+\infty} \cdots \int e^{-y' A^{-1} y} dY = |L| \int_{-\infty}^{+\infty} \cdots \int e^{-\sum_{\nu=1}^{f} z_\nu^2} dZ$$

$$= |L| \prod_{\nu=1}^{f} \int_{-\infty}^{+\infty} e^{-z_\nu^2} dz_\nu = |A|^{\frac{1}{2}} \pi^{\frac{f}{2}}.$$

This yields

$$a_m = \frac{\pi^{\frac{f}{2}}}{|A|^{\frac{1}{2}}} e^{-\pi^2 m' A m}, \qquad m \in \mathbb{Z}^f. \tag{4}$$

If we now replace A by $\pi t A$, $t > 0$, which is obviously permitted, then we obtain

$$\sum_{m \in \mathbb{Z}^f} e^{-\pi t (m+x)' A (m+x)} = \frac{1}{(\sqrt{t})^f |A|^{\frac{1}{2}}} \sum_{m \in \mathbb{Z}^f} e^{-\frac{\pi}{t} m' A^{-1} m + 2\pi i m' x}, \qquad \sqrt{t} > 0,$$

from (1), (2) and (3). Since each series of the above equation converges absolutely and uniformly for complex t lying in compact subsets of $\operatorname{Re} t > 0$, each represents a holomorphic function. Thus the equation which we have just derived for $t > 0$ is also valid for complex t with $\operatorname{Re} t > 0$ if we choose $\operatorname{Re} \sqrt{t} > 0$. Finally we set $t = -i\tau$ and obtain the transformation formula

$$\sum_{m \in \mathbb{Z}^f} e^{\pi i \tau (m+x)' A (m+x)} = \frac{1}{(\sqrt{-i\tau})^f |A|^{\frac{1}{2}}} \sum_{m \in \mathbb{Z}^f} e^{-\frac{\pi i}{\tau} m' A^{-1} m + 2\pi i m' x}, \tag{5}$$

where $\operatorname{Im} \tau > 0$ and the root is chosen so that it has positive real part.

The series in (5) represent functions holomorphic in $\mathscr{H}$. The series

$$\sum_{\boldsymbol{m}\in\mathbb{Z}^f} e^{\pi i\tau \boldsymbol{m}'\boldsymbol{A}\boldsymbol{m}}, \qquad \tau\in\mathscr{H},$$

is the *theta series associated with* $\boldsymbol{A}$.

3. A class of operators. Besides the functions (1), we also consider the functions which may be obtained from them by applying the linear operator

$$\mathscr{L} := \sum_{v=1}^{f} \ell_v \frac{\partial}{\partial x_v}, \qquad \ell_v\in\mathbb{C},$$

and its powers $\mathscr{L}^n$, $n\in\mathbb{N}$. This operator has the following properties:

If the functions $u(\boldsymbol{x})$ and $v(\boldsymbol{x})$ have partial derivatives with respect to $x_1,\dots,x_f$, then

$$\mathscr{L}(u+v)=\mathscr{L}(u)+\mathscr{L}(v), \qquad \mathscr{L}(uv)=\mathscr{L}(u)v+u\mathscr{L}(v),$$

$$\mathscr{L}(cu)=c\mathscr{L}(u) \quad \text{for} \quad c\in\mathbb{C}, \quad \text{and} \quad \mathscr{L}(e^u)=e^u\mathscr{L}(u).$$

Here, and in what follows, we set

$$\boldsymbol{\ell}'=(\ell_1,\dots,\ell_f).$$

Then we have

$$\mathscr{L}(\boldsymbol{n}+\boldsymbol{x})'\boldsymbol{A}(\boldsymbol{n}+\boldsymbol{x})=2\boldsymbol{\ell}'\boldsymbol{A}(\boldsymbol{n}+\boldsymbol{x}),$$

and by repetition

$$\mathscr{L}^2(\boldsymbol{n}+\boldsymbol{x})'\boldsymbol{A}(\boldsymbol{n}+\boldsymbol{x})=2\boldsymbol{\ell}'\boldsymbol{A}\boldsymbol{\ell}.$$

Thus the application of $\mathscr{L}$ to (1) yields

$$\mathscr{L}(\theta_{\boldsymbol{A},\boldsymbol{x}})=-2\sum_{\boldsymbol{n}\in\mathbb{Z}^f}\boldsymbol{\ell}'\boldsymbol{A}(\boldsymbol{n}+\boldsymbol{x})e^{-(\boldsymbol{n}+\boldsymbol{x})'\boldsymbol{A}(\boldsymbol{n}+\boldsymbol{x})},$$

and

$$\mathscr{L}^2(\theta_{\boldsymbol{A},\boldsymbol{x}})=-2\sum_{\boldsymbol{n}\in\mathbb{Z}^f}\{\boldsymbol{\ell}'\boldsymbol{A}\boldsymbol{\ell}-2[\boldsymbol{\ell}'\boldsymbol{A}(\boldsymbol{n}+\boldsymbol{x})]^2\}e^{-(\boldsymbol{n}+\boldsymbol{x})'\boldsymbol{A}(\boldsymbol{n}+\boldsymbol{x})}.$$

For what follows we assume that $\boldsymbol{\ell}$ *satisfies*

$$\boldsymbol{\ell}'\boldsymbol{A}\boldsymbol{\ell}=0. \tag{7}$$

Then

$$\mathscr{L}^k(\theta_{\boldsymbol{A},\boldsymbol{x}})=(-2)^k\sum_{\boldsymbol{n}\in\mathbb{Z}^f}\{\boldsymbol{\ell}'\boldsymbol{A}(\boldsymbol{n}+\boldsymbol{x})\}^k e^{-(\boldsymbol{n}+\boldsymbol{x})'\boldsymbol{A}(\boldsymbol{n}+\boldsymbol{x})}, \qquad k=0,1,2,\dots.$$

Since $\mathscr{L}(\boldsymbol{n}'\boldsymbol{x})=\boldsymbol{n}'\boldsymbol{\ell}$, application of $\mathscr{L}^k$ to (5) gives

$$\sum_{\boldsymbol{n}\in\mathbb{Z}^f}\{\boldsymbol{\ell}'\boldsymbol{A}(\boldsymbol{n}+\boldsymbol{x})\}^k e^{\pi i\tau(\boldsymbol{n}+\boldsymbol{x})'\boldsymbol{A}(\boldsymbol{n}+\boldsymbol{x})}$$

$$=\frac{(-i)^k}{(\sqrt{-i\tau})^{f+2k}|\boldsymbol{A}|^{\frac{1}{2}}}\sum_{\boldsymbol{n}\in\mathbb{Z}^f}(\boldsymbol{\ell}'\boldsymbol{n})^k e^{-\frac{\pi i}{\tau}\boldsymbol{n}'\boldsymbol{A}^{-1}\boldsymbol{n}+2\pi i(\boldsymbol{n}'\boldsymbol{x})}, \tag{8}$$

where the root is choosen so that it has positive real part.

§ 2. Special Theta Series

In connection with the transformation formula (8) we now construct functions which prove to be entire modular forms of a certain level.

1. Quadratic forms. Let A be the matrix of a positive definite quadratic form of even order $f=2r$, whose entries satisfy the following conditions:

$$A=(a_{\mu,\nu}),\quad a_{\mu,\nu}\in\mathbb{Z},\quad a_{\mu,\nu}=a_{\nu,\mu},\quad \text{and } a_{\mu,\mu}\equiv 0 \bmod 2\,. \tag{9}$$

Such matrices A and quadratic forms $x'Ax$ are called *even*. Then the quadratic form $\mathcal{Q}$, where

$$\mathcal{Q}(x):=\tfrac{1}{2}x'Ax=\tfrac{1}{2}\sum_{\mu,\nu=1}^{f} a_{\mu,\nu}x_\mu x_\nu=\sum_{1\le\mu\le\nu\le f} b_{\mu,\nu}x_\mu x_\nu \tag{10}$$

has rational integral coefficients $b_{\mu,\nu}$, where

$$b_{\mu,\nu}=a_{\mu,\nu}\quad \text{for } \mu<\nu \quad \text{and } b_{\mu,\mu}=\tfrac{1}{2}a_{\mu,\mu}\,.$$

We call $\mathcal{Q}$ and A *primitive* when the greatest common divisor of all coefficients $b_{\mu,\nu}$ is 1. Let D be the (positive) determinant of A; one calls $\Delta:=(-1)^r D$ the *discriminant* of the quadratic form $\mathcal{Q}$. If $A_{\mu,\nu}$ is the cofactor of $a_{\mu,\nu}$ in the matrix A, then

$$A^{-1}=\left(\frac{A_{\mu,\nu}}{D}\right).$$

The *level of the quadratic form* $\mathcal{Q}$ is the smallest natural number N such that

$$A^*:=N\,A^{-1}=\frac{N}{D}(A_{\mu,\nu}) \tag{11}$$

is an even matrix. We say that A^* is the *adjoint* of A (note the slight variation with standard matrix theory usage).

We now prove

Theorem 1. *The level of the quadratic form* $\mathcal{Q}(x)=\frac{1}{2}x'Ax$ *is*

$$N=\frac{D}{K}\quad \textit{with}\quad K=\text{g.c.d.}\left\{A_{\mu,\nu},\ \frac{A_{\mu,\mu}}{2}\right\}_{\mu,\nu=1,\dots,f}.$$

Proof. First we show that $A_{\mu,\mu}\equiv 0 \bmod 2$. Without restriction we may in fact take $\mu=f$. Then

$$A_{f,f}=\sum_{\pi}\ (\text{signum }\pi)\ a_{1,\pi_1}\cdots a_{f-1,\pi_{f-1}},$$

where the summation is over all permutations of the first $f-1$ natural numbers. A summand either contains a diagonal element of $\boldsymbol{A}$ and is therefore even, or it appears twice in the above sum due to the symmetry and odd order of $A_{f,f}$. Thus $A_{f,f} \equiv 0 \bmod 2$. Hence

$$\frac{1}{K}(A_{\mu,\nu}) = \frac{D}{K}\boldsymbol{A}^{-1}$$

is even and this is the maximal K for which this remains true. Since $K|D$ (cf. the remarks following the proof) and N is minimal, it follows that $N \left| \frac{D}{K} \right.$ and, since K is maximal, it follows that $N = \frac{D}{K}$. ▯

Obviously, the quadratic form $\mathcal{Q}^*(\boldsymbol{x}) = \frac{1}{2}\boldsymbol{x}'\boldsymbol{A}^*\boldsymbol{x}$ is primitive. One verifies easily that $\boldsymbol{A}$ is the adjoint of $\boldsymbol{A}^*$ when $\boldsymbol{A}$ is primitive, i.e. $(\boldsymbol{A}^*)^* = \boldsymbol{A}$.

We remark:

1) N and D have the same prime divisors. It follows from $|(A_{\mu\nu})| = D^{f-1}$ that $K^f | D^{f-1}$, and hence that each prime divisor of K divides D and does so in fact with greater multiplicity. Thus $N = \frac{D}{K}$ contains each prime factor of D.

2) $\boldsymbol{n} \in \mathbb{Z}^f$ and $\boldsymbol{A}\boldsymbol{n} \equiv \boldsymbol{0} \bmod N$ imply that

$$\frac{1}{2}\,\frac{\boldsymbol{n}'\boldsymbol{A}\boldsymbol{n}}{N} \in \mathbb{Z}.$$

Indeed, setting $\boldsymbol{A}\boldsymbol{n} = \boldsymbol{n}_1$, it follows from $\boldsymbol{n}_1 \equiv \boldsymbol{0} \bmod N$, i.e. from, $\boldsymbol{n}_1 = \boldsymbol{g}N$, $\boldsymbol{g} \in \mathbb{Z}^f$, that

$$\frac{1}{2}\,\frac{\boldsymbol{n}'\boldsymbol{A}\boldsymbol{n}}{N} = \frac{1}{2}\,\frac{\boldsymbol{n}_1'\boldsymbol{A}^{-1}\boldsymbol{n}_1}{N} = \frac{1}{2}\boldsymbol{g}'(\boldsymbol{A}^{-1}N)\boldsymbol{g}.$$

Now by the definition of N we see that the right hand side is an integer. It follows that $\frac{1}{2}\,\frac{\boldsymbol{n}'\boldsymbol{A}\boldsymbol{n}}{N^2} \bmod 1$ depends only on $\boldsymbol{n}$ modulo N, if $\boldsymbol{A}\boldsymbol{n} \equiv \boldsymbol{0} \bmod N$.

2. Division values. To construct entire modular forms we start with transformation formula (8) which we repeat:

$$\sum_{\boldsymbol{n} \in \mathbb{Z}^f} \{\boldsymbol{\ell}'\boldsymbol{A}(\boldsymbol{n}+\boldsymbol{x})\}^k e^{\pi i\tau(\boldsymbol{n}+\boldsymbol{x})'\boldsymbol{A}(\boldsymbol{n}+\boldsymbol{x})}$$
$$= \frac{(-i)^k}{(\sqrt{-i\tau})^{f+2k}|\boldsymbol{A}|^{\frac{1}{2}}} \sum_{\boldsymbol{n} \in \mathbb{Z}^f} (\boldsymbol{\ell}'\boldsymbol{n})^k e^{-\frac{\pi i}{\tau}\boldsymbol{n}'\boldsymbol{A}^{-1}\boldsymbol{n} + 2\pi i(\boldsymbol{n}'\boldsymbol{x})}.$$

Naturally we still assume (7), i.e. $\boldsymbol{\ell}'\boldsymbol{A}\boldsymbol{\ell} = 0$. $\boldsymbol{A}$ is the matrix of a positive definite quadratic form $\mathcal{Q}$ of even order $f = 2r$ that satisfies the conditions (9). In place of $\boldsymbol{x}$ we consider the *division value* $\boldsymbol{x}=\frac{\boldsymbol{h}}{N}$ with fixed $\boldsymbol{h}\in\mathbb{Z}^f$; N is the level of $\mathcal{Q}$. In addition we replace the variable τ by $-\frac{1}{\tau}$. Then (8) becomes

$$\sum_{\substack{\boldsymbol{n}\in\mathbb{Z}^f\\ \boldsymbol{n}\equiv\boldsymbol{h}\bmod N}} (\boldsymbol{\ell}'A\boldsymbol{n})^k e^{-\frac{\pi i}{\tau}\frac{\boldsymbol{n}'A\boldsymbol{n}}{N^2}} = \frac{(-i)^{2k+r}\tau^{r+k}}{\sqrt{D}} \sum_{\substack{\boldsymbol{n}_1\in\mathbb{Z}^f\\ A\boldsymbol{n}_1\equiv\boldsymbol{0}\bmod N}} (\boldsymbol{\ell}'A\boldsymbol{n}_1)^k e^{\pi i\tau\frac{\boldsymbol{n}_1'A\boldsymbol{n}_1}{N^2}+2\pi i\frac{\boldsymbol{n}_1'A\boldsymbol{h}}{N^2}}, \tag{12}$$

where $\boldsymbol{n}$ on the right hand side was replaced by $\boldsymbol{n}=\frac{1}{N}A\boldsymbol{n}_1$. *Concerning* $\boldsymbol{h}$ *we now assume that*

$$A\boldsymbol{h}\equiv\boldsymbol{0}\bmod N.$$

Then the rational number $\frac{\boldsymbol{n}_1'A\boldsymbol{h}}{N^2}\bmod 1$ depends only on $\boldsymbol{n}\bmod N$ and we may write equation (12) in the form

$$\sum_{\substack{\boldsymbol{n}\in\mathbb{Z}^f\\ \boldsymbol{n}\equiv\boldsymbol{h}\bmod N}} (\boldsymbol{\ell}'A\boldsymbol{n})^k e^{-\frac{\pi i}{\tau}\frac{\boldsymbol{n}'A\boldsymbol{n}}{N^2}} = \frac{(-i)^{2k+r}\tau^{k+r}}{\sqrt{D}} \sum_{\substack{\boldsymbol{g}\bmod N\\ A\boldsymbol{g}\equiv\boldsymbol{0}\bmod N}} e^{2\pi i\frac{\boldsymbol{g}'A\boldsymbol{h}}{N^2}} \sum_{\substack{\boldsymbol{n}\in\mathbb{Z}^f\\ \boldsymbol{n}\equiv\boldsymbol{g}\bmod N}} (\boldsymbol{\ell}'A\boldsymbol{n})^k e^{\pi i\tau\frac{\boldsymbol{n}'A\boldsymbol{n}}{N^2}}. \tag{13}$$

On the left side is a function which is similar to the function represented by the inner sum on the right side, however, with τ replaced by $-\frac{1}{\tau}$. In view of remark 2) of this section all these functions satisfy

$$\sum_{\substack{\boldsymbol{n}\in\mathbb{Z}^f\\ \boldsymbol{n}\equiv\boldsymbol{h}\bmod N}} (\boldsymbol{\ell}'A\boldsymbol{n})^k e^{\pi i(\tau+1)\frac{\boldsymbol{n}'A\boldsymbol{n}}{N^2}} = e^{\pi i\frac{\boldsymbol{h}'A\boldsymbol{h}}{N^2}} \sum_{\substack{\boldsymbol{n}\in\mathbb{Z}^f\\ \boldsymbol{n}\equiv\boldsymbol{h}\bmod N}} (\boldsymbol{\ell}'A\boldsymbol{n})^k e^{\pi i\tau\frac{\boldsymbol{n}'A\boldsymbol{n}}{N^2}}. \tag{14}$$

The transformation formulas (13) and (14) hold for every $\boldsymbol{\ell}$ with $\boldsymbol{\ell}'A\boldsymbol{\ell}=\boldsymbol{0}$. They obviously remain true if we replace the expression $(\boldsymbol{\ell}'A\boldsymbol{n})^k$ by one of the form $P_k^A(\boldsymbol{n})$ with

$$P_k^A(\boldsymbol{x}):=\sum_{\boldsymbol{\ell}} c_{\boldsymbol{\ell}}(\boldsymbol{\ell}'A\boldsymbol{x})^k,\qquad c_{\boldsymbol{\ell}}\in\mathbb{C}, \tag{15}$$

summed over finitely many $\boldsymbol{\ell}$ with the property $\boldsymbol{\ell}'A\boldsymbol{\ell}=0$.

Let A be a matrix of a positive definite quadratic form in an even number of variables that satisfies the conditions (9). Let $\boldsymbol{h}\in\mathbb{Z}^f$ be a vector satisfying $A\boldsymbol{h}\equiv\boldsymbol{0}\bmod N$. Finally let P_k^A be a function of the form

(15). We now define the *theta function* ϑ_{A,h,P_k^A} by

$$\vartheta_{A,h,P_k^A}(\tau) := \sum_{\substack{n\in\mathbb{Z}^f\\ n\equiv h \bmod N}} P_k^A(n)\, e^{\frac{2\pi i\tau}{N}\frac{1}{2}\frac{n'An}{N}}, \quad \tau\in\mathscr{H}, \quad k=0,1,2,\ldots. \tag{16}$$

We shall write P_k instead of P_k^A when there can be no confusion. Equations (13) and (14) imply

Theorem 2. *The behavior of the theta function* ϑ_{A,h,P_k} *under the generators* U *and* T *of the modular group is given by*

$$\left.\begin{aligned} &\vartheta_{A,h,P_k}(U\tau) = e^{\frac{2\pi i}{N}\frac{1}{2}\frac{h'Ah}{N}}\,\vartheta_{A,h,P_k}(\tau), \quad \text{and} && \text{I}\\ &\vartheta_{A,h,P_k}(T\tau) = \frac{(-i)^{r+2k}\tau^{r+k}}{\sqrt{D}} \sum_{\substack{g \bmod N\\ Ag\equiv 0 \bmod N}} e^{\frac{2\pi i}{N}\frac{g'Ah}{N}}\,\vartheta_{A,g,P_k}(\tau). && \text{II}\end{aligned}\right\} \tag{17}$$

3. First properties of theta functions.

1. The theta functions defined by (16) are holomorphic in $\mathscr{H}$, and at least in the case $P_k=1$ they are not identically 0. The exponents in the series expansion in powers of $e^{2\pi i\tau/N}$ are non-negative rational integers.

2. The following equations hold:

$$\vartheta_{A,h_1,P_k} = \vartheta_{A,h_2,P_k}, \quad \text{if } h_1 \equiv h_2 \bmod N,$$

and

$$\vartheta_{A,-h,P_k} = (-1)^k\,\vartheta_{A,h,P_k}.$$

3. In order to determine the behavior of the theta function ϑ_{A,h,P_k} under arbitrary modular transformations in **§3** we need the following decomposition. Let $c>0$ be integral, then

$$\begin{aligned}\vartheta_{A,h,P_k}(\tau) &= \sum_{\substack{n\in\mathbb{Z}^f\\ n\equiv h \bmod N}} P_k(n)\, e^{\frac{2\pi i\tau}{N}\cdot\frac{1}{2}\frac{n'An}{N}}\\ &= \sum_{\substack{g \bmod cN\\ g\equiv h \bmod N}} \sum_{\substack{n\in\mathbb{Z}^f\\ n\equiv g \bmod cN\\ (Ag\equiv 0 \bmod N)}} P_k(n)\, e^{\frac{2\pi i\tau}{N}\cdot\frac{1}{2}\frac{n'An}{N}}\\ &= \sum_{\substack{g \bmod cN\\ g\equiv h \bmod N}} \sum_{\substack{n\in\mathbb{Z}^f\\ n\equiv g \bmod cN\\ (cAg\equiv 0 \bmod cN)}} P_k(n)\, e^{\frac{2\pi i c\tau}{cN}\cdot\frac{1}{2}\frac{n'cAn}{cN}}.\end{aligned}$$

The congruences in parentheses are not additional summation conditions. Here the level of the quadratic form $\frac{1}{2}x'cAx$ is cN. Indeed in order to apply Theorem **1** to the matrix cA we must replace D by $c^f D$

and K by $c^{f-1}K$, we thus obtain cN as the level. The inner sum above is $\frac{1}{c^k}\vartheta_{cA,g,P_k^{cA}}(c\tau)$ with $P_k^{cA}(x)=\sum_{\ell} c_\ell(\ell' cAx)^k$. In conclusion we have

$$\vartheta_{A,h,P_k^A}(\tau)=\frac{1}{c^k}\sum_{\substack{g \bmod cN\\ g\equiv h \bmod N\\ (cAg\equiv 0 \bmod cN)}} \vartheta_{cA,g,P_k^{cA}}(c\tau),\quad c>0. \tag{18}$$

4. We now study the dependence of our theta function on the particular quadratic form of its class. We replace $P_k(x)$ in (16) by its representation (15), then

$$\vartheta_{A,h,P_k^A}(\tau)=\sum_{\substack{n\in\mathbb{Z}^f\\ n\equiv h \bmod N\\ (Ah\equiv 0 \bmod N)}}\Bigg\{\sum_{\substack{\ell\\ (\ell' A\ell=0)}} c_\ell(\ell' An)^k\Bigg\} e^{\frac{2\pi i\tau}{N}\cdot\frac{1}{2}\frac{n'An}{N}}.$$

For an integral matrix F of order f and determinant ± 1 we set

$$A_1=F'AF,\qquad n=Fn_1,\qquad \ell=F\ell_1,\qquad \text{and } h=Fh_1,$$

and obtain

$$\vartheta_{A,h,P_k^A}(\tau)=\sum_{\substack{n_1\in\mathbb{Z}^f\\ n_1\equiv h_1 \bmod N\\ (A_1h_1\equiv 0 \bmod N)}}\ \sum_{\substack{\ell_1\\ (\ell_1' A_1\ell_1=0)}} c_{F l_1}(l_1' A_1 n_1)^k e^{\frac{2\pi i\tau}{N}\cdot\frac{1}{2}\frac{n_1'A_1n_1}{N}}.$$

The level of both A_1 and A is equal to N. In fact both matrices $N^*A_1^{-1}$ and N^*A^{-1}, $N^*\in\mathbb{Z}$, are simultaneously entire and even, since $N^*A_1^{-1}=F^{-1}(N^*A^{-1})F'^{-1}$. Therefore we have

$$\begin{aligned}&\vartheta_{A,h,P_k^A}=\vartheta_{A_1,h_1,P_k^{A_1}},\quad \text{with}\\ &P_k^{A_1}(x)=\sum_{\ell_1} c_{\ell_1}^1(\ell_1' A_1 x)^k,\qquad c_{\ell_1}^1=c_\ell.\end{aligned} \tag{19}$$

4. Spherical functions. The functions

$$P_k^A(x)=\sum_{\ell} c_\ell(\ell' Ax)^k,\qquad \ell' A\ell=0,$$

as defined in (15) by a finite sum, are known to be spherical functions. We do not wish to elaborate this connection, however, we recall the definition: A homogeneous polynomial $S_k(x)$ of degree k in the variables $x_1,\dots,x_f$ is called a spherical function if it satisfies the differential equation

$$\frac{\partial^2 S_k}{\partial x_1^2}+\cdots+\frac{\partial^2 S_k}{\partial x_f^2}=0.$$

A homogeneous polynomial $S_k(x)$ is called *a spherical function with respect to the positive definite quadratic form* $x'Ax$, if it is a spherical function in $\boldsymbol{y}$, where V is a linear transformation which takes $\boldsymbol{x}'A\boldsymbol{x}$ into a sum of squares:

$$\boldsymbol{x} = V\boldsymbol{y}, \qquad \boldsymbol{x}'A\boldsymbol{x} = \boldsymbol{y}'\boldsymbol{y} = \sum_1^f y_\nu^2 .$$

Here we have the following theorem:

For a fixed k the set of spherical functions $S_k(\boldsymbol{x})$ with respect to the quadratic form $\boldsymbol{x}'A\boldsymbol{x}$ coincides with the set of functions $P_k(\boldsymbol{x})$ defined by (15).

For proofs and further details we refer to E. Hecke [4].

§ 3. Behavior of the Theta Series under Modular Transformations

By multiplying the series in (16) by $\omega_2^{-(r+k)}$ we obtain the series

$$\vartheta_{A,\boldsymbol{h},P_k}\left(\begin{pmatrix}\omega_1\\ \omega_2\end{pmatrix}\right) := \omega_2^{-(r+k)}\,\vartheta_{A,\boldsymbol{h},P_k}\left(\frac{\omega_1}{\omega_2}\right), \qquad \operatorname{Im}\left(\frac{\omega_1}{\omega_2}\right) > 0, \tag{16'}$$

which is homogeneous in ω_1, ω_2. For this series we obtain the following transformation formulas

$$\text{I} \quad \vartheta_{A,\boldsymbol{h},P_k}\left(U\begin{pmatrix}\omega_1\\ \omega_2\end{pmatrix}\right) = \vartheta_{A,\boldsymbol{h},P_k}\left(\begin{pmatrix}\omega_1\\ \omega_2\end{pmatrix}\right) e^{\frac{2\pi i}{N}\frac{1}{2}\frac{\boldsymbol{h}'A\boldsymbol{h}}{N}}, \qquad U = \begin{pmatrix}1 & 1\\ 0 & 1\end{pmatrix},$$

$$\text{II} \quad \vartheta_{A,\boldsymbol{h},P_k}\left(T\begin{pmatrix}\omega_1\\ \omega_2\end{pmatrix}\right) = \frac{(-i)^{r+2k}}{\sqrt{D}} \sum_{\substack{\boldsymbol{g} \bmod N\\ A\boldsymbol{g}\equiv 0 \bmod N}} e^{\frac{2\pi i}{N}\frac{\boldsymbol{g}'A\boldsymbol{g}}{N}}\,\vartheta_{A,\boldsymbol{g},P_k}\left(\begin{pmatrix}\omega_1\\ \omega_2\end{pmatrix}\right), \tag{17'}$$

$$T = \begin{pmatrix}0 & -1\\ 1 & 0\end{pmatrix}.$$

Since U and T generate Γ it follows: for fixed A and P_k the vector space over $\mathbb{C}$ generated by the $\vartheta_{A,\boldsymbol{h},P_k}\left(\begin{pmatrix}\omega_1\\ \omega_2\end{pmatrix}\right)$ is mapped into itself by the homogeneous modular substitutions S, i.e.

$$\vartheta_{A,\boldsymbol{h},P_k}|_{r+k}S = \sum_{\boldsymbol{g} \bmod N} \psi_{\boldsymbol{h},\boldsymbol{g}}^S\,\vartheta_{A,\boldsymbol{g},P_k}, \tag{20}$$

where the $\psi_{\boldsymbol{h},\boldsymbol{g}}^S$ are certain constants—which are not uniquely determined. According to our agreement concerning the symbol $|_{r+k}$ the system of formulas (20) also holds for the inhomogeneous series (16).

1. Determination of a system of solutions for (20). We use inhomogeneous notation and assume $S=\begin{pmatrix} a & b \\ c & d \end{pmatrix}$ with $c>0$. The case $c=0$ was already considered in (17) I and taking $c>0$ is no restriction. We note that

$$\frac{a\tau+b}{c\tau+d}=\frac{a}{c}-\frac{1}{c(c\tau+d)} \quad \text{for } c\neq 0,$$

and, applying (18), (17) I and (17) II, we derive

$$\vartheta_{A,h,P_k^A}\left(\frac{a\tau+b}{c\tau+d}\right)=\vartheta_{A,h,P_k^A}\left(\frac{a}{c}-\frac{1}{c(c\tau+d)}\right)=\sum_{\substack{g \bmod cN \\ g\equiv h \bmod N}} \frac{1}{c^k}\,\vartheta_{cA,g,P_k^{cA}}\left(a-\frac{1}{c\tau+d}\right)$$

$$=\frac{(-i)^{r+2k}(c\tau+d)^{r+k}}{\sqrt{D}\,c^{r+k}} \sum_{\substack{g \bmod cN \\ g\equiv h \bmod N}} e^{\frac{2\pi i}{N}\cdot\frac{1}{2}\cdot\frac{g' A g}{cN}a} \sum_{\substack{q \bmod cN \\ Aq\equiv 0 \bmod N}} e^{\frac{2\pi i}{N}\frac{q' A g}{cN}}\,\vartheta_{cA,q,P_k^{cA}}(c\tau+d).$$

We again use (17) I and, collecting our results, we obtain

$$\vartheta_{A,h,P_k}\left(\frac{a\tau+b}{c\tau+d}\right)=\frac{(-i)^{r+2k}(c\tau+d)^{r+k}}{\sqrt{D}\,c^{r+k}} \sum_{\substack{q \bmod cN \\ Aq\equiv 0 \bmod N}} \varphi_{h,q}^S\,\vartheta_{cA,q,P_k^{cA}}(c\tau) \tag{21}$$

with the coefficients

$$\varphi_{h,q}^S := \sum_{\substack{g \bmod cN \\ g\equiv h \bmod N}} \exp\left\{\frac{2\pi i}{cN^2}\left(a\frac{1}{2}\,g' A g+g' A q+d\frac{1}{2}\,q' A q\right)\right\}. \tag{22}$$

After an easy computation we arrive at

$$\varphi_{h,q}^S=\varphi_{h+dq,0}^S \exp\left\{-\frac{2\pi i}{N^2}\left(\frac{1}{2}\,q' A q b d+h' A q b\right)\right\}. \tag{23}$$

The exponential in (23) depends on q only modulo N because $Aq\equiv Ah\equiv 0 \bmod N$, and by (22) the second factor likewise depends on q only modulo N. Hence (23) implies that $\varphi_{h,q}^S$ depends on q only modulo N. Therefore (21) takes the form

$$\begin{aligned}\vartheta_{A,h,P_k^{cA}}\left(\frac{a\tau+b}{c\tau+d}\right)&=\frac{(-i)^{r+2k}(c\tau+d)^{r+k}}{\sqrt{D}\,c^{r+k}} \sum_{q_1 \bmod N} \varphi_{h,q_1}^S \sum_{\substack{q \bmod cN \\ q\equiv q_1 \bmod N \\ Aq\equiv 0 \bmod N}} \vartheta_{cA,q,P_k^{cA}}(c\tau) \\ &=\frac{(-i)^{r+2k}(c\tau+d)^{r+k}}{\sqrt{D}\,c^{r+k}} \sum_{\substack{q \bmod N \\ Aq\equiv 0 \bmod N}} \varphi_{h,q}^S\,\vartheta_{A,q,P_k^{cA}}(\tau),\end{aligned} \tag{24}$$

where we again used equation (18). We have now reached our first goal.

2. Behavior of theta series under special modular transformations. Our formulas (23) and (24), valid in general, take a very simple form if we assume that $d \equiv 0 \bmod N$ and, as earlier, $c>0$. Then with $S=\begin{pmatrix} a & b \\ c & d \end{pmatrix}$ we have

$$\frac{\vartheta_{A,h,P_k}\left(\dfrac{a\tau+b}{c\tau+d}\right)}{(c\tau+d)^{r+k}} i^{r+2k} c^r \sqrt{D} = \varphi^S_{h,0} \sum_{q \bmod N} e^{-\frac{2\pi i}{N}\frac{h' A q b}{N}} \vartheta_{A,q,P_k}(\tau).$$

In both sides of the above equation we now subject τ to the transformation $\tau \mapsto -\dfrac{1}{\tau}$ and then apply formula (17) II to the right hand side. We obtain

$$\frac{\vartheta_{A,h,P_k}\left(\dfrac{b\tau-a}{d\tau-c}\right)}{(d\tau-c)^{r+k}} = \frac{\varphi^S_{h,0}}{(-c)^r D} \sum_{\substack{q,g \bmod N \\ Aq \equiv Ag \equiv 0 \bmod N}} e^{\frac{2\pi i}{N}\frac{(g-bh)' A q}{N}} \vartheta_{A,g,P_k}(\tau). \tag{25}$$

However, we prove next that

$$\sum_{\substack{q \bmod N \\ Aq \equiv 0 \bmod N}} e^{\frac{2\pi i}{N}\frac{(g-bh)' A q}{N}} = \begin{cases} 0, & \text{if } g-bh \not\equiv 0 \bmod N, \text{ and} \\[2ex] D, & \text{if } g-bh \equiv 0 \bmod N. \end{cases}$$

For this we denote the sum by σ and set $g_1 := g - bh$. Then with $q_0 \in \mathbb{Z}^f$, $Aq_0 \equiv 0 \bmod N$, we have

$$\sigma = \sum_{\substack{q \bmod N \\ Aq \equiv 0 \bmod N}} e^{\frac{2\pi i}{N}\frac{g_1' A(q+q_0)}{N}} = e^{\frac{2\pi i}{N}\frac{g_1' A q_0}{N}} \sigma.$$

Here if $g_1 \not\equiv 0 \bmod N$, then there is a choice of q_0, namely $q_0 = A^{-1} N q^0$ with suitable $q^0 \in \mathbb{Z}^f$, for which $(g_1' A q_0)/N$ is not divisible by N. This proves the first part of the assertion. In order to prove the second part we note from (25) that

$$\frac{\vartheta_{A,h,P_k}\left(\dfrac{b\tau-a}{d\tau-c}\right)}{(d\tau-c)^{r+k}} = \frac{Z}{D}\,\frac{\varphi^S_{h,0}}{(-c)^r}\,\vartheta_{A,bh,P_k}(\tau),$$

where Z denotes the number of solutions q incongruent modulo N of $Aq \equiv 0 \bmod N$. In particular, this equation holds for $S = T = \begin{pmatrix} 0 & -1 \\ 1 & 0 \end{pmatrix}$ and $k=0$. Since $\vartheta_{A,h,1} = \vartheta_{A,-h,1} \neq 0$ and since $\varphi^T_{h,0} = 1$ by (22), it follows that $Z = D$. We change notation and write $\begin{pmatrix} a & b \\ c & d \end{pmatrix}$ in place of $\begin{pmatrix} b & -a \\ d & -c \end{pmatrix}$.

Our result is then stated in the form:

The behavior of the theta function $\vartheta_{A,\boldsymbol{h},P_k}$ under the modular substitution

$$S=\begin{pmatrix} a & b \\ c & d \end{pmatrix}, \qquad c\equiv 0 \bmod N, \qquad d>0$$

is given by

$$\left.\begin{aligned} &\frac{\vartheta_{A,\boldsymbol{h},P_k}\left(\dfrac{a\tau+b}{c\tau+d}\right)}{(c\tau+d)^{r+k}}=\frac{\varphi_{\boldsymbol{h}}^{b,d}}{d^r}\,\vartheta_{A,a\boldsymbol{h},P_k}(\tau),\\ &\varphi_{\boldsymbol{h}}^{b,d}:=\sum_{\substack{\boldsymbol{g} \bmod dN\\ \boldsymbol{g}\equiv\boldsymbol{h} \bmod N}} e^{\frac{2\pi i}{N}\frac{b}{d}\frac{1}{2}\frac{\boldsymbol{g}'A\boldsymbol{g}}{N}}. \end{aligned}\right\} \tag{26}$$

Here $\varphi_{\boldsymbol{h}}^{b,d}$ is a new notation for $\varphi_{\boldsymbol{h},\boldsymbol{0}}^{S}$.

§ 4. Behavior of the Theta Series under Congruence Groups. Gaussian Sums

In this section we investigate more closely the multiplier

$$\Phi_{\boldsymbol{h}}^{b,d}:=\varphi_{\boldsymbol{h}}^{b,d}/d^r \tag{27}$$

of the transformation equation (26). Among other things we will find that the theta series $\vartheta_{A,\boldsymbol{h},P_k}$ are entire modular forms of level N where, as before, N is the level of the quadratic form $\frac{1}{2}\boldsymbol{x}'A\boldsymbol{x}$.

1. Residue class characters $\bmod N$. In the sum (26) for $\varphi_{\boldsymbol{h}}^{b,d}$ we write the summation condition for $\boldsymbol{g}$ in the form $\boldsymbol{g}=ad\boldsymbol{h}+\boldsymbol{g}_1N$, $\boldsymbol{g}_1 \bmod d$ as we may, since $ad\equiv 1 \bmod N$. Then from $A\boldsymbol{h}\equiv\boldsymbol{0} \bmod N$ it follows that

$$\varphi_{\boldsymbol{h}}^{b,d}=e^{\frac{2\pi i}{N}\frac{1}{2}\frac{\boldsymbol{h}'A\boldsymbol{h}}{N}ab}\sum_{\boldsymbol{g} \bmod d} e^{2\pi i\frac{1}{2}\frac{\boldsymbol{g}'A\boldsymbol{g}}{d}b}=e^{\frac{2\pi i}{N}\frac{1}{2}\frac{\boldsymbol{h}'A\boldsymbol{h}}{N}ab}\varphi_{\boldsymbol{0}}^{b,d}. \tag{28}$$

Here, because of the definition (26) of $\varphi_{\boldsymbol{h}}^{b,d}$, it is necessary that $d>0$ and g.c.d. $(d,bN)=1$. The following considerations hold for all such $\varphi_{\boldsymbol{h}}^{b,d}$, since there is a matrix $\begin{pmatrix} a & b \\ c & d \end{pmatrix}\in\Gamma$ with $c\equiv 0 \bmod N$. For the further investigation of the $\varphi_{\boldsymbol{0}}^{b,d}$ we look at the function theoretic behavior of the theta functions. Since $\varphi_{\boldsymbol{0}}^{b,d}$ is independent of P_k we take $k=0$. In (26) we replace τ by $\tau+n$ and, since $\vartheta_{A,\boldsymbol{0},1}\neq 0$, we conclude that

$$\Phi_{\boldsymbol{0}}^{b,d}=\Phi_{\boldsymbol{0}}^{b+na,d+nc} \quad \text{for } n\in\mathbb{Z} \text{ and } d+nc>0. \tag{29}$$

This says that $\Phi_0^{b,d}$ lies in the field of $(d+nc)^{th}$ roots of unity over the rational number field $\mathbb{Q}$ for all natural n for which $d+nc>0$. Hence $\Phi_0^{b,d}$ already lies in $\mathbb{Q}$. Because $(b,d)=1$, the map $e^{2\pi i\frac{b}{d}}\mapsto e^{2\pi i\frac{1}{d}}$ defines an automorphism of the dth roots of unity. This map takes $\Phi_0^{b,d}$ to $\Phi_0^{1,d}$ and, since these numbers are rational, $\Phi_0^{b,d}=\Phi_0^{1,d}$. On the other hand, since $(b+na,d+nc)=1$, (29) holds and

$$\Phi_0^{1,d}=\Phi_0^{b+na,d+nc}=\Phi_0^{1,d+nc}.$$

Here we are permitted to set $c=N$ since, if $\begin{pmatrix}a & b\\ c'N & d\end{pmatrix}$ is admissible, then so is $\begin{pmatrix}a & bc'\\ N & d\end{pmatrix}$. Thus if we define

$$S(d):=\Phi_0^{1,d},\qquad d>0,\qquad (d,N)=1, \tag{30}$$

the preceding remarks show that

$$S(d)=S(d')\quad \text{if } d\equiv d' \bmod N,\ d,d'>0\quad \text{and } (dd',N)=1.$$

Moreover, if $(dd',N)=1$ we have

$$S(dd')=S(d)S(d').$$

This can be seen by applying an admissible transformation $\begin{pmatrix}a' & b'\\ c' & d'\end{pmatrix}$ in (26), where we take $P_k=1$ and $\boldsymbol{h}=\boldsymbol{0}$. Since $S(d)$ is defined for all positive d relatively prime to N and since $S(1)=1$, it now follows that

$$S(d)=\frac{1}{d^r}\sum_{\boldsymbol{g} \bmod d} e^{2\pi i\frac{1}{2}\frac{\boldsymbol{g}'A\boldsymbol{g}}{d}} \tag{31}$$

is a residue class character modulo N for all positive d and that it takes only the values ± 1.

If we define $S(d)$ for negative d with $(d,N)=1$ by

$$S(d):=(-1)^r S(-d), \tag{32}$$

then for all $\begin{pmatrix}a & b\\ c & d\end{pmatrix}\in\Gamma$ with $c\equiv 0 \bmod N$ we obtain instead of (26) the following formula:

$$\frac{\vartheta_{A,\boldsymbol{h},P_k}\left(\dfrac{a\tau+b}{c\tau+d}\right)}{(c\tau+d)^{r+k}}=e^{\frac{2\pi i}{N}\frac{1}{2}\frac{\boldsymbol{h}'A\boldsymbol{h}}{N}ab}\,S(d)\,\vartheta_{A,a\boldsymbol{h},P_k}(\tau). \tag{33}$$

We immediately have

Theorem 3. *The function S defined for rational integers d relatively prime to N by*

$$S(d) = \frac{1}{d^r} \sum_{\boldsymbol{g} \bmod d} e^{2\pi i \frac{1}{2} \frac{\boldsymbol{g}' \boldsymbol{A} \boldsymbol{g}}{d}} \quad \text{if } d>0,$$

and by (34)

$$S(d)=(-1)^r S(-d) \qquad \text{if } d<0,$$

is a residue class character $\bmod N$ *which only assumes the values* ± 1.

Proof. The right side of (33) is invariant under the transformation $\tau \mapsto \tau + nN$, $n \in \mathbb{Z}$. On the other side $S(d)$ is replaced by $S(d+ncN)$. Since $d+ncN$ can be both positive and negative, the equation

$$S(d) = S(d') \quad \text{if } d \equiv d' \bmod N, \qquad (d\,d', N) = 1,$$

also holds for d, d' of opposite sign, since it holds when they have the same sign. By the same reasoning we also obtain the extension

$$S(d)\,S(d') = S(d\,d'), \qquad (d\,d', N) = 1. \quad \square$$

Now (33) implies

$$\vartheta_{\boldsymbol{A},\boldsymbol{h},P_k}(\tau)\,|_{r+k} M = \vartheta_{\boldsymbol{A},\boldsymbol{h},P_k}(\tau) \quad \text{for } M \in \Gamma[N]. \tag{35}$$

The functions $\vartheta_{\boldsymbol{A},\boldsymbol{h},P_k}$ are holomorphic in $\mathcal{H}$. At $i\infty$ they have an expansion in powers of $e^{2\pi i \frac{\tau}{N}}$ with non-negative exponents. Likewise, because of (24), they have expansions at the rational cusps as is required of entire modular forms. Thus by (33) we have

Theorem 4. *The series* $\vartheta_{\boldsymbol{A},\boldsymbol{h},P_k}(\tau)$, *if not identically* 0, *represent entire modular forms of level N and dimension* $-(r+k)$. *If* $k>0$ *they are cusp forms.*

For the transition to homogeneous forms one must observe that $S(-1)=(-1)^r$.

2. Evaluation of the residue class characters $S(d)$. In order to determine the value of $S(d)$ for $(d, N)=1$ (and so $(d, D)=1$), we assume $d>0$ and choose an odd prime $p_d \equiv d \bmod N$. Then $S(d)=S(p_d)$ and the quadratic form $\frac{1}{2}\boldsymbol{x}'\boldsymbol{A}\boldsymbol{x}$ can be transformed modulo p_d to a diagonal form:

$$\tfrac{1}{2}\boldsymbol{x}'\boldsymbol{A}\boldsymbol{x} = \boldsymbol{z}'\boldsymbol{F}\boldsymbol{z}, \qquad \boldsymbol{F} \equiv (\delta_{\mu,\nu} a_\mu) \bmod p_d, \qquad a_\mu \in \mathbb{Z},$$
$$\boldsymbol{x} = T\boldsymbol{z}, \quad \det T \equiv t \not\equiv 0 \bmod p_d.$$

By (31), if $z'=(z_1,\dots,z_f)$, then

$$S(d)=\frac{1}{p_d^r}\sum_{z \bmod p_d} e^{\frac{2\pi i}{p_d}\sum_{\nu=1}^{2r} a_\nu z_\nu^2}=\frac{1}{p_d^r}\prod_{\nu=1}^{2r}\sum_{z_\nu \bmod p_d} e^{\frac{2\pi i}{p_d}a_\nu z_\nu^2}.$$

As is known and as we will prove in the next subsection,

$$\sum_{z \bmod p} e^{\frac{2\pi i}{p}az^2}=\left(\frac{a}{p}\right)\sqrt{\left(\frac{-1}{p}\right)p}\,, \tag{36}$$

with positive or positive imaginary square root, where p is an odd prime, $a\in\mathbb{Z}$, $(a,p)=1$ and where $\left(\frac{a}{p}\right)$ denotes Legendre's symbol. Hence

$$S(d)=\frac{1}{p_d^r}\sqrt{\left(\frac{-1}{p_d}\right)p_d}^{\,2r}\prod_{\nu=1}^{2r}\left(\frac{a_\nu}{p_d}\right)=\left(\frac{\prod_{\nu=1}^{2r}a_\nu}{p_d}\right)\left(\frac{-1}{p_d}\right)^r.$$

Since $\prod_{\nu=1}^{2r} a_\nu\equiv 2^{-2r}t^2 D \bmod p_d$, it follows that

$$S(d)=\left(\frac{(-1)^r D}{p_d}\right).$$

This equation holds for all odd primes $p_d\equiv d \bmod N$ and hence for all odd primes $p_d\equiv d \bmod D$. This, however, is impossible if $(-1)^r D\equiv 2$ or 3 mod 4. In these cases the generalized Jacobi symbol $\left(\frac{(-1)^r D}{n}\right)$ as a function of odd natural n is a character modulo $4D$, not mod D (see E. Hecke [1], A. Scholz—B. Schoeneberg [1], § 24 for this). Hence

There are no even quadratic forms with $(-1)^r D\equiv 2$ or $\equiv 3$ mod 4.

In case $(-1)^r D\equiv 0$ or $\equiv 1$ mod 4, the residue symbol $\left(\frac{(-1)^r D}{n}\right)$ represents a character mod D. Consequently

$$S(d)=\left(\frac{(-1)^r D}{d}\right)\quad \text{if } d>0.$$

We collect these results in

Theorem 5. *If $M=\begin{pmatrix} a & b\\ c & d\end{pmatrix}\in\Gamma$, with $c\equiv 0 \bmod N$, then*

$$\vartheta_{\boldsymbol{A},\boldsymbol{h},P_k}(\tau)|_{r+k}M=\chi(d)\,e^{\frac{2\pi i}{N}\frac{1}{2}\frac{\boldsymbol{h}'\boldsymbol{A}\boldsymbol{h}}{N}ab}\,\vartheta_{\boldsymbol{A},a\boldsymbol{h},P_k}(\tau),$$

$\chi(d)=\left(\frac{(-1)^r D}{d}\right)$ *if $d>0$, and $\chi(d)=(-1)^r\chi(-d)$ if $d<0$.*

3. Gaussian sums. We now prove formula (36) for the so-called *Gaussian sums*,

$G_p(a) := \sum_{z \bmod p} \zeta_p^{az^2}$, with a prime $p \neq 2$, $a \in \mathbb{Z}$, $a \not\equiv 0 \bmod p$, and $\zeta_p = e^{\frac{2\pi i}{p}}$,

which already appeared in the previous subsection. For our purposes we only need the definition of the Legendre symbol

$$\left(\frac{a}{p}\right) := 1, \qquad \text{if } x^2 \equiv a \bmod p \text{ has a solution, and}$$

$$\left(\frac{a}{p}\right) := -1, \quad \text{if } x^2 \equiv a \bmod p \text{ is not solvable,}$$

and the properties

$$\left(\frac{a}{p}\right)\left(\frac{b}{p}\right) = \left(\frac{ab}{p}\right), \quad \text{and} \quad \left(\frac{-1}{p}\right) = (-1)^{\frac{p-1}{2}} \quad \text{if } p \neq 2.$$

We assume that the 'numerator' of the symbol is not divisible by p.

First,

$$\begin{aligned} G_p(a) &= 2 \sum_{\substack{n \bmod p \\ \left(\frac{na}{p}\right)=1}} \zeta_p^n + 1 = 2 \sum_{\substack{n \bmod p \\ \left(\frac{na}{p}\right)=1}} \zeta_p^n - \sum_{\substack{n \bmod p \\ n \not\equiv 0 \bmod p}} \zeta_p^n \\ &= \sum_{\substack{n \bmod p \\ n \not\equiv 0 \bmod p}} \left(\frac{na}{p}\right) \zeta_p^n = \left(\frac{a}{p}\right) G_p(1). \end{aligned} \tag{37}$$

It is therefore sufficient to prove (36) with $a=1$. Here we use transformation formula (5) with $A=(1)$:

$$\sum_{m \in \mathbb{Z}} e^{\pi i \tau (m+x)^2} = \frac{1}{\sqrt{-i\tau}} \sum_{m \in \mathbb{Z}} e^{-\frac{\pi i}{\tau} m^2 + 2\pi i m x}, \qquad \operatorname{Im} \tau > 0, \operatorname{Re}(\sqrt{-i\tau}) > 0.$$

If $x=0$ we thus obtain

$$\sum_{m \in \mathbb{Z}} e^{\pi i \tau m^2} = \frac{1}{\sqrt{-i\tau}} \sum_{m \in \mathbb{Z}} e^{-\frac{\pi i}{\tau} m^2}. \tag{38}$$

If $x = \frac{\alpha}{q}$; $\alpha, q \in \mathbb{Z}$; $q > 0$, then after the substitution $\tau \mapsto \tau q^2$ it follows that

$$\sum_{m \equiv \alpha \bmod q} e^{\pi i \tau m^2} = \frac{1}{q\sqrt{-i\tau}} \sum_{m \in \mathbb{Z}} e^{-\frac{\pi i m^2}{\tau q^2} + \frac{2\pi i}{q} m\alpha}.$$

From this we conclude that

$$\lim_{\tau\to 0}\sqrt{-i\tau}\sum_{m\equiv\alpha \bmod q} e^{\pi i\tau m^2}=\frac{1}{q}, \tag{39}$$

where τ goes to 0 in such a way that $\operatorname{Im}\left(-\frac{1}{\tau}\right)$ tends to $+\infty$. If we set

$$\tau=\frac{2}{q}+iw^2, \qquad w>0,$$

then the left side of (38) becomes

$$\sum_{m\in\mathbb{Z}} e^{\pi i\tau m^2}=\sum_{\alpha \bmod q}\zeta_q^{\alpha^2}\sum_{m\equiv\alpha \bmod q} e^{-\pi m^2 w^2}, \qquad \zeta_q=e^{\frac{2\pi i}{q}},$$

and thus by (39)

$$\lim_{w\to 0}\left(w\sum_{m\in\mathbb{Z}} e^{\pi i\tau m^2}\right)=\sum_{\alpha \bmod q}\zeta_q^{\alpha^2}\cdot\lim_{w\to 0}\left(w\sum_{m\equiv\alpha \bmod q} e^{-\pi m^2 w}\right)=\frac{1}{q}\sum_{\alpha \bmod q}\zeta_q^{\alpha^2}.$$

Now, since

$$-\frac{1}{\tau}=-\frac{q}{2}+\frac{w^2q^2i}{4+2iqw^2},$$

the right side of (38) becomes

$$\sqrt{\frac{qi}{2}+\frac{w^2q^2}{4+2qiw^2}}\sum_{\alpha \bmod 4} e^{-\frac{\pi i\alpha^2 q}{2}}\sum_{m\equiv\alpha \bmod 4} e^{-\frac{w^2q^2m^2\pi}{4+2iqw^2}}.$$

As $w\to 0$ the root converges to $\sqrt{\frac{qi}{2}}$, where the root is taken to have positive imaginary part, and where by (39) the inner sum when multiplied by w has the limit $\frac{2}{q}\cdot\frac{1}{4}$. Hence

$$\frac{1}{q}\sum_{\alpha \bmod q}\zeta_q^{\alpha^2}=\left|\sqrt{\frac{q}{2}}\right| e^{\frac{\pi i}{4}}\frac{1}{2q}\sum_{\alpha \bmod 4}\zeta_4^{-\alpha^2 q}.$$

For odd q,

$$\sum_{\alpha \bmod 4}\zeta_4^{-\alpha q^2}=2(1+(-i)^q),$$

and since $e^{\frac{\pi i}{4}}=\frac{\sqrt{2}}{2}(1+i)$,

$$\sum_{\alpha \bmod q}\zeta_q^{\alpha^2}=\sqrt{(-1)^{\frac{q-1}{2}}q},$$ where the root is positive or has positive imaginary part.

This proves (36) for $a=1$ and odd primes p, because $\left(\frac{-1}{p}\right)=(-1)^{\frac{p-1}{2}}$. (36) is proved in general by (37).

§ 5. Examples and Applications

1. Quadratic forms of level 1. Even positive definite quadratic forms $\boldsymbol{x}'\boldsymbol{A}\,\boldsymbol{x}$ of level $N=1$ have determinant $D=|\boldsymbol{A}|=1$, since N and D have the same prime divisors (cf. § **2**, **1**). Here we only consider forms of this kind. In this case equation (24) with $k=0$ and $S=T=\begin{pmatrix}0 & -1\\ 1 & 0\end{pmatrix}$ becomes

$$\vartheta_{\boldsymbol{A},\boldsymbol{0},1}\left(-\frac{1}{\tau}\right)=(-i)^r\tau^r\,\vartheta_{\boldsymbol{A},\boldsymbol{0},1}(\tau).$$

Since $\vartheta_{\boldsymbol{A},\boldsymbol{0},1}$ is a modular form, this equation only holds if $r\equiv 0 \bmod 4$, i.e.

The order $f=2r$ of an even positive definite quadratic form of determinant 1 is divisible by 8.

An example of such a form with $f=8$ is $\mathcal{Q}_8$:

$$2\,\mathcal{Q}_8(\boldsymbol{x})=\boldsymbol{x}'\boldsymbol{A}_8\,\boldsymbol{x},\quad \boldsymbol{A}_8=\begin{pmatrix}2 & 1 & 0 & & & \dots & & 0\\ 1 & 2 & 1 & 0 & & \dots & & 0\\ 0 & 1 & 4 & 3 & & \dots & & 0\\ 0 & 0 & 3 & 4 & 5 & \dots & & 0\\ 0 & \dots & & 5 & 20 & 3 & & 0\\ 0 & \dots & & & 3 & 12 & 1 & 0\\ 0 & \dots & & & & 1 & 4 & 1\\ 0 & \dots & & & & & 1 & 2\end{pmatrix}.$$

$\vartheta_{\boldsymbol{A}_8,\boldsymbol{0},1}$ is an entire modular form of dimension -4 which up to a constant factor is thus the Eisenstein series G_4. From **III**, Theorem **2**, we obtain

$$\sum_{\boldsymbol{n}\in\mathbb{Z}^8} e^{2\pi i\tau\frac{1}{2}\boldsymbol{n}'\boldsymbol{A}_8\boldsymbol{n}}=1+240\sum_{n=1}^{\infty}\sigma_3(n)e^{2\pi i\tau n} \tag{40}$$

by comparing the constant terms in the expansions in powers of $e^{2\pi i\tau}$. Analogously, for $f=16$ and with

$$2\,\mathcal{Q}_{16}(\boldsymbol{x})=\boldsymbol{x}'\boldsymbol{A}_{16}\,\boldsymbol{x},\qquad \boldsymbol{A}_{16}=\begin{pmatrix}\boldsymbol{A}_8 & \boldsymbol{0}\\ \boldsymbol{0} & \boldsymbol{A}_8\end{pmatrix},$$

we see that

$$\sum_{\boldsymbol{n}\in\mathbb{Z}^{16}} e^{2\pi i\tau\frac{1}{2}\boldsymbol{n}'\boldsymbol{A}_{16}\boldsymbol{n}}=1+480\sum_{n=1}^{\infty}\sigma_7(n)e^{2\pi i\tau n}. \tag{41}$$

Thus the number of different solutions $\boldsymbol{n}$ of

$$\tfrac{1}{2}\boldsymbol{n}'\boldsymbol{A}_8\,\boldsymbol{n}=n \quad\text{and}\quad \tfrac{1}{2}\boldsymbol{n}'\boldsymbol{A}_{16}\,\boldsymbol{n}=n,\qquad n\in\mathbb{Z},\quad n>0,$$

is

$$240\,\sigma_3(n) \quad\text{and}\quad 480\,\sigma_7(n),$$

respectively.

Since up to a constant factor there is only one entire modular form of dimension -4 or -8, it follows that equation (40) holds except for a constant factor for each even positive definite matrix of determinant 1 and order 8, and that (41) likewise holds for all such matrices of order 16.

If the matrix A has order $f=2r\equiv 0 \bmod 8$, then **III**, Theorem **2**, implies that

$$\vartheta_{\boldsymbol{A},\boldsymbol{0},1}(\tau)=\frac{1}{2\zeta(r)}\,G_r(\tau)+H_{\boldsymbol{A}}(\tau), \tag{42}$$

where G_r is the Eisenstein series of dimension $-r$ and where

$$H_{\boldsymbol{A}}(\tau):=\sum_{n=1}^{\infty} h_{\boldsymbol{A}}(n)\,e^{2\pi i n\tau}$$

is a cusp form of dimension $-r$. It follows from **III**, Theorem **2** and **III**, (6) that

$$\frac{1}{2\zeta(r)}\,G_r(\tau)=1-\frac{2r}{B_r}\sum_{n=1}^{\infty}\sigma_{r-1}(n)\,e^{2\pi i n\tau},$$

where B_r is the rth Bernoulli number. Hence for the matrix A of order $f=2r$ the number of solutions of

$$\tfrac{1}{2}\,\boldsymbol{n}'\boldsymbol{A}\boldsymbol{n}=n,\qquad n\in\mathbb{Z},\qquad n>0,$$

is

$$a_{\boldsymbol{A}}(n)=-\frac{2r}{B_r}\,\sigma_{r-1}(n)+h_{\boldsymbol{A}}(n), \tag{43}$$

where $h_{\boldsymbol{A}}(n)$, $n\geqq 1$, is the nth Fourier coefficient of the cusp form $H_{\boldsymbol{A}}$. If A is of order 24, then $H_{\boldsymbol{A}}$ is up to a constant factor equal to the discriminant Δ. For

$$\boldsymbol{A}=\boldsymbol{A}_{24}=\begin{pmatrix}\boldsymbol{A}_8 & \boldsymbol{0} & \boldsymbol{0}\\ \boldsymbol{0} & \boldsymbol{A}_8 & \boldsymbol{0}\\ \boldsymbol{0} & \boldsymbol{0} & \boldsymbol{A}_8\end{pmatrix}$$

one obtains the equation

$$a_{\boldsymbol{A}_{24}}(n)=\tfrac{1}{691}\,(65\,520\,\sigma_{11}(n)+432\,000\,\tau(n)), \tag{44}$$

where $\tau(n)$ is the nth Fourier coefficient of Δ.

For this derivation we used $a_{\boldsymbol{A}_{24}}(1)=720$ and $B_{12}=-\frac{691}{2730}$.

We now consider the functions $P_k^{\boldsymbol{A}}$, defined in (15), for $\boldsymbol{A}=\boldsymbol{A}_8$. For suitable $c_k(n)$ we have

$$\vartheta_{\boldsymbol{A}_8,\boldsymbol{0},P_k}(\tau)=\sum_{\boldsymbol{n}\in\mathbb{Z}^8} P_k(\boldsymbol{n})\,e^{2\pi i\frac{1}{2}\boldsymbol{n}'\boldsymbol{A}_8\boldsymbol{n}}=\sum_{n=1}^{\infty} c_k(n)\,e^{2\pi i\tau n}.$$

These series are entire modular forms of dimension $-(4+k)$ and thus are trivially 0 for odd k. They are also identically 0 for even $k, 0<k<8$, since there are no cusp forms for these dimensions. These results hold for all such P_k. If $k=8$ there is a P_8 for which $c_8(1)=1$, and with this choice of P_8 we have

$$\vartheta_{\boldsymbol{A}_8,\boldsymbol{0},P_8}(\tau)=\Delta(\tau). \tag{45}$$

For the construction of such a P_8 we refer to E. Hecke [4] who makes use of a basis for the spherical functions $P_k(\boldsymbol{x})$.

2. The order of magnitude of $a_{\boldsymbol{A}}(n)$. We now study formula (43) in greater depth and generalize it to the case $|\boldsymbol{A}|\geqq 1$. Suppose $\boldsymbol{A}$ is a matrix of order $f=2r$, $r\geqq 1$, and $|\boldsymbol{A}|\geqq 1$. We put

$$\vartheta_{\boldsymbol{A}}(\tau):=\vartheta_{\boldsymbol{A},\boldsymbol{0},1}(\tau)$$

and derive from (16) that

$$\vartheta_{\boldsymbol{A}}(\tau)=\sum_{\boldsymbol{n}\in\mathbb{Z}^f} e^{2\pi i\tau\frac{1}{2}\boldsymbol{n}'\boldsymbol{A}\boldsymbol{n}}=\sum_{n=0}^{\infty} a_{\boldsymbol{A}}(n)e^{2\pi i n\tau}, \tag{46}$$

where $a_{\boldsymbol{A}}(n)$ is the number of different representations of n by $\frac{1}{2}\boldsymbol{n}'\boldsymbol{A}\boldsymbol{n}$. $\vartheta_{\boldsymbol{A}}(\tau)$ is an entire modular form of level N and dimension $-r$. Thus by Theorems **4**, **9** and **10** of Chapter **VI**,

$$\vartheta_{\boldsymbol{A}}(\tau)=L_{\boldsymbol{A}}(\tau)+H_{\boldsymbol{A}}(\tau)$$

if $r\geqq 1$. Here $L_{\boldsymbol{A}}$ is a linear combination of Eisenstein series of dimension $-r$ and level N for $r>2$ and $r=1$. For $r=2$, $L_{\boldsymbol{A}}$ is a linear combination of $\wp$-division values. $H_{\boldsymbol{A}}$ is a cusp form. If

$$L_{\boldsymbol{A}}(\tau)=\sum_{n=0}^{\infty} l_{\boldsymbol{A}}(n)e^{2\pi i\tau n}, \qquad H_{\boldsymbol{A}}(\tau)=\sum_{n=1}^{\infty} h_{\boldsymbol{A}}(n)e^{2\pi i\tau n},$$

then

$$a_{\boldsymbol{A}}(n)=l_{\boldsymbol{A}}(n)+h_{\boldsymbol{A}}(n). \tag{47}$$

Next we determine the order of $l_{\boldsymbol{A}}(n)$. For this we prove the following

Lemma. *Let*

$$\sigma_s(n):=\sum_{d\mid n,\, d>0} d^s, \qquad s\in\mathbb{Z}, \qquad s\geqq 0,$$

be the sum of the sth powers of the divisors of n and ζ the Riemann zeta function. Then we have the estimates

$$n^s<\sigma_s(n)<n^s\zeta(s) \quad \textit{if } s>1,$$
$$n<\sigma_1(n)<n(1+\log n), \quad \textit{and}$$
$$\sigma_0(n)=O(n^\delta) \quad \textit{as } n\to\infty \textit{ for arbitrary } \delta>0.$$

[Here the O-assertion has its usual meaning — $\sigma_0(n)/n^\delta$ is bounded.]

Proof. If $s \geqq 2$, then

$$n^s < \sigma_s(n) = n^s \sum_{d|n, d>0} \frac{1}{d^s} < n^s \zeta(s).$$

If $s=1$ then

$$n < \sigma_1(n) = n \sum_{d|n, d>0} \frac{1}{d} < n \sum_{\nu=1}^{n} \frac{1}{\nu} < n(1+\log n).$$

For $s=0$, i.e. for the number of divisors of n, we refer to G.H. Hardy and G.H. Wright [1]. There it is also shown that the estimate

$$\sigma_0(n) = O(\log^k n) \quad \text{is false for each } k>0. \quad \square$$

Since $L_A(\tau)$ is a finite sum of Eisenstein series or of $\wp$-divison values, it follows from the representations (5), (25) and (27) of Chapter **VII** that

$$l_A(n) = \begin{cases} n^{r-1}\mathscr{S}(n) & \text{with } |\mathscr{S}(n)| < C \text{ for suitable } C \text{ if } r>2, \text{ and} \\ O(n^{r+\varepsilon-1}) & \text{for each } \varepsilon>0 \text{ if } r=1 \text{ or } r=2. \end{cases} \tag{48}$$

Further for $|A|=1, r>2$ the lemma implies that there is a $\gamma>0$ such that $|\mathscr{S}(n)| \geqq \gamma$. Whether such a γ exists for $|A|>1$ and $r>2$ can only be decided by a deeper investigation of $\mathscr{S}$.

We now determine the order of the Fourier coefficient $h_A(n)$ of the cusp form $H_A(\tau)$. First we prove the general

Lemma. *Let f be a cusp form of dimension $-k<0$ for an arbitrary subgroup Γ_1 of finite index μ in Γ. Then*

$$f(x+iy) = O(y^{-k/2}) \quad \text{as } y \to 0, y>0, \tag{49}$$

uniformly in x.

Proof. As in **V**, (10) we set

$$f\left(\frac{a\tau+b}{c\tau+d}\right)(c\tau+d)^{-k} = (f|_k S)(\tau), \qquad S = \begin{pmatrix} a & b \\ c & d \end{pmatrix} \in \Gamma.$$

By Theorem **6** of Chapter **V**, $f|_k S$ is a modular form for $S^{-1}\Gamma_1 S$, and is in fact a cusp form. If

$$\Gamma = \bigcup_{\nu=1}^{\mu} \Gamma_1 S_\nu, \qquad \mu = [\Gamma : \Gamma_1],$$

is a coset decomposition for Γ modulo Γ_1, then for the function

$$F(\tau) := \sum_{\nu=1}^{\mu} |(f|_k S_\nu)(\tau)|^2$$

we have

$$F(S\tau) = |c\tau+d|^{2k} F(\tau),$$

because the transformation $S\in\Gamma$ permutes the terms $f|_k S_\nu$. Now we note that the function

$$g(\tau)=|\tau-\bar{\tau}|, \qquad \bar{\tau} \text{ the conjugate of } \tau,$$

satisfies the equation

$$g(S\tau)=|c\tau+d|^{-2}g(\tau).$$

Hence the function

$$F(\tau)\,|\tau-\bar{\tau}|^k$$

is invariant under the transformations of Γ. Since the functions $f|_k S_\nu$ are cusp forms, the function $F(\tau)\,|\tau-\bar{\tau}|^k$ is bounded in a neighborhood of $i\infty$ and is continuous in $\mathscr{F}-\{i\infty\}$, where $\mathscr{F}$ is the fundamental region for the modular group Γ. Thus this function is bounded in $\mathscr{H}$. This implies the lemma. □

We now easily derive

Theorem 6. *Let f be a cusp form of dimension $-k<0$ for Γ_1, $[\Gamma:\Gamma_1]<\infty$, and let*

$$f(\tau)=\sum_{n=1}^{\infty} a_n e^{\frac{2\pi i\tau}{N}n}$$

be its Fourier expansion. Then

$$a_n=O(n^{k/2}). \tag{50}$$

Proof. We set $\tau_0=\frac{i}{n}$ in the integral representation

$$a_n=\frac{1}{N}\int_{\tau_0}^{\tau_0+N} f(\tau)e^{-\frac{2\pi i\tau}{N}n}\,d\tau=e^{-\frac{2\pi i\tau_0}{N}n}\int_0^N f(\tau_0+x)e^{-\frac{2\pi ix}{N}n}\,dx$$

which is valid for all $\tau_0\in\mathscr{H}$. Then

$$a_n=e^{\frac{2\pi}{N}}\cdot\frac{1}{N}\int_0^N f\left(x+\frac{i}{n}\right)e^{-\frac{2\pi ix}{N}n}\,dx,$$

and the assertion follows from the Lemma. □

One may conjecture that the estimate (50) may be strengthened for $\Gamma(1)$ to

$$a_n=O\left(n^{\frac{k}{2}-\frac{1}{2}+\varepsilon}\right) \quad \text{for each } \varepsilon>0. \tag{51}$$

S. Ramanujan made the yet sharper conjecture

$$|a_p|=|\tau(p)|\leqq 2p^{\frac{11}{2}} \quad \text{for primes } p$$

for the Fourier coefficients of the discriminant Δ.

P. Deligne has proved the conjecture of Ramanujan: IHES No. 43, to appear.

The investigations of this section were motivated by the question as to the order of the number of representations of n by a positive definite quadratic form. However, these results are true for all modular forms for the group $\Gamma(N)$. We state these results in

Theorem 7. *If f is an entire modular form of dimension $-k$ for the group $\Gamma(N)$ with the Fourier expansion*

$$f(\tau)=\sum_{n=0}^{\infty} a_n e^{\frac{2\pi i\tau}{N}n},$$

then the Fourier coefficients a_n have the estimate

$$a_n = n^{k-1}\mathcal{S}_k(n)+O(n^{k/2}).$$

Here

$$\begin{aligned}&\mathcal{S}_k(n)=O(1) \quad \textit{if } k\geqq 3,\\ &\mathcal{S}_2(n)=O(\log n), \quad \textit{and}\\ &\mathcal{S}_1(n)=O(n^{\varepsilon}) \quad \textit{for each } \varepsilon>0.\end{aligned}$$

If we return once more to the number $a_{\boldsymbol{A}}(n)$ of representations of n by the quadratic form $\frac{1}{2}\boldsymbol{x}'\boldsymbol{A}\boldsymbol{x}$ of order $2r$ we have the equation (47)

$$a_{\boldsymbol{A}}(n)=l_{\boldsymbol{A}}(n)+h_{\boldsymbol{A}}(n),$$

where

$$\begin{aligned}&l_{\boldsymbol{A}}(n)=O(n^{r-1}) && \text{for } r>2,\\ &l_{\boldsymbol{A}}(n)=O(n^{r-1+\varepsilon}) && \text{for } r=1 \text{ and } r=2 \text{ with arbitrary } \varepsilon>0, \text{ and}\\ &h_{\boldsymbol{A}}(n)=O\big(n^{\frac{r}{2}}\big) && \text{for } r\geqq 1.\end{aligned}$$

Cf. B. Schoeneberg [1].

Literature

Behnke, H. and Sommer, F.: 1. Theorie der analytischen Funktionen einer komplexen Veränderlichen.
Reprint of the 3rd ed. Berlin-Heidelberg-New York: Springer 1972.

Blij, van der, F.: 1. The function $\tau(n)$ of S. Ramanujan. J. Indian Math. Soc. **XVIII**, 83—99 (1950).

Carathéodory, C.: 1. Conformal representation. 2nd ed. Cambridge: Univ. Press 1952.

Davenport, H.: 1. Multiplicative number theory. Chicago: Markham 1967.

Deuring, M.: 1. Die Klassenkörper der komplexen Multiplikation. Enzyklopädie d. math. Wissensch. I 2, 23. Stuttgart: B. G. Teubner Verlagsgesellschaft 1958.

Gunning, R. C.: 1. Lectures on modular forms. Princeton, New Jersey: Ann. of Math. Studies **48**, 1962.

Hardy, G. H. and Wright, E. M.: 1. An introduction to the theory of numbers. 4th ed. London: Oxford University Press 1960.

Hasse, H.: 1. Vorlesungen über Zahlentheorie. 2nd ed. Berlin-Heidelberg-New York: Springer 1964.

Hecke, E.: 1. Vorlesungen über die Theorie der algebraischen Zahlen. Leipzig: Akademische Verlagsgesellschaft 1923. 2nd ed. 1954. Reprint, Chelsea Publ. Comp. 1970.
2. Zur Theorie der elliptischen Modulfunktionen. Math. Ann. **97**, 210—242 (1926) = Math. Werke, 428—460.
3. Theorie der Eisensteinschen Reihen höherer Stufe und ihre Anwendung auf Funktionentheorie und Arithmetik. Abh. Math. Sem. Univ. Hamburg **5**, 199—224 (1927) = Math. Werke, 461—486.
4. Analytische Arithmetik der positiven quadratischen Formen. Danske Vidensk. Selsk. Math.-fys. Meddel. **XVII**, 12, Copenhagen 1940 = Math. Werke, 789—918.
5. Mathematische Werke. 2nd ed. Göttingen: Vandenhoeck und Ruprecht 1970.

Klein, F.: 1. Vorlesungen über das Ikosaeder und die Auflösung der Gleichung vom 5. Grade. Leipzig: Teubner 1884. Rev. transl., 2nd ed. London: Stechert 1913.

Klein, F.—Fricke, R.: 1. Vorlesungen über die Theorie der elliptischen Modulfunktionen. Ausgearbeitet und vervollständigt von Robert Fricke. 2 vol. Leipzig: Teubner 1890 and 1892. Reprint 1966.

Lehner, J.: 1. Discontinuous groups and automorphic functions. Providence, Rhode Island: Amer. Math. Soc. 1964.

Le Veque, W. J.: 1. Topics in number theory. 2 vol. Reading, Mass.: Addison-Wesley 1956.

Maak, W.: 1. Elliptische Modulfunktionen. Unter Benutzung einer Vorlesung von E. Hecke aus dem Jahre 1935. Lecture notes, Göttingen Univ. 1955/56.

Maass, H.: 1. Lectures on modular functions of one complex variable. Bombay: Tata Institute of Fundamental Research 1964.

Niven, I. and Zuckermann, H. S.: 1. An introduction to the theory of numbers. 3rd ed. New York: Wiley 1972.

Ogg, A.: 1. Modular forms and Dirichlet Series. Benjamin 1969.

Petersson, H.: 1. Theorie der automorphen Formen beliebig reeller Dimension und ihre Darstellung durch eine neue Art Poincaréscher Reihen. Math. Ann. **103**, 369—436 (1930).
2. Konstruktion der sämtlichen Lösungen einer Riemannschen Funktionalgleichung durch Dirichlet-Reihen mit Eulerscher Produktentwicklung I, II, III. Math. Ann. **116**, 401—412 (1939); **117**, 39—64, 277—300 (1940/41).
3. Über die systematische Bedeutung der Eisenstein Reihen. Abh. Math. Sem. Univ. Hamburg **16**, 104—126 (1949).

Pfetzer, W.: 1. Die Wirkung der Modulsubstitutionen auf mehrfache Thetareihen zu quadratischen Formen ungerader Variablenzahl. Arch. Math. **VI**, 448—454 (1953).

Ramanujan, S.: 1. Collected papers. Ed. by G. H. Hardy, P. V. Seshu Aiyar and B. M. Wilson. Cambridge: Univ. Press 1927.

Rankin, F. K. C. and Swinnerton-Dyer, H. P. F.: 1. On the zeros of Eisenstein Series, Bull. London Math. Soc. **2**, 169—170 (1970).

Riemann, B.: 1. Gesammelte mathematische Werke. 2. Aufl. Leipzig 1892. Reprints.

Rudin, W.: 1. Real and complex analysis. New York: Mc Graw-Hill 1966.

Schoeneberg, B.: 1. Das Verhalten von mehrfachen Thetareihen bei Modulsubstitutionen. Math. Ann. **116**, 511—523 (1939).
2. Verhalten von speziellen Integralen 3. Gattung bei Modultransformationen und verallgemeinerte Dedekindsche Summen. Abh. Math. Sem. Univ. Hamburg **30**, 1—10 (1967).
3. Zur Theorie der verallgemeinerten Dedekindschen Modulfunktionen. Nachr. Akad. Wiss. Göttingen 1969, 119—128.

Scholz, A.—Schoeneberg, B.: 1. Einführung in die Zahlentheorie. 5th ed. Berlin-New York: Walter de Gruyter 1973.

Shimura, G.: 1. Introduction to the arithmetic theory of automorphic finctions. Iwanami Shoten Publ. and Princeton Univ. Press. Publ. Math. Soc. Japan **11**, 1971.

Siegel, C. L.: 1. Gesammelte Abhandlungen. 3 vol. Berlin-Heidelberg-New York: Springer 1966.
2. Topics in complex function theory. New York: Wiley-Interscience. Vol. 1, 1969, Vol. 2, 1971.

Springer, G.: 1. Introduction to Riemann surfaces. Reading, Mass.: Addison-Wesley 1957.

Titchmarsh, E. C.: 1. The theory of functions. 2nd ed. London: Oxford University Press 1939. Many corrected reprints.
2. Introduction to the theory of Fourier integrals. 2nd ed. London: Oxford University Press 1948.

Watson, G. N.: 1. A table of Ramanujan's function $\tau(n)$. Proc. London Math. Soc. (2) **51**, 1—13 (1949).

Whittacker, E. T. and Watson, G. N.: 1. A course of modern analysis, 4th ed. Cambridge: Univ. Press 1927. Many reprints.

Wohlfahrt, K.: 1. Über Dedekindsche Summen und Untergruppen der Modulgruppe. Abh. Math. Sem. Univ. Hamburg **23**, 5—10 (1959).

Zuckermann, H. S.: 1. The computation of the smaller coefficients of $J(\tau)$. Bull. Amer. Math. Soc. **45**, 917—919 (1939).

Proceedings International Summer School, University of Antwerp, 1972, Modular Functions of One Variable. Part I (1973), Part II (1974), Part III (1973), Part IV to appear. Lecture Notes in Mathematics, Springer-Verlag.

Index of Definitions

Index of Notations

(in alphabetical order)

Die Grundlehren der mathematischen Wissenschaften in Einzeldarstellungen mit besonderer Berücksichtigung der Anwendungsgebiete

Eine Auswahl

---

23. Pasch: Vorlesungen über neuere Geometrie
41. Steinitz: Vorlesungen über die Theorie der Polyeder
45. Alexandroff: Topologie. Band 1
46. Nevanlinna: Eindeutige analytische Funktionen
63. Eichler: Quadratische Formen und orthogonale Gruppen
102. Nevanlinna/Nevanlinna: Absolute Analysis
114. Mac Lane: Homology
127. Hermes: Enumerability, Decidability, Computability
131. Hirzebruch: Topological Methods in Algebraic Geometry
135. Handbook for Automatic Computation. Vol. 1/Part a: Rutishauser: Description of ALGOL 60
136. Greub: Multilinear Algebra
137. Handbook for Automatic Computation. Vol. 1/Part b: Grau/Hill/Langmaack: Translation of ALGOL 60
138. Hahn: Stability of Motion
139. Mathematische Hilfsmittel des Ingenieurs. 1. Teil
140. Mathematische Hilfsmittel des Ingenieurs. 2. Teil
141. Mathematische Hilfsmittel des Ingenieurs. 3. Teil
142. Mathematische Hilfsmittel des Ingenieurs. 4. Teil
143. Schur/Grunsky: Vorlesungen über Invariantentheorie
144. Weil: Basic Number Theory
145. Butzer/Berens: Semi-Groups of Operators and Approximation
146. Treves: Locally Convex Spaces and Linear Partial Differential Equations
147. Lamotke: Semisimpliziale algebraische Topologie
148. Chandrasekharan: Introduction to Analytic Number Theory
149. Sario/Oikawa: Capacity Functions
150. Iosifescu/Theodorescu: Random Processes and Learning
151. Mandl: Analytical Treatment of One-dimensional Markov Processes
152. Hewitt/Ross: Abstract Harmonic Analysis. Vol. 2: Structure and Analysis for Compact Groups. Analysis on Locally Compact Abelian Groups
153. Federer: Geometric Measure Theory
154. Singer: Bases in Banach Spaces I
155. Müller: Foundations of the Mathematical Theory of Electromagnetic Waves
156. van der Waerden: Mathematical Statistics
157. Prohorov/Rozanov: Probability Theory
159. Köthe: Topological Vector Spaces
160. Agrest/Maksimov: Theory of Incomplete Cylindrical Functions and their Applications
161. Bhatia/Shegö: Stability Theory of Dynamical Systems
162. Nevanlinna: Analytic Functions
163. Stoer/Witzgall: Convexity and Optimization in Finite Dimensions
164. Sario/Nakai: Classification Theory of Riemann Surfaces
165. Mitrinović/Vasić: Analytic Inequalities
166. Grothendieck/Dieudonné: Eléments de Géométrie Algébrique I

167. Chandrasekharan: Arithmetical Functions
168. Palamodov: Linear Differential Operators with Constant Coefficients
170. Lions: Optimal Control Systems Governed by Partial Differential Equations
171. Singer: Best Approximation in Normed Linear Spaces by Elements of Linear Subspaces
172. Bühlmann: Mathematical Methods in Risk Theory
173. F. Maeda/S. Maeda: Theory of Symmetric Lattices
174. Stiefel/Scheifele: Linear and Regular Celestial Mechanics. Perturbed Two-body Motion—Numerical Methods—Canonical Theory
175. Larsen: An Introduction of the Theory of Multipliers
176. Grauert/Remmert: Analytische Stellenalgebren
177. Flügge: Practical Quantum Mechanics I
178. Flügge: Practical Quantum Mechanics II
179. Giraud: Cohomologie non abélienne
180. Landkoff: Foundations of Modern Potential Theory
181. Lions/Magenes: Non-Homogeneous Boundary Value Problems and Applications I
182. Lions/Magenes: Non-Homogeneous Boundary Value Problems and Applications II
183. Lions/Magenes: Non-Homogeneous Boundary Value Problems and Applications III
184. Rosenblatt: Markov Processes. Structure and Asymptotic Behavior
185. Rubinowicz: Sommerfeldsche Polynommethode
186. Wilkinson/Reinsch: Handbook for Automatic Computation II. Linear Algebra
187. Siegel/Moser: Lectures on Celestial Mechanics
188. Warner: Harmonic Analysis on Semi-Simple Lie Groups I
189. Warner: Harmonic Analysis on Semi-Simple Lie Groups II
190. Faith: Algebra: Rings, Modules, and Categories I
192. Mal'cev: Algebraic Systems
193. Pólya/Szegö: Problems and Theorems in Analysis. Vol. 1
194. Igusa: Theta Functions
195. Berberian: Baer $*$-Rings
196. Athreya/Ney: Branching Processes
197. Benz: Vorlesungen über Geometrie der Algebren
198. Gaal: Linear Analysis and Representation Theory
200. Dold: Lectures on Algebraic Topology
203. Schoeneberg: Elliptic Modular Functions
206. André: Homologie des algèbres commutatives
207. Donoghue: Monotone Matrix Functions and Analytic Continuation
209. Ringel: Map Color Theorem